PLANCHETTE;

OR,

THE DESPAIR OF SCIENCE.

BEING A

FULL ACCOUNT OF MODERN SPIRITUALISM, ITS PHENOMENA, AND THE VARIOUS THEORIES REGARDING IT.

WITH A

SURVEY OF FRENCH SPIRITISM.

By E Sargent.

"Search where thou wilt, and let thy reason go
To ransom Truth, even to the abyss below."

BOSTON:
ROBERTS BROTHERS.
1869.

CAMBRIDGE:
PRESS OF JOHN WILSON AND SON.

A PREFACE IN A LETTER.

To the Rev. W. M.:—

MY dear Friend, — More than twenty years ago, we ventured to cross the border of what Ennemoser calls "the great ill-famed land of the marvellous." Certain manifestations arrested our notice. Repelled and, for a long time, baffled by what seemed merely grotesque or trivial, we did not abandon inquiry. Our interest in the proscribed phenomena has not yet abated. We have lived to see the smile of derision with which the spiritual hypothesis that accompanied them was at first saluted, grow fainter and fainter, until it now rarely appears on the lips of well-informed persons; and the question is put seriously, even by doctors of divinity and veterans of science, "What do these things mean?"

I cannot presume to answer dogmatically; but having kept trace of the so-called spiritual movement that began at Hydesville in 1847, and having, long before that period, investigated the kindred phemomena of somnambulism, independent and mesmeric, I have hoped to offer such a survey of the facts and theories as would be acceptable to earnest and uncommitted seekers after truth, always excepting those who, like Mr. Herbert Spencer, decline, "on *à priori* grounds," to look into the subject. Recently, attention has been directed to it anew by the wooden trifle known as the Planchette; and I have chosen the name of this mysterious toy as the title of my book, rather as a convenient sign-post, pointing to one little phase of the complex whole, than as indicating fully the character of the facts here collected; for these are, I am persuaded, of supreme importance, embracing, as they do, in their relations, most of the authentic marvels in the pneumatology of ancient and modern times.

Without undervaluing the tributary services of Planchette in certain rare cases, I cannot doubt that its eccentricities are often explicable by unconscious nervous movement or by wanton deception. But, after making allowance for all that is unprofitable, trifling, and tedious in the experiments,—for all that ought to be deducted as giving no conclusive evidence of supersensual knowledge or power,—there is a remainder of well-attested results, which cannot be explained by any theory of imposture, hallucination, or unexplored nervous action; and these results belong to the class here considered.

I regret that the circumstances under which the present work was written did not permit me to shape it in nearer accordance with my own notions of completeness and of the far-reaching significance of the developments; but this earth-life is so brief and uncertain, that to have deferred my task, in order to accomplish more, might have been to accomplish nothing; and when one has something to say, he may leave it for ever unspoken if he is over-nice in his choice of modes of presentation. May I not tell the public that to your pen, my friend, they may look for something more in keeping with the amplitude of the theme; something of which I may venture to announce, —

> "'Tis not the hasty product of a day,
> But the well-ripened fruit of sage delay?"

In treating of the anti-supernaturalism of the age, you remark, —

"Every now and then comes forth some one who says aloud, after this manner: 'I know it, and also every man living knows, by his own eyes and ears, that there has nothing ever been known of the spiritual world, not a word from it even, not a miracle. . . . That anybody knows, or ever has known, more about it than anybody else, is nonsense. I am myself the standard by which you may measure Abraham, the patriarch; and as to his visions, they were merely dreams, such as I have myself. I am the measure of the man Paul. And, you may believe me, as to voice or light from heaven ever having come to him at the time of his conversion, that it was not so. Simply, at that time, he had an attack of vertigo, such as we all know something about. Oh, this glorious clearing of the mind, by which now, in my view, there is nothing higher anywhere than the level of my own experience! Oh, what a comfort it is to have miracles shrink into common earthly things, and to know that nobody has ever seen them, any more than I have!' This would seem to be odd comfort; but there are persons whose needs it would seem to meet."

You anticipate that the child born this year will see in his generation our men of science become reverent believers in the supernatural of the Scriptures. In the facts I here present, you will find some reasons for this opinion; but you will also learn that there are persons who admit the phenomena, but denounce Spiritualism as a "nervous epidemic, based on a gigantic assumption, and propping an ancient superstition," namely, that of individual spirits! So, you see, there will be work still for the Spiritualists, even when the facts are generally accepted, as they are likely to be within the next quarter of a century.

For portions of this volume, I can claim no other merit than that of a compilation. Some of it, however, is compiled from past publications of my own. For what I have adopted from others I have endeavored to give credit, except where the purely narrative form of the matter seemed not to require it, or its source could not be traced.

Of late years, our public journalists have generally been not only tolerant, but liberal, toward investigators into the modern phenomena. In some instances, however, holders of the spiritual hypothesis have been met in a way which would be insulting, if it were possible for a pedantic arrogance to insult. To some of their speculations a learned critic, whose audience is rather select than large, replies with impertinent personal detraction, — his ready relief when confronted by a *non-ego* that will not fit into his pigeon-holes. Then, in all dissentients, he sees at once a base congenital defect; by the easy and polite imputation of which, without the foreign aid of argument, he would explain the audacious phenomenon of a way of thinking, different from his own.

Less amusing, but equally supercilious, is the regard which a somewhat higher authority bestows on the spiritual theory.

At this time, when the extreme materialism that denies a soul and a future life lifts its head with so assured an air, and assumes the tone of scientific certainty, claiming the latest discoveries of physiology and biology in its support, it is not for one who accepts substantially the leading facts of this volume, to be deterred by the ignorant misconceptions and the *de-haut-en-bas* affectations of any would-be dictators in the world of letters and philosophy, from handing on to the next willing hand the torch which may help to illumine the occult places of truth. These enterprising critics aspire to put a stop to the great phenomena of Spiritualism by means of powerful leading articles and ingenious feats of irony. A sneer at gravitation would have about as much effect as their clever writing has had in arresting progress.

Thanks to that Providence who ever proportions our natural supply to our needs, physical and spiritual, the means of opposing to the hypothesis of the extreme materialist an array of positive facts are now widely familiar; so much so indeed, that the time for sarcasms and condescensions towards them is past, except among the loiterers in the present quick march of mind. An accumulation of facts, supported by the most respectable contemporaneous testimony, is presented in this book, such as no free, sincere intelligence can dismiss with contumely or flippant unconcern. To neutralize their force, or to induce any person, who does his own thinking, to be blind to their importance, it will require something more than a sprightly critique or even a crushing "editorial,"— disgraceful as such insensibility may seem to the able editor himself.

What are we to do with these facts? Criticism has done its worst, and they are still irrepressible. May we not hope that what is now the *despair* of Science may one day be its key to much that is obscure in the duplex nature of man; its clew to a complete rational assurance of his immortal destiny?

E. S.

BOSTON, January, 1869.

TABLE OF CONTENTS.

CHAPTER I.

CHAPTER II.

CHAPTER III.

CHAPTER IV.

CHAPTER V.

CHAPTER VI.

CHAPTER VII.

CHAPTER VIII.

CHAPTER XII.

CHAPTER XIII.

CHAPTER XIV.

PLANCHETTE.

CHAPTER I.

WHAT SCIENCE SAYS OF IT.

"Have any of the rulers or of the Pharisees believed on Him?" *St. John.*

THE future historian of the marvellous cannot well avoid some mention of the planchette or "little plank." For his benefit, we will remark that the year 1868 witnessed the appearance of the planchette, in great numbers, in the booksellers' shops of the United States.

Why so sudden a demand for it should have sprung up, nobody could explain. Planchette was nothing new. For twelve or fifteen years it had been common in France, where it received its name. It was simply an improvement on some ruder instrument that had been in use among the original American investigators, of the year 1848, into the rapping and table-tipping phenomena.

The planchette is a little heart-shaped table with three legs, one of which is a pointed lead-pencil, that can be slipped in and out of a socket, and by means of which marks can be made on paper. The other two legs have casters attached, which can be easily moved in any direction. The size of this table is usually about seven inches long and five wide. At the apex of the heart is the socket, lined with rubber, through which the pencil is thrust.

Not improbably, some future antiquarian will discover that this mystic toy was in use long before the days of Pythagoras.

The phenomenon of the tipping tables was known twenty centuries ago.

The form of the planchette is of little consequence, and may be regulated by the caprice of the manufacturer. The instrument is made light, so that the slightest application of force will move it. As for the insulated casters and other "patent" contrivances, they are of no account, except to give novelty to an advertisement.

When the modern rapping phenomena began to be investigated, communications were received by the tedious process of calling over the alphabet, and noting down the letters at which the rap was given. Then, when the movements of tables took place, it was suggested that by arranging a pencil at the foot of a light table, and placing a sheet of paper under it, the intelligent force that was operating might produce written sentences.

The device was tried, and found successful. The table, once set in motion by the passivè influence of a medium, began to trace characters, then words and sentences. This method was finally simplified by substituting little tables, the size of a hand; then small baskets, pasteboard boxes, and finally the flat piece of wood, running on little wheels, and called Planchette.

Here we have the genealogy of the planchette. It is, you see, the direct offspring of the tipping table. The phenomena in which it is made instrumental are, for the most part, the same.

And now, what will Planchette do?

Place it on the smooth wood of a table, and let one person, or two or more, *of a particular organization*, rest the fingers on it lightly, and it will soon begin to move; and this without any *conscious* intent or action on the part of any individual present, as there is reason to believe.

Then, by placing a sheet of white paper under the pencil, it will be found that intelligible sentences will be written out by these movements.

There would be nothing curious in all this, were it not for the character of these sentences in many instances. Expressions wholly foreign to the mental habits of the operators will be found on the paper. Thus, the pious will be made to write profanely; and the profane will be suddenly made instrumental in the production of messages which might do credit to Madame Guyon or to Vincent de Paul. But the results are as various as the idiosyncrasies of individuals.

Frequently, answers to mental questions will be given with a directness that leaves no doubt as to the intelligence of the operating force.

For example: the other day an affectionate father put a mental inquiry, to which the instantaneous reply, under the hands of a child, was "A husband." The question had been, "What does Miss Susan want?" The inquirer then asked what sum he had paid for repairing a certain garment, and the answer was correctly given, "Three dollars and seventy-five cents."

What wonder that the planchette should be getting to be a puzzle and a study to thousands of intelligent inquirers, for whom the great problems of psychology and physiology have a not irrational interest?

It must not be supposed that the "little plank" will be equally communicative under the fingers of all. In the majority of cases it obstinately refuses to move. The failures are very numerous. Probably not more than ten out of a hundred persons in a mixed assemblage would be found, through whom the phenomena would take place; and in these hundred there might possibly be one who would prove a good medium. Such a one will soon discard the planchette as of no use, in the production of phenomena far more extraordinary than any got by its aid.

The editor of the "Boston Journal of Chemistry," Dr. James R. Nichols, with a candor somewhat rare among men of science, remarks (September, 1868), of the phenomena of Planchette and the tipping tables: "The position assumed by a majority of scientific men towards this class of phenomena is that of entire disbelief. They do not separate the physical disturbances, the outward show of force by unseen agencies, from the spiritual interpretation mixed up with, or inseparably connected, as they suppose, with the phenomena. The whole matter is regarded as a sham and a delusion, unworthy of thought or investigation.

"A considerable number, however, have reached a different conclusion. They only direct attention to a single point, and first clear away all the rubbish with which it is incumbered. The great question is, *Whether these alleged physical disturbances actually occur or not, independent of direct and palpable human agency.* Is it mischief, or is it not? Is it delusion, or is it not? These questions they have settled in their own minds; and the conclusion is, that the *phenomena are undeniably real.*

"Not a step further will they go; beyond this all is misty and dark. Many occupy this position who hesitate to admit it, *as there is in scientific circles a peculiar sensitiveness upon the subject;* and odium and disgrace are liable to rest on any one, no matter how high his position may be, who cherishes a belief even in the reality of the physical disturbances. We incline to think the popularity of Planchette may serve to break a link in the chain of prejudice that binds fast honest convictions, and permit a little more freedom in thought and investigation."

If the "little plank" shall accomplish as much as this, it will not have been wholly unproductive of good; but science must put off its dictatorial attitude, and take facts as they present themselves, before it can hope to make any progress in the path of interpretation and induction. The writer adds:—

"We are asked to explain Planchette. To do this would be to explain a most remarkable and extensive class of physical phenomena, beginning with the antics of the little heart-shaped table, and running up through parlor table-tippings, rappings, writing, &c., to the more astounding physical disturbances, noises, and hub-bub, witnessed in so many dwellings in this country and in Europe. There are probably a dozen or more families disturbed in this mysterious manner in the United States at the present moment; but every effort is made at concealment, as but few people of respectability feel that they can bear up under the public odium attached to such proceedings.

"We once, for several hours, listened to the recital of what occurred in the dwelling of Rev. Dr. Phelps, of Stratford, Conn., from the lips of the venerable man himself. We were reduced to the alternative, from listening to his statement, of regarding him, his family, and a wide circle of intelligent friends, as the most egregiously duped, deluded, cheated circle of men and women, the greatest liars and impostors, that ever lived, or of believing in the reality of phenomena, which human reason and human science were incompetent to explain. We felt compelled to adopt the latter alternative.

"Thousands, from the strange and unusual character of the phenomena, have been driven to a belief in their supernatural origin, and the *unfortunate delusion* has spread throughout the civilized world. We incline to think exaggerated views are entertained respecting the competency of scientific men to shed light upon the subject. The *key* to the mystery must be found before any reliable solution is reached. We will not weary the reader with details of what the writer has seen. Suffice it to say, that enough has been observed to lead to the conclusion, that there is *one power, impulse, or force, in nature, regarding the character of which, mankind are totally in the dark.*

"It has proved, so far as our experiments extend, a most difficult and baffling subject to investigate. The nature of this difficulty is illustrated in Planchette. Why cannot one person cause it to move as well as another? Why does it sometimes utterly and ignominiously fail when those are present who have the strongest desire to witness its movements, and when those who are supposed to influence its movements share in this desire? The attempt, or design, to carefully and methodically investigate and study the phenomenon appears to arrest it. In some families, a lady, or a child even, stands in such relations to the instrument as to cause it to move by passing it at a considerable distance. It seems full of impatience to work when such persons are in the house; and it will write, leap, and run about as if impelled by an irresistible impulse. It has occurred, when such a family has invited one or more ladies or gentlemen to an investigation of its performances, and they have come, that the results have been frivolous and unsatisfactory. A calm, philosophical, careful man is not likely to become convinced of the reality of this class of phenomena from such exhibitions.

"Several years ago we invited a friend — a highly distinguished professor in one of our largest Universities — to visit a house where certain extraordinary physical disturbances were alleged to be taking place, apparently in connection with a girl about twelve years of age belonging to the family. In this instance, the power was uncommonly demonstrative, the force being brought to bear upon several articles of furniture, but more particularly upon a parlor-table, which danced and tumbled about the room, entirely regardless of the professor's cool investigations and ingenious tests to 'discover the trick.' This he entirely failed to accomplish. There were no conducting wires, springs, pulleys, or levers to be found; and the little girl and family were manifestly as ignorant of what produced the phenomena as ourselves. A large number of theories were propounded and discussed, not one of which was in the least satisfactory; and the whole affair remains a mystery.

"In explanation, we hear it often stated that it is due to animal magnetism. Of course, such declarations must come from the unlearned or unscientific, as science recognizes no such force or principle in nature as animal magnetism. Some kinds of fishes possess electrical power, and can impart shocks; but then they carry about with them a little arrangement of cells or batteries, which is the source of the electrical force. Human beings are not supposed to possess any such endowment. It is very convenient to have a term to apply in explanation of the phenomena among the crowd, although it may be entirely unmeaning and empirical. Electricity offers no explanation; neither does magnetism, as at present understood. Chemical laws and principles are appealed to in vain for a solution; and as regards 'odic force,' we have not the slightest knowledge of what that is.

"In conclusion, we venture the opinion, that if the phenomena are ever explained, they will be found to be due to a blending of the psychological and physical endowments of the human organization, acting under certain laws *entirely dissimilar to any now known or understood.* Who will produce the key that will unlock the mystery?"

Such are the conclusions of an educated chemist in regard to these phenomena which have been attracting so large a share of public attention for the last twenty years. Instead of being "put down" and "exploded," as we have been repeatedly told, they are now extorting from men of science a reluctant recognition, after years of bitter hostility and denunciation on their part, to which, however, there have been conspicuous exceptions.

Perhaps Dr. Nichols is a little hasty in pronouncing the spiritual hypothesis "an unfortunate delusion." This, to say the least, is as yet an open question. But it is something gained to have the phenomena admitted. We may honestly differ as to their *cause.* Having agreed upon the *facts,* — whatever our theory as to the origin may be, whether we decide in favor of some unknown force not spiritual, or conclude, for want of a better term, that the force is spiritual, — let us not stigmatize those who differ from us on this point, as under "an unfortunate delusion."

The "Scientific American," the principal scientific journal of the United States, has had its attention attracted by Planchette to these despised phenomena; and, in one of its issues, of July, 1868, manfully makes the admission, that "a peculiar class of phenomena have manifested themselves within the last quarter of a century, which seem to indicate that the human body may become the medium for the transmission of force to inert and dead matter, either in obedience to the will of others, or by the action of the nervous power upon the muscular system, in such a way that those through whom or from whom it emanates are totally unconscious of any exercise of volition, or of any muscular movement, as acts of their own wills."

The only expression here that we would modify, is in the

remark that these phenomena have manifested themselves within "the last quarter of a century." The annals of the race are full of them, back to the first dawn of authentic history. They have been interrogated and examined in a different spirit during the last quarter of a century; and that is the only respect in which they can be said to differ essentially from many of the phenomena of witchcraft, necromancy, somnambulism, mesmerism, &c., so long known and disputed.

The same journal remarks, "The spirit with which scientific men have looked upon these phenomena has been unfortunately such as has retarded their solution. Skepticism as to their reality, although they were corroborated by evidence that would be convincing upon any other subject; refusal to investigate, except upon their own conditions; and ridicule not only of the phenomena themselves, but of those who believe in them,—have marked the course of scientific men ever since these manifestations have laid claim to public credence. Such a spirit savors of bigotry. The phenomena of table-tipping, spirit-rapping so called, and the various manifestations which many have claimed to be the effect of other wills acting upon and through the medium of their persons, are exerting an immense influence, good or bad, throughout the civilized world. They should, therefore, be candidly examined; and if they are purely physical phenomena, as has been claimed, they should be referred to their true cause."

Dr. J. Ray, well known in the United States for his works on Medical Jurisprudence, contributes to the "American Journal of Insanity," of October, 1867, a paper, in which he admits that many of the facts of Spiritualism "are susceptible of proof, and are attested by evidence that places them beyond a reasonable doubt." "They indicate," he says, "the existence of agencies, certainly, that have not yet been admitted into the philosophy of the schools. It is to be regretted, that the present tendency is to ignore them entirely, rather than to make them a subject of scientific investigation. It is surprising that physicians, especially, with such well-recognized affections before them as catalepsy, somnambulism, ecstasies, and double consciousness, should jump to the conclusion that all the facts of Spiritualism and animal magnetism are utterly anomalous and impossible."

The first elaborate attempt to give a bad name to these phenomena was the "knee-joint theory" of Drs. Lee and Flint, in 1849. It was declared that the raps were made by a slipping of the knee-joint, and a pamphlet was published to prove it. Innumerable were the denunciations from scientific quarters, that then followed the contemned phenomena. The testimony of thousands of competent witnesses was set aside as worthless. They did not know "how to observe." They had not had the advantage of a "thorough scientific training;" and they could not use their eyes and ears and other senses in a manner to afford any guaranty whatever that they were not under an hallucination.

Such was the language of the late Professor Felton, of Harvard College; and in England, of the celebrated Faraday and of Sir David Brewster.

Every now and then paragraphs would appear in the newspapers, headed "The humbug exploded at last," "Spiritualism exposed," &c. And then we would be told that some "medium" had turned State's evidence, and had revealed how the "tricks" were accomplished. There have been many such mediums, who, having failed to attract attention by genuine phenomena, have hoped to reach the public ear and the public purse by undertaking to disclose how the manifestations were brought about. But, like Balaam, they could not curse whom God would not curse.

All such attempts on the part of deserters have resulted in little that was satisfactory; although they have had a good effect in making investigators more wary, by showing that some of the phenomena of the dark circles, especially the rope-tying experiments, may be adroitly simulated.

In the year 1857, a reward having been offered by the publishers of the "Boston Courier" for the production of certain phenomena, a well-known investigator, Dr. H. F. Gardner, of

Boston, undertook to exhibit them before a committee of professors of Harvard University, composed of Benjamin Peirce, Louis Agassiz, B. A. Gould, and E. N. Horsford, all of them gentlemen of the highest scientific distinction.

The result of the rash experiment may be read in the following report, made by this committee, and dated Cambridge, Mass., June 29, 1857: —

"The Committee award that Dr. Gardner having failed to produce before them an agent or medium who 'communicated a word imparted to the spirits in an adjoining room,' 'who read a word in English written inside a book or folded sheet of paper,'* who answered any question 'which the superior intelligence must be able to answer,' who 'tilted a piano without touching it, or caused a chair to move a foot,' and having failed to exhibit to the committee any phenomenon which, under the widest latitude of interpretation, could be regarded as equivalent to either of these proposed tests, or any phenomenon which required for its production, or in any manner indicated a force which could technically be denominated spiritual, or which was hitherto unknown to science, or a phenomenon of which the cause was not palpable to the committee, is, therefore, not entitled to claim from the 'Boston Courier' the proposed premium of five hundred dollars.

"It is the opinion of the committee, derived from observation, that any connection with spiritualistic circles, so called, corrupts the morals and degrades the intellect. They therefore deem it their solemn duty to warn the community against this contaminating influence, which surely tends to lessen the truth of man and the purity of woman.

* That there has been some progress since the Cambridge professors set down this phenomenon of seeing through opaque substances as one of their impossibilities, may be seen from the "Edinburgh Review" (July, 1868), where that highly conservative authority admits the fact, as follows: "Sleep-walkers have been known, who could not only walk, and perform all ordinary acts in the dark as well as in the light, but who went on writing or reading without interruption, though an opaque substance — a book or a slate — was interposed, and would dot the *i's* and cross the *t's* with unconscious correctness, without any use of their eyes."

"The committee will publish a report of their proceedings, together with the results of additional investigations and other evidence, independent of the special case submitted to them, but bearing upon the subject of this stupendous delusion."

The promised report has not yet (1868) seen the light.

The solemn admonitions of the Cambridge professors against the "stupendous delusion" seem to have been of little effect in repressing inquiry or checking belief in the manifestations; inasmuch as Spiritualists who could then be reckoned by thousands must now be estimated at millions.

Dr. Gardner, on his side, reported that the four learned gentlemen insisted upon prescribing conditions that were fatal to the production of the subtle and evasive phenomena, obtained, independently of the will, from various mediums.

On this occasion, the mediums were Miss Kate Fox, Mrs. Brown (of the Fox family), Mr. J. V. Mansfield, Mr. Kendrick, the Davenport Brothers, and Dr. G. A. Redman, since deceased.

Raps were produced, but the committee were not satisfied that this now common manifestation, which no intelligent person questions, was not some mechanical trick.

At the first sitting, Mr. Agassiz and others refused to sit at the table. The committee had agreed to make the conditions harmonious, as far as they could. Here, at the outset, was a deviation which discomposed the mediums. Mr. Redman, in his "Mystic Hours," states, that, on being importuned to join the circle, Mr. Agassiz averred that he had sworn never to sit in a circle, and meant to adhere to his oath.

Redman significantly asks, "For what was he present? Receiving no manifestations of any consequence, Dr. Gardner and myself retired to an ante-room to inquire of the operating intelligence what next should be done? Scarcely were we seated at the table, when it moved violently; and a communication was written, from right to left, to the purport, that unless all present were willing to receive, and shaped their actions accordingly, nothing could be done. We announced the substance of the message to the party. Mr. Agassiz desired to see the manuscript: it was shown to him; when, without hesitation, he declared *I* had written it, and 'that it was sheer humbug.' . . .

"I now politely invited Mr. Agassiz to join me in the anteroom, and we would try alone; that no doubt he would be more successful. 'Sit with *you?*' said Mr. A., 'No! 'I have resolved to sit with no one. I made up my mind before coming here, that nothing would come of it; and I am only the more convinced it is all deception.' I could say no more."

The experiments with the Brothers Davenport were reserved for the last. The following is the account given by Dr. T. L. Nichols, their English biographer (1865): —

"At the beginning, they were submitted to a cross-examination. The professors exercised their ingenuity in proposing tests. 'Would they submit to be handcuffed?' — 'Yes.' — 'Would they allow men to hold them?' — 'Yes.' A dozen propositions were made, accepted, and then rejected by those who made them. If any test was accepted by the brothers, that was reason enough for not trying it. They were supposed to be prepared for that, so some other must be found. It was of no use to put them to any test to which they were ready and apparently eager to submit.

"At last the ingenious professors fell back upon rope, — their own rope and plenty of it. They brought five hundred feet of new rope, selected for the purpose. They bored the cabinet, set up in one of their own rooms, and to which they had free access, full of holes. They tied the two boys in the most thorough and the most brutal manner. They have, as any one may see, or feel, small wrists, and hands large in proportion, — good, solid hands, which cannot be slipped through a ligature which fits even loosely on the wrists. When they were tied hand and foot, arms, legs, and in every way, and with every kind of complicated knotting, the ropes were drawn through the holes bored in the cabinet, and firmly knotted outside so as to make a network over the boys. After all, the knots were tied with linen thread.

"Professor Peirce then took his place in the cabinet between the two brothers, who could scarcely breathe, so tightly were they secured. As he entered, Professor Agassiz was seen to put something in his hand. The side doors were closed and fastened. The centre door was no sooner shut than the bolt was shot on them inside, and Professor Peirce stretched out both hands to see which of the two firmly bound boys had done it. The phantom hand was shown, the instruments were rattled: the professor felt them about his head and face, and at every movement kept pawing on each side with his hands, to find the boys both bound as firm as ever. Then the mysterious present of Professor Agassiz became apparent. The professor ignited some phosphorus by rubbing it between his hands, and half-suffocated himself and the boys with its fumes in trying to see the trick or the confederate.

"At last, both boys were untied from all the complicated fastenings without and within the cabinet; and the ropes were found twisted around the neck of the watchful Professor Peirce! Well, and what came of it all? Did the professors of Harvard tell what they had seen? Not in the least. To this day they have made no report whatever of the result of their investigation, and are probably, to this day, denouncing it all as humbug, imposture, delusion, &c. What can a man of science do with a fact he cannot account for, except deny it? It is the simplest way of overcoming a difficulty, and avoiding the confession that there is something in the world which he does not understand. Of all men in the world, men of science, and especially scientific professors, are the last to acknowledge that 'there are more things in heaven and earth, than are dreamt of in their philosophy.'"

Thus ended the famous investigation into the phenomena by the Cambridge professors. As appropriate to the subject, we quote the following remarks from a letter by the late Dr. William Gregory, of Edinburgh, a well-known writer and physician: —

"The rational inquirer will soon find that there are innumerable causes of failure, — such as the state of health of the subject; the state of the weather; the state of body or mind of the experimenter; and last, not least, the influence of the bystanders, above all if they be skeptical, prejudiced, or excited by controversy. Whether in magnetism, in clairvoyance, or spiritual manifestations, we who have experimented know these things; but the scientific committees never do, and hence they most unreasonably expect, and indeed some observers as unreasonably promise, uniform success, as the test of truth.

"For many years past I have never accepted any such challenge or test; nor have I made any attempt to convince, in this way, men who are capable of expressing decided opinions previous to their having examined the subject. All that I ever consent to do is to make the trial, on the express understanding that failure proves nothing as to the disputed truth. And even then I reject all dictation as to conditions, as I will only experiment under the conditions presented by Nature, to whom the skeptics have no right to dictate. Our duty is to study Nature as she presents herself, and to take the facts as we find them. We may alter the conditions if we please; but we have no right to insist that the facts shall be produced under such altered conditions as the uneducated judgment may dictate or fancy suggest."

In England the *savans* have been quite as intractable as their American brethren. Mr. Herbert Spencer settles the question on *à priori* grounds, as glibly as if Bacon had not long since shown the absurdity of *à priori* objections to attested facts. Professor W. D. Gunning, of Boston, who lately (1868) had an interview with Mr. Spencer, writes: "In the course of the conversation, he referred to a great naturalist. 'Mr. Spencer,' said I, 'do you know that Mr. ——— has become a Spiritualist?' 'Yes,' he said, 'and I am greatly surprised.' 'Did you ever look at the phenomena?' 'No,' he said, 'I never did. I have settled this question in my own mind on *à priori* grounds'! Now, Herbert Spencer, for whose power as a thinker no one has a higher respect than myself, is writing a great work on psychology; and he settles these questions of odyle, trance, and of obsession — involving the very nature of the soul and its powers — on *à priori* grounds. The *savans* had settled the impossibility of meteoric stones *à priori*. But things settled in that way won't stay settled."

Everybody has heard of the philosopher who refused to look through a microscope, on being told it would unsettle a favorite theory.

In England, the late Professor Faraday committed himself, at an early period, against the *possibility* of the "spiritual" phenomena. His declaration that, "before we proceed to consider any question involving physical principles, we should set out with clear ideas of the naturally possible and impossible," was severely handled by Professor A. De Morgan, the distinguished mathematician.

The whole assumption on which Faraday based his objection to facts of supposed spiritual agency was a misconception. Neither in table-moving nor any other of these phenomena is the *creation of force* implied, as he imagined, but simply the employment of existing forces by invisible intelligences; a view which, whether it be true or false, is at least not manifestly impossible.

The only practical suggestion on this subject by Faraday was the employment of an instrument to test whether the alleged table-movements were, or were not, caused by the muscular pressure of the sitters around it; but, apart from other considerations, this suggestion was at once disposed of by the fact that these movements frequently occurred without the slightest contact with the table.

In 1865, Faraday wrote, "They who say they see these things are not competent witnesses of facts."

The facts which Faraday had so unhesitatingly pronounced to proceed from *involuntary muscular action*, and from no other cause, had become so unruly, that, after some fruitless attempts to right himself, he gave up the subject in disgust.

At length, however, the numerous and circumstantial descriptions given by men of high note, and dinned into his ears, had their effect; and he signified his desire to see for himself. A meeting was accordingly arranged for him by Sir Emerson Tennant, at which Mr. Home, the medium, was to be present. But, lo! the day before the sitting was to have been held, Faraday demanded *a programme of what was to take place!*

In a letter dated June 14, 1861, he says, "It would be a *condescension on my part* to pay any more attention to them [the occult phenomena] now." He asks, "Does Mr. Home wish me to go? Is he willing to investigate as a philosopher, and, as such, to have no concealments, no darkness? . . . Does he make himself responsible for the effects, and identify himself more or less with their cause? Would he be glad if their delusive character were established and exposed; and would he gladly help to expose it, or would he be annoyed and personally offended? Does he consider the effects natural or supernatural? If natural, what are the laws which govern them? or does he think they are not subject to laws? If supernatural, does he suppose them to be miracles or the work of spirits? If the work of spirits, *would an insult to the spirits be considered as an insult to himself?* If the effects are miracles, or the work of spirits, does he admit the utterly contemptible character, both of them and their results, up to the present time, in respect either of yielding information or instruction, or supplying any force or action of the least value to mankind?"

Such was the spirit in which the great scientist approached the subject. And Mr. John Tyndall, the eulogist of Faraday, and hardly his inferior as a man of science, makes the following announcement, under date of May 8, 1868: "I hold myself in readiness to witness and investigate, *in the spirit of the foregoing letter*, such phenomena as Mr. Home may wish to reveal to me during the month of June."

Mr. Tyndall, echoing Faraday, calls upon Mr. Home, as preliminary to the "condescension" of an investigation by Mr. Tyndall, to "admit the utterly contemptible character of the manifestations and their results"!

In his reply to Mr. Tyndall, Mr. Home writes, "I would ask

if this is the tone of a humble student and inquirer, prepared to analyze and ascertain facts, or whether it be not the sign of a mind far gone in prejudging the question at issue? . . . When these matters first engaged public attention, Professor Faraday had, unfortunately, publicly decided they were due to involuntary muscular action; and, as time went on, every development of them which proved the incorrectness of his explanation was received almost as a personal affront by him. . . . Mr. Tyndall says he is ready to witness and investigate in the spirit of Mr. Faraday's letter. From the attitude he takes up, I fully believe it; and as such spirit is not that of logic, nor according to the true scientific method, I will wait until he can approach the subject in a more humble frame of mind."

Mr. Tyndall having introduced into his correspondence the name of Mr. W. M. Wilkinson, that gentleman addressed a letter to the "Pall Mall Gazette," of London, in which, referring to Mr. Tyndall's proposition to retain Faraday's preliminary tests, he says, if conditions are to be the order of the day, he would like answers from Mr. Tyndall on certain preliminaries: —

"If he insist on having an answer to the question whether what he is about to investigate 'can be of any use or value to mankind,' I shall require him to answer whether the *cui bono* has been introduced into science *as a bar to inquiry, and, if so, when?* The history of science is full of instances in which centuries have elapsed between the observation of phenomena and their application to useful purposes. More than a thousand years the world had to wait before the known qualities of conic sections were applied in carpentry; and it was many years before the first experiments in electricity ended in the electric telegraph."

When the Davenport Brothers visited England in 1864, yet another opportunity was offered Professor Faraday to set himself right on the great question. Twenty-four gentlemen met at the house of Mr. Dion Boucicault, and, after a most searching and thorough investigation of the manifestations, unanimously agreed, that they "could arrive at no other conclusion than that there was no trace of trickery in any form, and certainly there were neither confederates nor machinery; and that, so far as their investigations enabled them to form an opinion, the phenomena which had taken place in their presence were not the product of legerdemain."

Professor Faraday was one of those invited; and, on this occasion, he might have had, what on the former occasion he had thought so necessary, a programme; inasmuch as with the Davenports the same general order of phenomena, and in the same sequence, usually take place. This time, however, the demand was not repeated; but, while acknowledging the courteous invitation of the Brothers, he expressed himself "disappointed" with the "manifestations," and therefore left them "in the hands of the professors of legerdemain."

"If," he wrote, "spirit communications, not utterly worthless, should happen to start into activity, I will trust the spirits to find out for themselves *how they can move my attention. I am tired of them.*"

It is barely possible that the spirits did not regard it as a matter of supreme importance to find out how they might move Mr. Faraday's attention.

Among the gentlemen present at the meeting which Faraday declined to attend, and all of whom testified to the good faith with which the experiments were conducted, were Lord Bury; Sir Charles Nicholson; Sir John Gardiner; Rev. E. H. Newenham; Charles Reade, author of "Foul Play," &c.; Rev. W. Ellis; Captain Inglefield, the Arctic explorer; Robert Bell, the author; Robert Chambers, publisher and author; Dr. E. Tyler Smith; and other well-known persons.

The late Sir David Brewster was almost as unfortunate as Faraday in his relations to this perplexing subject. In the early part of 1855, on the invitation of Mr. William Cox, of Jermyn Street, London, Brewster was at a *séance*, where Mr. Home was the medium. The late Lord Brougham, the late Mrs. Trollope, her son Mr. Thomas Trollope, and Mr. Benjamin Coleman, were also present. Seated in a private room, in the open light of day, the party saw, among other extraordinary things, a heavy table rise from the floor; a phenomenon which Faraday had asserted the "undeviating truth" of Newton's law would not permit, and which, to believe in, was proof of "deficiency of judgment."

In a letter to Mr. Coleman (Oct. 9, 1855), Brewster writes, "It is true that, at Mr. Cox's house, Mr. Home, Mr. Cox, Lord Brougham, and myself sat down to a small table, Mr. Home having previously requested us to examine if there was any machinery about his person; an examination, however, which we declined to make. When all our hands were upon the table, noises were heard, rappings in abundance; and finally, when we rose up, the table actually rose, *as appeared to me*, from the ground. This result I do not attempt to explain."

In a conversation afterwards with Mr. Coleman, as the latter testifies, Sir David Brewster scouted the idea of there having been trick or delusion in the matter; but said, "*Spirit is the last thing I will give in to.*"

At first stunned and surprised, Sir David seems to have subsequently been laughed out of his profound impressions, and to have joined the scoffers. He wrote a letter to the "Morning Advertiser," in which he affected to cast ridicule on the subject. Mr. Cox, Mr. Trollope, and Mr. Coleman, each wrote to refute Sir David, and placed him in a position before the public not the most honorable to his consistency and courage.

Goaded by these confutations, he afterwards, in a published address, dismissed the stupendous amount of testimony confirming the phenomena of Spiritualism, with the following words: "All such beliefs *are the result of an imperfect education*, of the want of general knowledge. They are the observations of ill-trained faculties, the cravings of morbid and mystic temperaments that have been suckled on the husks and garbage of literature," &c.

And yet his own letter is in existence, in which he says, "This result [the rising of the table] I do not attempt to explain"!

He sees a table under his nose rise from the ground; does not attempt to explain it; will, on reflection, rather distrust his senses than admit that the fact was other than an *appearance;*

and contents himself with referring the belief of others, seeing a similar thing, and believing that they see it, to "ill-trained faculties"!

If such is to be the last word of science on the subject, is it to be wondered that science has been told not to block the way?

It is pleasant to turn from these instances of arrogance and illiberality on the part of men of science to others of a very different character, from men who are their peers. The late Professor Hare (born, in Philadelphia, 1781, died 1858) was eminent both as a chemist and electrician. For twenty-nine years he was professor of chemistry in the medical school of the university of Pennsylvania. After first maintaining, in a published letter, dated July 27, 1853, the mechanical view of the phenomena taken by Faraday, Dr. Hare instituted a series of scientific tests and experiments, the result of which was, that he became convinced there was a new order of facts which could not be explained on Faraday's theory. Though he may be charged with credulity in accepting much that came by supposed spiritual communications, no one can deny that he investigated the physical phenomena with a rare amount of patience and skill. He was thoroughly convinced of the genuineness of the manifestations.

Dr. Hare had been an unbeliever in deity and in the immortality of the soul. Shortly before his death, he avowed himself not only a Spiritualist, but "a believer in revelation, and in a revelation through Jesus of Nazareth."

Somewhat similar to the experience of Dr. Hare was that of Dr. John Elliotson, F.R.S., president of the Royal Medical and Chirurgical Society of London, and who died July, 1868, at the age of eighty. He had been a fearless investigator of the phenomena of mesmerism, but had rejected some of the higher marvels of somnambulism and clairvoyance. A materialist in his belief, he had written an elaborate treatise denying the existence of an immortal soul. He denounced all mediums as impostors, and regarded as mere delusions the facts claimed by modern Spiritualism.

In the year 1863, being at Dieppe, he was introduced to Mr. D. D. Home, and spent some time in investigating, with the aid of the sons of his friend, Dr. Symes, the phenomena attributed to Spiritualism. The result was, in the language of the "London Morning Post," of Aug. 3, 1868, "that he expressed his conviction of the truth of the phenomena, and became a sincere Christian, whose handbook henceforth was the Bible. Some time after this, he said he had been living all his life in darkness, and had thought there was nothing in existence but the material."

Professor A. De Morgan, of London, as contemporary encyclopædias will show, is of the first eminence as a mathematician. In 1863, a volume of some four hundred pages, from the pen of his wife, appeared, bearing the title, "From Matter to Spirit: the Result of Ten Years' Experience in Spirit Manifestations." The preface is by the professor himself; and in it he plainly admits his belief in the reality of the phenomena, although incredulous as to their spiritual origin.

"I have no acquaintance," he says, "either with P. or Q.; but I feel sure that the decided conviction of all who can see both sides of the shield must be, that it is more likely that P. has seen a ghost than that Q. *knows* he cannot have seen one."

"I am perfectly convinced," he says, "that I have both seen and heard, in a manner which should make unbelief impossible, things *called* spiritual, which cannot be taken by a rational being to be capable of explanation by imposture, coincidence, or mistake. So far I feel the ground firm under me. But when it comes to what is the *cause* of these phenomena, I find I cannot adopt any explanation which has yet been suggested. . . . Spirit or no spirit, there is at least a reading of one mind by something out of that mind. . . .

"The Spiritualists, beyond a doubt, are in the track that has led to all advancement in physical science: their opponents are the representatives of those who have striven against progress. . . .

"There is a higher class of obstructives who, without jest or sarcasm, bring up principles, possibilities, and the nature of things. These most worthy and respectable opponents are, if wrong, to be reckoned the lineal descendants of those who proved the earth could not be round, because the people on the other side would then tumble off. . . .

"I have said that the deluded spirit-rappers are on the right track: they have the spirit and the method of the grand time when those paths were cut through the uncleared forest in which it is now the daily routine to walk. What was that spirit? It was the spirit of universal examination, wholly unchecked by fear of being detected in the examination of nonsense. . . .

"I hold those persons to be incautious who give in at once to the spirit doctrine, and never stop to imagine the possibility of unknown power other than disembodied intelligence. But I am sure that this calling in of the departed spirit, because they do not know what else to fix it on, may be justified by those who do it, upon the example of the philosophers of our own day. . . .

"My state of mind, which refers the whole either to unseen intelligence, or something which man has never had any conception of, proves me to be out of the pale of the Royal Society. . . .

"What I reprobate is, not the wariness which widens and lengthens inquiry, but the assumption which prevents or narrows it; the imposture theory, which frequently infers imposture from the assumed impossibility of the phenomena asserted, and then alleges imposture against the examination of the evidence. . . .

"It is now [1863] twelve or thirteen years since the matter began to be everywhere talked about; during which time there have been many announcements of the total extinction of the spirit-mania. But in several cases, as in Tom Moore's fable, the extinguishers have caught fire."

The late Daniel Davis, of Boston, well known as an electrical instrument-maker, was so thoroughly persuaded of the genuineness of the phenomena, as manifested in raps, movements of tables, &c., that, after exhausting all his practical knowledge in testing them, he offered a reward of a thousand dollars to any one who would produce them independently of any medium. It is needless to say that his offer was never accepted.

Another electrician, Mr. C. F. Varley, at the trial of the celebrated case of Lyon *versus* Home, in London, April, 1868, made oath as follows: "I have been a student of electricity, chemistry, and natural philosophy, for twenty-six years, and a telegraphic engineer by profession, for twenty-one years; and I am the consulting electrician of the Atlantic Telegraph Company, and of the Electric and International Company.

"About nine or ten years ago, having had my attention directed to the subject of Spiritualism, by its spontaneous and unexpected development in my own family, in the form of clairvoyant visions and communications, I determined to test the truth of the alleged physical phenomena, to the best of my ability, and to ascertain, if possible, the nature of the force which produced them.

"Accordingly, about eight years ago, I called on Mr. Home, and stated that I had not yet witnessed any of the physical phenomena, but that I was a scientific man, and wished to investigate them carefully.

"He immediately gave me every facility for the purpose, and desired me to satisfy myself in every possible way; and I have been with him on divers occasions when the phenomena have occurred. I have examined and tested them with him and with others, under conditions of my own choice, under a bright light, and have made the most jealous and searching scrutiny. I have been since then, for seven months, in America, where the subject attracts great attention and study, and where it is cultivated by some of the ablest men; and having experimented with, and compared the forces with electricity and magnetism, and after having applied mechanical and mental tests, I entertained no doubt whatever that the manifestations which I have myself examined were not due to the operation of any of the recognized physical laws of nature, and that there has been present on the occasions above mentioned some intelligence other than that of the medium and observers."

Mr. J. H. Simpson, another English electrician, the inventor of electrical apparatus, including one for printing at a distance by the telegraph, writes (1868) to "Human Nature," a monthly magazine, published in London, as follows: "That the physical effects are, in Mr. Home's case, produced without aid from electricity, ferro-magnetism, or apparatus of any kind, I am well satisfied. They are *bonâ fide*. Of that no one who witnesses them can have a doubt." He adds, however, "I believe that nine-tenths of the phenomena produced through Mr. Home will some day be shown to have nothing to do with aid lent by disembodied spirits."

With regard to the one-tenth remaining, Mr. Simpson suggests no theory as to their origin.

One of the earliest and most accomplished inquirers into these phenomena was William Martin Wilkinson, of 44, Lincoln's Inn Fields, London, Solicitor. He is a brother of the distinguished J. Garth Wilkinson. In his affidavit (1868) in the Home case, already referred to, he says, —

"Such phenomena have been carefully observed by several of the most powerful sovereigns of Europe, and by persons of eminence in the leading professions, and in literature and science, and by practical men of business, under conditions when any thing like fraud or contrivance was impossible. Various theories have been suggested, by way of explanation, connected with the abstrusest problems in biology and metaphysics. My own views on this subject are probably unimportant; but as charges and insinuations are made against me, and the subject of Spiritualism is so misunderstood by the public, I have the right to say, that having had my attention drawn to certain remarkable occurrences, about eighteen years ago, in the house of a relative, and which continued for nearly twelve years, I have since that time occupied a portion of my leisure in inquiring into the subject, and in arranging the various phenomena, and comparing them with historical statements of similar occurrences.

"I have very seldom been at any *séances*, and that not for

many years, having entirely satisfied myself years ago of the truth of most of the phenomena, — that is, of their actual happening; and I have at the same time and for many years formed and constantly expressed the opinion that it was wrong to believe in, or act upon, what might appear to be communications from the unseen, on their own evidence merely. I have invariably inculcated that no such communication should be received as of so much value as if it were told by a friend in this world, inasmuch as you know something of your friend here, and cannot know the identity or origin of the communicant.

"I have frequently referred to the passage in the Old Testament, in which it is said that God sent a lying spirit, and to the directions given us in the New Testament, to try or test the spirits. I have pursued the inquiry under great misrepresentations and obloquy, and I intend to continue it as long as I can; and I believe that the subjects of spiritual visions, trances, ecstasies, prophecies, angelic protection, and diabolic possession, anciently recorded, have already had light thrown upon them, and will have much more. I submit that I have a right to pursue an inquiry into psychological laws, without being subjected to ridicule or abuse, and that the proof of supernatural occurrences is valuable in both a scientific and religious point of view. The mere physical phenomena which the public erroneously fancies to be the whole of Spiritualism, and which, of course, afford room for spurious imitation and fraud, are in my belief the most unimportant part of the subject, and have not for years engaged my attention."

Mr. G. H. Lewes, author of a "History of Philosophy," took part in the Tyndall-Home controversy, in a long letter, (May, 1868), the burden of which was, that men of science were quite right to refuse to waste their valuable time in investigating the pretensions of mediums, and that "had the tone of Faraday's letter been ten times more offensive, it would have been no excuse for Mr. Home's declining his investigation."

Upon which the London "Spiritual Magazine" remarks, "Let the matter as to examination be put on its right ground; namely, that scientific and literary men have the same opportunities of examination of the question as any one else, and that these opportunities are so open, easy, and common, that many millions of people have already examined and satisfied themselves, many of them men of the highest science, learning, and ability. It would be stepping out of the way now to ask any scientific man in. We protest against conceited, and, on this question, profoundly ignorant men, treating it *as some novelty just discovered in a corner*, because they wilfully keep themselves uninformed of it. Spiritualism is a great fact, as much past the mere day of testing and proving as even the law of gravitation. When as many men and women have accepted it as would people Scotland several times over, it is surely ridiculous for such as Professor Tyndall and Mr. Lewes *to ask for some scientific nob to settle the point for them. If he wishes, let the nob do it on his own account, or stand out of the way.*"

We have said enough to show the attitude of science, past and present, with some honorable exceptions, towards the great facts re-asserted by modern Spiritualism.

The reality of the alleged facts, supposed to be spiritual, must be tried by the same tests as any other class of alleged facts; that is, by testimony and experiment. It is believed that they have been so tried. Whether they are caused by spiritual agency is another and separate question; and whether scientific men are the best qualified to decide this point may well admit of doubt. They have no instruments to lay hold of spirits; no chemical tests by which to detect their presence. Retorts and galvanic batteries are here of no avail. A simple woman, like Joan of Arc or the Seeress of Prevorst, may be the true experts here.

The complaint is often made that science has outrun religious belief; that as men have acquired more knowledge, they have become more and more unsettled in their opinions as to their inner life, and in the existence even of the spiritual world. The facts of modern Spiritualism present themselves no sooner than they are needed to meet the want which this tendency has created.

There has long been a vague notion that the discoveries of the age have so far enlightened men, that they are better qualified to form accurate opinions in regard to certain occult phenomena than were the great intellects of antiquity, or of three centuries since. Many persons quietly accept it as something not to be questioned, that such men as Pythagoras, Socrates, Plato, Plutarch, Origen, Augustine, Luther, Baxter, and Mather, were mere children, compared with the college professors of our own day, in their ability to judge of the genuineness of these phenomena. Because science has invented a few chemical tricks, and has made great discoveries in electricity and magnetism, it is assumed that the ancients must have been more easily imposed on than we, in regard to psychological marvels. There is no evidence whatever that such was the fact.

The phenomena on which the ancients based their belief in gods or spirits, and the Blackstones and Glanvils their belief in witchcraft, were, with unimportant exceptions, experiences analogous to those to which thousands of persons are now bearing testimony. Science has not made us one jot better able to dispute the genuineness of these phenomena than were the men of former ages.

"We refuse," says a recent writer, "to believe assertions without evidence: we decline to reject testimony merely because it vouches what is new or strange. It is not in the least impossible, it is not even improbable, it is probable, reasoning from the past it is even certain, that real phenomena should reveal themselves totally inexplicable by any known law, apparently a violation of physical laws, perhaps new principles, pregnant with marvels to which the fictions of the past are prosaic. What Paul ever thought of making the sun paint? What Joseph or Elisha could ever converse with a friend three thousand miles across the ocean? Talk of prophecy! Why, Halley predicted the very day and minute of the appearance of a comet which was myriads of miles away at the time he died! There is no event better authenticated in history than Swedenborg's vision of the great fire of Stockholm. The perfectly ascertained facts of mesmerism, clairvoyance, and electricity, prepare us to wait with reverence and candor upon the unfolding of such phenomena as are attested by Bell, Gully, and Collier; and we shall never be ashamed to own, that as truth in all ages has owed very much more to credulity than to conceited skepticism and self-sufficient prejudice, so there is no phenomenon, however marvellous, that we should *à priori* reject as impossible, in the face of cognate facts, and accumulated, intelligent, and unexceptionable testimony."

It is the duty of Science to wait upon Nature, to reverently listen to what she chooses to tell, and in the way it pleases her to utter it, and deal with the facts that are manifested without ignoring them because others are not manifested. We must be glad to learn her lessons on the conditions she chooses to prescribe, thankful to accept such insight into her arcana as she vouchsafes to grant.

We can readily understand why timid sectarians should denounce the investigation of these phenomena as dangerous; but how, in this nineteenth century, a committee of intelligent gentlemen, renowned for their attainments in their respective departments of science, and whose province it was to consider the subject from a purely scientific stand-point, should think to frighten grown men and women from pursuing an inquiry into certain remarkable facts of nature, by raising the cry of "immorality," and talking of their "solemn duty to warn the community," &c., would indeed be astounding, did we not remember that history repeats itself, and that there were "professors" before the year of grace 1857; even among those wise ones who denounced the researches and revelations of Copernicus and Galileo as immoral, pernicious, diabolical, tending "to lessen the truth of man and the purity of woman."

CHAPTER II.

THE PHENOMENA OF 1847.

"There is in nature nothing interpolated or without connection, as in a bad tragedy." — *Aristotle*.

IN the little village of Hydesville, Wayne County, New York, there stood, in 1847, a small house, which had been occupied by Mr. Michael Weekman. He had been troubled by certain rappings, of which he could give no explanation. But they attracted little attention, and may have had no connection with subsequent developments. It was reserved for the family of Mr. John D. Fox, of Rochester, a respectable farmer, to have their names inseparably associated with the first development of the modern spiritual movement, based on the phenomena now challenging the regards of all thoughtful persons.

Mr. Fox moved into the house the 11th of December, 1847. His family consisted of himself, his wife, and six children; but only the two youngest were staying with them at the time of the manifestations, — Margaret, twelve years old, and Kate, nine years old. The former of these sisters subsequently became the wife of the celebrated Captain Kane, the Arctic explorer.

From the first, the family were disturbed by noises in the house; but these they attributed for a time to rats and mice. In January, 1848, however, the sounds became loud and startling. Knocks, so violent as to produce a tremulous motion in the furniture and floor, were heard. Occasionally there would be a patter of footsteps. The bed-clothes would be pulled off; and Kate would feel a cold hand passed over her face.

Throughout February, and to the middle of March, the disturbances increased. Chairs and the dining-table were moved from their places. Mr. and Mrs. Fox, night after night, with a lighted candle, explored the house, but in vain. While they stood close to the door, raps would be made on it; and on their opening it no one would be found.

On the night of March 31st, having been broken of their rest for several nights previous, they retired to bed earlier than usual, hoping to sleep without disturbance. The sounds, however, were resumed. They occurred near the bed occupied by Kate and Margaret. Kate attempted to imitate the sounds by the snapping of her fingers. There was the same number of raps in response. She then said, "Now do as I do; count one, two, three, four, five, six," at the same time striking her hands together. The same number of raps responded, at similar intervals. The mother of the girls then said, "Count ten!" and ten distinct raps were heard. "Count fifteen!" and that number of sounds followed. She then said, "Tell us the age of Katie" (the youngest daughter), "by rapping one for each year;" and the number of years was rapped correctly. "How many children have I?" There were seven raps in reply. "Ah!" she thought, "it can blunder sometimes." "Try again." Still the number of raps was seven. Mrs. Fox was surprised. "Are they all alive?" she asked. No answer. "How many are dead?" There was a single rap. She had lost one child.

"Do as I do," said Kate Fox. Such was the commencement. "Who can tell," asks Owen, "where the end will be?"

"A Yankee girl, but nine years old, following up, more in sport than earnest, a chance observation, became the instigator of a movement, which, whatever its true character, has had its influence throughout the civilized world. The spark had been several times ignited, — once, at least, two centuries ago; but it had died out each time without effect. It kindled no flame till the middle of the nineteenth century."

The instances here referred to are the answers by knocks elicited by Mr. Mompesson in 1661, and by Glanvil and the Wesley family.

The Rev. Joseph Glanvil, chaplain in ordinary to Charles II., was a writer of great erudition and ability. In his "Sadducismus Triumphatus," written to show that the phenomena of witchcraft were genuine occurrences, he gives an account of Mr. Mompesson's haunted house at Tedworth, where it was observed that, on beating or calling for any tune, it would be exactly answered by drumming. When asked by some one to give three knocks, if it were a certain spirit, it gave three knocks, and no more. Other questions were put, and answered by knocks exactly. Glanvil himself says, that, being told it would imitate noises, he scratched, on the sheet of the bed, five, then seven, then ten times; and it returned exactly the same number of scratches each time.

Melancthon relates that at Oppenheim, in Germany, in 1620, the same experiment of rapping, and having the raps exactly answered by the spirit which haunted a house, was successfully tried; and he tells us that Luther was visited by a spirit who announced his coming by "a rapping at his door."

In the famous Wesley case, the haunting of the house of John Wesley's father, the Parsonage at Epworth, Lincolnshire, in 1716, for a period of two months, the supposed spirit used to imitate Mr. Wesley's knock at the gate. It responded to the Amen at prayers. Emily, one of the daughters, knocked; and it answered her. Mr. Wesley knocked a stick on the joists of the kitchen; and it knocked again, in number of strokes and in loudness exactly replying. When Mrs. Wesley stamped, it knocked in reply.

It is not surprising that John Wesley was a Spiritualist. "With my latest breath," he writes, "will I bear my testimony against giving up to infidels one great proof of the invisible world; I mean that of witchcraft, confirmed by the testimony of all ages."

A writer in the "Encyclopædia Metropolitana" (London, 1861), referring to these and similar phenomena, observes: "It is, to say the least, a remarkable fact, that such occurrences are to be found in the histories of all ages, and, if inquiries are but sincerely made, in the traditions of nearly all living families. The writer can testify to several monitions of this kind portending death; and the authentic records of such things would make a volume."

In the "Life of Frederica Hauffé, the Seeress of Prevorst, by Dr. Justinus Kerner, chief physician at Weinsberg" (who died in 1859), almost every phase of the recent spiritual phenomena is described as pertaining to her experience. To these more than twenty credible witnesses testify. They consisted in repeated knockings, noises in the air, a tramping up and down stairs by day and night, the moving of ponderable articles, &c.

But we must return to the experiences of the Fox family. Startled and somewhat alarmed by the manifestations of intelligence, Mrs. Fox asked if it was a human being that was making the noise, and, if it was, to manifest it by making the same noise. There was no sound. She then said, "If you are a spirit, make two distinct sounds." Two raps were accordingly heard.

The members of the family by this time had all left their beds, and the house was again thoroughly searched, as it had been before, but without discovering any thing that could explain the mystery; and, after a few more questions and responses by raps, the neighbors were called in to assist in tracing the phenomenon to its cause. But the neighbors were no more successful than the family had been, and confessed themselves thoroughly confounded.

For several subsequent days, the village was in a turmoil of excitement; and multitudes visited the house, heard the raps, and interrogated the apparent intelligence which controlled them, but without obtaining any clew to the discovery of the agent, further than its own persistent declaration that it was a spirit.

About three weeks after these occurrences, David, a son of Mr. and Mrs. Fox, went alone into the cellar, where the raps were then being heard, and said, "If you are the spirit of a human being who once lived on the earth, can you rap to the letters that will spell your name? and if so, now rap three times." Three raps were promptly given, and David proceeded to call the alphabet, writing down the letters as they were indicated; and the result was the name, "Charles B. Rosma," a name quite unknown to the family, and which they were afterward unable to trace. The statement was in like manner obtained from the invisible intelligence, that he was the spirit of a peddler, who had been murdered in that house some years previous. According to Mr. David Fox, the floor was subsequently dug up, to the depth of more than five feet, when the remains of a human body were found.

Soon after these occurrences, the family removed to Rochester, at which place the manifestations still accompanied them; and here it was discovered, by the rapping of the letters of the alphabet in the manner before described, that different spirits were apparently using this channel of communication; and that, in short, almost any one, in coming into the presence of the two girls, could get a communication from what purported to be the spirits of his departed friends, the same often being accompanied by tests which satisfied the interrogator as to the spirits' identity.

A new phenomenon was also observed in the frequent moving of tables and other ponderable bodies, without appreciable agency, in the presence of these two girls. These manifestations, growing more and more remarkable, attracted numerous visitors, some from long distances; and the phenomenon began, as it were, to propagate itself, and to be witnessed in other families in Rochester and vicinity; while, as coincident therewith, susceptible persons would sometimes fall into apparent trances, and become clairvoyant, and re-affirm these raps and physical movements to be the production of spirits.

In November, a public meeting was called; and a committee appointed to examine into the phenomena. They reported that they were unable to trace the phenomena to any known mundane agency. Of course, the large majority of persons pronounced the whole thing an imposture; and the public press was against it, almost without an exception. There were stories that the Fox girls produced the sounds by their knees and toe-joints; and one of their relations, a Mrs. Culver, declared that Kate Fox had told her how it was done. If the young and mischief-loving Kate had ever told her so, it must have been in sport; for Mrs. Culver's explanation was soon rejected as not covering the phenomena.

The girls were subjected to the examination of a committee of ladies, who had them divested of their clothes, laid on pillows, and watched; still the sounds took place on walls, doors, tables, ceilings, and at quite a distance from the mediums.

We have before us a letter, received by us, dated Rochester, N.Y., Feb. 16, 1850. It is from the pen of a friend, an English gentleman of high culture, who, at our request, availed himself of a brief stay in Rochester to look into the subject of the mysterious knockings. He made two calls on the Misses Fox, to hear the rappings, and wrote us as follows in regard to them: —

"My opinion of the rappings is that they are human, very human, sinfully human, made to get money by. If really there is a ghost in the matter, then quite certainly he is very fickle, something of a liar, very clumsy, very trifling, and altogether wanting in good taste. It would indeed be painful to me, exceedingly, if I thought that any man on this earth, on dying, had ever turned into such a paltry, contemptible ghost.

"Yet at a distance from this place, as I understand, there are men affecting philosophy, and even a skeptical philosophy, who are ready to believe, and who do believe, that these Rochester knockings are a spirit. A very ridiculous spirit! An untrue ghost, a very pretending ghost! a ghost of no reverence or awe whatever! Indeed, a ghost that is no ghost at all!

"Here, now, I have written what will satisfy your curiosity about this absurd business. My experience in it will be useful to me, in regard to superstition as a disease of the human mind. I have learned something from the errand I have been on. But to me the knockings themselves are not nearly so wonderful as the echoes they make in the city of New York."

The gentleman who wrote this letter subsequently made a very careful investigation of the phenomena, as manifested through the mediumship of the late G. A. Redman, and became fully convinced of their genuineness. He accepted the spiritual hypothesis as to their origin, and is now (1868) — after years of examination and reflection, both in this country and in Europe — an unwavering believer,* and one who can give solid reasons for his belief; thus justifying that remark of Novalis, who says, "To become properly acquainted with a truth, we must first have disbelieved it, and disputed against it."

It was soon found that the marvellous phenomena could be produced through numerous persons of either sex. Mediums for the manifestations began to spring up on all sides; and, as a matter of course, spurious phenomena began to be mixed with the genuine.

The raps were soon superseded by more astonishing and inexplicable experiences. Tables, chairs, and other furniture would be moved about, raised from the floor, and, in some cases, so powerfully, that six full-grown men have been known to be carried about a room on a table, the feet of which did not touch the floor, and which no other person touched. Hand-bells would be rung, guitars floated about the room and played on, tambourines played on, and moved about with marvellous force; and at last spirit-hands would be both seen and felt. Although these phenomena would be generally produced in the dark, there were enough of them produced in the light to satisfy inquirers that the effects were not imaginary or spurious.

Mediums were developed with various powers. There rapidly sprang into notice musical, writing, speaking, drawing, and healing mediums. The press and the pulpit sneered and fulminated; but the work went on with amazing celerity, until millions were not ashamed to admit their belief in the phenomena.

At the rooms of J. Koon, Athens County, Ohio, in February,

* See on page 125, of this volume, an account, from his pen, of certain phenomena for which Miss Lord was the medium, which he witnessed in our company the latter part of the year 1860.

1854, musical instruments were played on with astonishing force. Five witnesses, whose names are published,* testify to seeing spirit-hands on these occasions. They say, "They [the spirits] beat a march on the drum, and carried the tambourine all around over our heads, playing on it the while. They then dropped it on the table, took the triangle from the wall, and carried it all around, as they did the other instruments, for some time. We could only hear the dull sound of the steel; then would peal forth the full ring of the instrument. They let this fall on the table also. After this, they spoke through the trumpet to all, stating that they were glad to see them. Then they went to a gentleman who was playing on the violin, and took it out of his hand up into the air, all around, thrumming the strings, and playing as well as mortals can do. They played on the trumpet, then took the harp, and played on both instruments; and, at the same time, sang with four voices, sounding like female voices, which made the room swell with melody.

"After this, they made their hands visible again, took paper, brought it out on the other table, and commenced writing slowly, when one of the visitors asked them if they could not write faster: the hand then moved so fast we could hardly see it go; but all could hear the pencil move over the paper for some five minutes or so. When done, the spirit took up the trumpet and spoke, saying the communication was for friend Pierce; and, at the same time, the hand came up to him, and gave the paper into his hand. Now, said the spirit, if friend Pierce would put his hand on the table, they would shake hands with him for a testimony to the world, as he could do much good with such a fact while on his spiritual mission. He then put his hand on the table by their request; the hand came up to him, took his fingers, and shook them. Then it went away, but soon came back, patted his hand some minutes, then left again. Now it came back the third time; and, taking his whole hand for some

* D. Hasteler, Pittsburg; A. P. Pierce, Philadelphia; H. F. Partridge, Wheeling, Va.; Lewis Dugdale, farmer, Ohio; Charles C. Stillman, Marion, Ohio.

five minutes, he examined it all over, and found it as natural as a human hand, even to the nails on the fingers. He traced the hand up as far as the wrist, and found nothing any further than that point."

Having, on some forty occasions, witnessed phenomena analogous to these, and quite as remarkable, we cannot doubt that this account is scrupulously true, so far as the *facts* are concerned.

We have already had something to say of the Davenport Brothers. In 1846, their family in Buffalo were disturbed by what they described as "raps, thumps, loud noises, snaps, cracking noises in the dead of night." In 1850, having read in the newspapers of the Rochester knockings, they sat round a table with their hands upon it, and waited further developments. These began by knockings and other noises, and table-tippings. Soon, the alphabet was called into use; then, through the hand of Ira, the elder boy, messages were written by an invisible scribe; and Ira was "floated in the air over the heads of all the people, and from one end of the room to the other, at a height of nine feet from the floor, every person in the room having the opportunity of seeing him as he floated in the air above them." To add to the wonder, William and Elizabeth (a sister) were also upborne; and other marvels took place.

On the fifth evening of their proceedings (according to Dr. Nichols), "in compliance with a direction rapped out on the table by the now familiar method of calling over the alphabet, a pistol was procured, and capped, but not loaded. One of the boys was then directed to go to a vacant corner of the room and fire it. At the instant that he fired, the pistol was taken from his hand; and, by its flash, it was plainly seen, by every person in the room, held by a human figure, looking smilingly at the company. The light and the form vanished together, as when we see a landscape in a flash of lightning; and the pistol fell upon the floor."

Under the directions of supposed spirits, the brothers were tied with all sorts of complicated knots, and then released in an inexplicably brief space of time. The news of what was taking place soon spread; and many eager inquirers came to the house. Such was the curiosity, that public exhibitions were given. The fact that the phenomena were produced for the most part in the dark, naturally gave rise to suspicion and dispute.

In the year 1868, at the Cleveland Convention of Spiritualists, a report was adopted, reprobating what were called "the *dark circle impostors*, who pretend to do physical impossibilities, claiming that spirits do them, while they give no proof of what they assert." "After a diligent and careful investigation of the subject," says the report, "we are irresistibly forced to the conclusion that darkness is not a necessary condition for physical manifestations; but that it *is* a condition assumed and insisted upon by tricksters, having no other use than to afford opportunities for deception."

These remarks are likely to mislead. They appear to be aimed principally at the class of manifestations for which the Davenports are celebrated. That the most remarkable of the manifestations produced in the dark have been produced in the light, will not be disputed; but it does not follow from this that darkness may not sometimes be more favorable to their production. Darkness, it is true, may offer more opportunity for fraud; but a little more trouble taken will soon satisfy the patient investigator. We do not doubt that genuine mediums are often tempted to "help on" the phenomena. But careful observers do not find it difficult to separate the true from the simulated. We must not expect to find all mediums persons of scrupulous integrity.

The Davenports were mere boys when they commenced their exhibitions; and it would not be surprising, if sometimes, impatient of the capriciousness or slowness of "the spirits," they tried to make them "hurry up," by some boyish acts that may properly be denounced as tricks. Indeed, Dr. John F. Gray, of New York, well known to American Spiritualists as identified from the first with the cause, and a thoroughly impartial, independent investigator, wrote us, under date of New York, June 7, 1864, as follows: —

"I have not seen the Davenports this time here; but I entertain no doubt of the genuineness of the manifestations made in their presence. When they were here some years ago, they were detected in making spurious manifestations when the genuine failed."

Surely the testimony of careful, scientific investigators, like Dr. Gray, thoroughly prepared against fraud, and anticipating it, is worth something in a case like this.

Dr. Loomis, professor of chemistry in the Medical College, Georgetown, has given a minute account of his investigations into the phenomena produced through the Davenports. His testimony will carry the more weight with the skeptic, when it is known that he does not admit the spiritual hypothesis, but attributes the thaumaturgic occurrences to some new, unknown force. From Dr. Loomis's report, we extract enough to indicate the thoroughness of his investigation, and the character of his conclusions: —

"At one end of Willard's Hall is a large platform about fifteen feet square, and three feet from the floor, carpeted. At the back side of this platform, resting on three horses, about eighteen inches high, with four legs, each one inch in diameter, was a box or cabinet, in which the phenomena occurred.

"I find the box seems to be made for two purposes only. 1st, to exclude the light; and 2d, to be easily taken apart and packed in a small space for transportation. It is made of black walnut boards, from one-fourth to one-half of an inch in thickness. The boards are mostly united by hooks and hinges, so as to be taken apart and folded up. The box is about seven feet high, six feet wide, and two feet deep; and the back was one inch in front of the brick wall of the building. It has three doors, each two feet wide and as high as the box; so that when the doors are open the entire interior of the box is exposed to the audience.

"Across each end and along the back are boards about ten inches wide, arranged for seats, firmly attached to the box. These are one-half inch walnut boards. At the middle and near the back edge of each of these seats are two half-inch holes, through which ropes may be passed for the purpose of tying the boys firmly to their seats. The entire structure is so light and frail as to utterly preclude the idea that any thing whatever could be concealed within or about its several parts, by which any aid could be given in producing the phenomena witnessed. The top and bottom of the box are of the same thin material, and not tongued and grooved; so that the joints were all open. The floor was carpeted with a loose piece of carpet, which was taken out. The entire inside of the box was literally covered with bruises and dents, from mere scratches to those of an eighth of an inch deep. I examined the box thoroughly in all its parts, and am satisfied that there was nothing concealed in it; nor was there any way by which any thing could be introduced into it to aid in producing the phenomena. The phenomena exhibited may be divided into several classes.

"*a*. Before the performance commenced, the audience chose a committee of three, of which I was one. The other two were strangers to each other and to myself. I never saw them before that evening, have never seen them since, and do not know their names. One of the committee — a stout, muscular man, over six feet in height, professionally a sea-captain, and who remarked to me as he was performing the operation, that he had pinioned many prisoners — tied one of the boys in the following manner: viz., a strong hemp rope was passed three times round the wrist, and tied. It was then passed three times round the other wrist, and tied again, the hands being behind the back. The rope was then passed twice around the body, and tied in front as tightly as possible. Before this was completed, the wrists had commenced swelling, so that the flesh between the cords was even with their outer surface, the hands puffed with blood and quite cool. The circulation was almost completely stopped in the wrists.

"The boy complained of pain, and said, 'Tie the rope as you wish; but I cannot stand it. I am in your power; but you must loosen the rope.' I remarked to the captain that it was cruel to let the rope remain so tight as it was, that security could be gained without being unnecessarily cruel. We examined his wrists again; and the captain decided not to loosen the rope. The whole work of tying the boy was closely watched by me during the entire progress, and thoroughly examined when done; and I must say that very little feeling was exhibited for the boy. No human being could be bound so tightly without suffering excruciating pain. His hands were released in about fifteen minutes. I then examined his wrists carefully. Every fibre of the rope had made its imprint on the wrists. I examined them a second time, one hour and thirty minutes after; and the marks of the rope were plainly visible. He was pinioned as tightly around the body. After being thus tied by his hands, he was seated at one end of the box; and a second rope being passed around his wrists, was drawn both ends through the holes in the seat, and firmly tied underneath. His legs were tied in a similar manner, so that movement of his body was almost impossible. All the knots were a peculiar kind of sailor knots, and entirely beyond reach of the boy's hands or mouth.

"The other Davenport boy was tied in a similar way by another member of the committee. After being tied, I carefully examined every knot, and particularly noticed the method in which he was bound. The knots were all beyond the reach of his hands or mouth. He was as securely bound as the other, the only difference being that the ropes were not as tight around the wrists. This one, as the other, was tied to his seat; the ropes being passed through the holes, and tied underneath to the ropes attached to his legs. Thus fastened, one at one end of the box and one at the other, they were beyond each other's reach.

"Thus far I was perfectly satisfied of three things. 1st, There was in the box no person except the boys, bound as above described; 2d, It was physically impossible for the boys to liberate themselves; 3d, There was introduced into the box nothing whatever besides the boys, and the ropes with which they were bound.

"These being the conditions, the right-hand door was closed;

then the left-hand door; and finally the middle door was closed. At the same time the gas-lights were lowered, so that it was twilight in the room. Within ten seconds, two hands were seen by the committee and by the audience, at an opening near the top of the middle door; and, one minute after, the doors opened of their own accord, and the boy bound so tightly walked out unbound, the ropes lying on the floor, every knot being untied. The other boy had not been released; and a careful examination showed every knot and every rope to be in the precise place in which the committee left it.

"The doors being closed as before, with nothing in the box besides one of the boys, bound as described, hand and foot, with all the knots beyond the reach of his hands or mouth, in less than one minute they opened without visible cause; and the boy walked out unbound, every knot being untied.

"*b*. The box being again carefully examined, and found to contain nothing but the seats, the boys were placed in them unbound, one seated at one end and one at another. Between them on the floor was thrown a large bundle of ropes. The doors were then closed. In less than two minutes, they opened as before; and the boys were bound hand and foot in their seats. The committee examined the knots and the arrangements of the ropes, and declared them more securely bound than when they had tied them themselves. I then made a careful examination of the manner in which they were tied, and found as follows: viz., a rope was tightly passed around each wrist and tied, the hands being behind the back; the ends were then drawn through the holes in the seat, and tied underneath, drawing the hands firmly down on the seat. A second rope was passed several times around both legs and firmly tied, binding the legs together. A third rope was tied to the legs and then fastened to the middle of the back side of the box. A fourth rope was also attached to the legs and drawn backward, and tied to the ropes underneath the seat, which bound the hands. This last rope was so tightened as to take the slack out of the others. Every rope was tight; and no movement of the body could make any rope slacken. They were tied precisely alike. I also examined the precise points where the ropes passed over the wrists, measuring from the processes of the radial, ulnar, and metacarpal bones. I also carefully arranged the ends of the ropes in a peculiar manner. This arrangement was out of reach and out of sight of the boys, and unknown to any one but myself. The examination being ended, the following facts were apparent: 1st, There was no one in the box with the boys; 2d, There was no thing in the box with the boys except the ropes; 3d, It was physically impossible for the boys to have tied themselves, every one of the knots being beyond the reach of their hands or mouths, and the boys being four feet apart; 4th, The time elapsing from the closing of the doors to their opening — less than two minutes by the watch — was altogether too short for any known physical power to have tied the ropes as they were tied.

"*c.* The boys being tied in this manner, one of the committee was requested to shut the doors. He stepped forward, closed the right-hand door, also the left-hand door, and was about closing the middle door, when two hands came out of the box, one of which hit him a severe blow on the right shoulder. The committee-man was partly in the box and felt the blow, but did not know what struck him. He immediately threw open the doors; but nothing could be found but the boys, tied as before. I carefully re-examined the positions of the ropes, and found them as I had left them. The hands were seen by the audience distinctly. The lights had not been turned down; and the hands were seen in the plain gas-light, and remained in sight several seconds. Having satisfied myself of the reality of the hands, having seen the blow given by one of them, which was sufficient to turn the committee-man partly round, I examined them with reference to their position in relation to the boys anatomically considered. The middle door had not been closed, and the committee-man had not left the box; both boys were firmly tied to their seats, and the gas was fully lighted. The hand that appeared to the left of the committee-man might have been, so far as position and anatomical relation were concerned, the right hand of the boy at the left side of the box; but the hand that struck the man could not have belonged to either boy. It was more than four feet from either one, and at least two feet high; and, had either boy been sufficiently near, it must have been a right hand on a left arm.

"*d.* The box was then carefully examined again; and nothing could be found except the boys, bound as described before. There were then placed on the floor, between the boys, a bell, a violin, a guitar, a tambourine, and a trumpet. This being done, the left door was closed, then the right door; and, as the committee-man was closing the middle door, the brass trumpet, weighing about two pounds, jumped up from the floor, struck the top of the box with great force, and fell out on the floor. This took place while the committee-man stood facing the box. The door was wide open; and the committee-man stood partly in the box. The boys were again carefully examined, and found to be tied as at first. I examined the ropes that I had carefully and privately arranged, as before described, and found them as I had left them.

"*e.* The trumpet was placed back, and all the doors closed. Within ten seconds the violin was tuned and began to play; at the same time the guitar, tambourine, and bell began to play, all joining in the same tune. Part of the time the bell was thrust out of the window in the upper part of the middle door, by an arm, and played in sight of the audience. While the music was being made, there were a multitude of raps, both light and heavy, on all parts of the box. The first tune was played and repeated; and a few seconds of comparative quiet followed, broken only by the instruments jumping about the box, and a few raps. Soon a second tune was begun, in which all the instruments joined as before. In the midst of this tune, the doors suddenly opened themselves; and the instruments tumbled about, some one way, some another; and part fell out on the floor. The time between the stopping of the music and the opening of the door was not a single second. I went at once to the box and found both boys bound, hand and foot, as I had left them. I examined the ropes particularly about the wrists, and found them in the precise position in which I had left them, measuring from the processes of the radial, ulnar, and metacarpal bones. I also found the ends of the ropes under the seats, which I had, as previously described, privately arranged in a peculiar manner, in precisely the same position as I had left them."

The late Professor Mapes, well known for his scientific attainments, described an exhibition witnessed by him through the Davenport Boys. These boys permitted themselves to be bound by cords, hand and foot, in any way the operator pleased; and in an instant they were liberated by the supposed spirits. The spirit of one John King claimed to be the chief actor of their band. With this spirit Professor Mapes said *he conversed for half an hour*. The voice was loud and distinct, spoken through a trumpet. He shook hands with him, the spirit giving a most powerful grasp; then taking his hand again, it was increased in size *and covered with hair*. The professor said he went, accompanied only by his friends, among whom were Dr. Warren and Dr. Wilson. They had a jocular sort of evening, into which King entered heartily, and at length played them a trick, for which they were not prepared, and which rather astonished them. Their hats and caps were suddenly whisked from their heads, and replaced in an instant. Turning on the lights, they found each hat and cap was turned inside out; and it took many minutes to replace them. Dr. Warren's gloves, which were in his hat, were also turned completely inside out. This exhibition took place in a large club-room at Buffalo, selected by the professor and his party, having but one place of entrance and exit. The boys sat on an elevated platform at a large table; and this table, in an instant of time, was carried over the heads of the auditors, and deposited at the most distant part of this large room.

It is unnecessary to multiply descriptions of the phenomena. After giving exhibitions in the principal cities of the United States, in the latter part of 1864, the Davenport Brothers went to England. Here their reception was of rather a mixed character. By some they were denounced or mobbed; by others they were treated with the attention which was due to the extraordinary manifestations produced in their presence. They were accompanied by Mr. William Fay, himself the medium for some inexplicable specimens of modern thaumaturgy.

Captain Richard F. Burton, the African traveller, in a letter, dated Nov. 10, 1864, gives a detailed description of a sitting with the brothers at his own lodgings. He says, "Mr. Fay's coat was removed whilst he was securely fastened, hand and foot; and a lucifer match was struck at the same instant, showing us the two gentlemen fast bound, and the coat in the air on its way to the other side of the room. Under precisely similar circumstances, the coat of another gentleman present was placed upon him.

"I have spent a great part of my life in Oriental lands, and have seen there many magicians. Lately, I have been permitted to see and be present at the performances of Messrs. Anderson and Tolmaque. The latter showed, as they profess, clever conjuring; but they do not even attempt what the Messrs. Davenport and Fay succeed in doing, — for instance, the beautiful management of the musical instruments. Finally, I have read and listened to every explanation of the Davenport 'tricks' hitherto placed before the English public; and, believe me, if any thing would make me take that tremendous jump 'from matter to spirit,' it is the utter and complete unreason of the reasons by which the manifestations are explained."

In France the Davenports were well received by the emperor; but a great clamor was raised against them by the press, and the unbelievers generally. Two experts, however, in the art of legerdemain, in Paris, — namely, M. Hamilton, a professor of the art, and M. Rhys, a manufacturer of conjuring implements, — fully exonerated, in published letters, the brothers from all suspicion of trick. M. Rhys is the maker of all the articles used by the well-known Robert Houdin, who is himself the inventor and originator of almost the whole of the tricks performed by the less accomplished jugglers, and who declared some time since that nothing in the magic art could account for the so-called spiritual phenomena which he had witnessed. The letters alluded to were published in the "Gazette des Etrangers" in Paris, on the 27th of September, 1865, and are as follows: —

"MESSRS. DAVENPORT, — Yesterday I had the pleasure of being present at the *séance* you gave; and I came away from it convinced that jealousy alone was the cause of the outcry raised against you. The phenomena produced surpassed my expectations; and your experiments were full of interest for me. I consider it my duty to add that those phenomena are inexplicable, and the more so by such persons as have thought themselves able to guess your supposed secret, and who are, in fact, far indeed from having discovered the truth. HAMILTON."

"MESSRS. DAVENPORT, — I have returned from one of your *séances* quite astonished. Like all other persons, I was admitted to examine your cabinet and instruments. I went through that examination with the greatest care, but failed to discover any thing that could justify legitimate suspicions. From that moment, I felt that the insinuations cast about you were but false and malevolent. I must also declare that, your cabinet being completely isolated, all participation in the manifestation of your phenomena by strangers is absolutely impossible; that the knots are made by persons selected indiscriminately, and that the public has been admitted to watch them; and I shall add that, under these conditions, no one has ever yet produced any thing similar to the phenomena I witnessed. RHYS."

The Davenports met with great success in Belgium, where the press treated them with unwonted candor and fairness. In St. Petersburg, they gave private *séances* before the emperor and the nobility, and were received with much attention.

On the 11th of April, 1868, they re-appeared in London, and drew a crowded audience. Their powers had not diminished. A gentleman who was present writes, "In the cabinet exhibition, hands, life-like in form and texture, were frequently seen before the doors were closed; and from the aperture two long, naked, femininely formed arms, and also a group of not less than five hands of various sizes, were protruded at the same instant."

Mr. Benjamin Coleman, of London, a gentleman personally known to us, and who has been an indefatigable investigator of the phenomena for many years, writes, under date of May, 1868, of the Messrs. Davenport and Mr. William Fay, "I desire to convey to those of my friends in America, who introduced them to me, the assurance of my conviction that the Brothers' mission to Europe has been of great service to Spiritualism. . . . I have had no reason whatever to change my opinion of the genuine and marvellous character of their mediumship, which is entirely free from the imputation of trickery and bad faith of any kind."

Mr. Robert Cooper, of London, a sincere and disinterested investigator, and who accompanied the Davenports to Ireland, Scotland, Belgium, and Germany, solely in the pursuit of truth, writes as follows: "I have been intimately associated with the Davenports for seven months. I have witnessed the manifestations under a variety of circumstances, — in the dark and in the light, in public and in private, — and I have never seen any indication whatever of the slightest approach to trickery. On the contrary, I have seen much to convince me of the absence of any thing of the kind. For instance, I have seen lights struck, contrary to regulations, when the instruments were sounding and floating in the air; but no one was discovered out of his place, the only result being the falling of the guitars to the ground.

"At Brussels, at a *séance* before the first literary society of the town, blue paint was placed on the instruments unknown to any of us; but, though the instruments were all played on, no trace of the paint was found on the hands of the brothers. At Antwerp, at the conclusion of the cabinet *séance*, a gentleman exhibited his hand covered with some black composition of a greasy nature. He said he had caught hold of the hands that appeared at the cabinet window, and fully expected, when the

Davenports came from the cabinet, to find their hands blackened, but, to his great surprise, such was not the case. I have also known black composition placed on the hands of the brothers during the dark *séance*, with the idea that the instruments would show traces of the pigment; but such was not the case. None of our party knew of these experiments being made till the termination of the *séances*."

Mr. Cooper has heard the "spirits" speak in an audible voice, and has held long conversations with them. He says, "It is obviously impossible for any one to be with the Davenports, as I have been, and not discover fraud, if any existed. I could multiply proofs in favor of the genuineness of these manifestations. If they are not a reality, then all creation is a myth, and our senses are nothing worth."

The occurrences in the family of the Rev. Dr. Phelps, of Stratford, Conn., which took place not long after the manifestations through the Fox family (1848-9), are of a character strictly analogous to those that were established as true, so far as human testimony can establish any thing, in the days of witchcraft.

For seven months, the phenomena were of the most unaccountable character. We took the pains to write to Dr. Phelps at the time, and have from him a letter confirming the facts in every particular. On returning one day from church, the family found the doors of rooms, which had been carefully locked, all thrown open; and the furniture tossed about in the utmost confusion. In one room were from eight to ten figures formed with articles of clothing, and arranged with singular skill. They were all kneeling, and each with an open Bible before it, as if in mockery of their own church-going. Nothing was missing. The family locked the door of this room, but only to find, on opening it again, the number of figures increased, and that with articles of dress which three minutes before they had seen in other parts of the house. Heavy tables were lifted up and let down again, strange noises were heard; and a boy of eleven years of age was lifted up and carried across the room. His clothes were carried away, and only discovered after a long and patient search. He was sent from home to a distant school, but had to be recalled, as his clothes there were cut to pieces repeatedly in a most extraordinary manner. The panes in the windows used to fly to pieces as Dr. Phelps and others stood looking at them.

In his letter, Dr. Phelps writes, "I have seen things in motion above a thousand times; and, in most cases, where no visible power existed by which the motion could be produced. There have been broken from my windows more than seventy-one panes of glass, more than thirty of which I have seen broken before my own eyes."

About the year 1850, the Hon. James F. Simmons, of Rhode Island, a well-known member of the United-States Senate, was the witness of some remarkable phenomena. In the autumn of 1852, Mr. Horace Greeley,* editor of the "New-York Tribune," received a letter which he published in his paper, with the following introduction: "The writer has received the following letter from Mrs. Sarah H. Whitman, in reply to one of inquiry from him as to her own experience in 'Spiritualism,' and especially with regard to a remarkable 'experience,' currently reported as having occurred to Hon. James F. Simmons, late United-States Senator from Rhode Island, and widely known as one of the keenest and clearest observers, most unlikely to be

* In his "Recollections of a Busy Life" (1868), Mr. Greeley admits that "the jugglery hypothesis utterly fails to account for occurrences which I have personally witnessed," and that "certain developments strongly indicate that they do proceed from departed spirits." But he complains that nothing of any value is obtained by the investigation; that the spirits "did not help to fish up the Atlantic cable nor find Sir John Franklin;" that Spiritualism has not made the body of believers "better men and women." Much the same kind of objection might be brought against the Copernican theory of the universe. Mr. Greeley admits that the phenomena may enable us "to answer with more confidence that old momentous question, If a man die, shall he live again?" Did it never occur to Mr. G. that *this is something;* a trifle, perhaps, compared with fishing up an old cable, but still something? We fear that Mr. G.'s life has been too "busy" to enable him to give to these matters the reflection they require.

the dupe of mystery or the slave of hallucination. Mrs. Whitman's social and intellectual eminence are not so widely known; but there are very many who know that her statement needs no confirmation whatever." Here is her letter:—

"DEAR SIR,—I have had no conversation with Mr. Simmons on the subject of your note until to-day. I took an early opportunity of acquainting him with its contents; and this morning he called on me to say that he was perfectly willing to impart to you the particulars of his experience in relation to the mysterious writing *performed under his very eyes, in broad daylight, by an invisible agent.*

"In the fall of 1850, several messages were telegraphed to Mrs. Simmons through the electric sounds, purporting to come from her step-son, James D. Simmons, who died some weeks before in California. The messages were calculated to stimulate curiosity, and lead to an observation of the phenomena. Mrs. Simmons, having heard that messages in the handwriting of deceased persons were sometimes written through the same medium, asked if her son would give her this evidence. She was informed, through the sounds, that the attempt should be made, and was directed to place a slip of paper in a certain drawer at the house of the medium, and to lay beside it her own pencil, which had been given her by the deceased. Weeks passed; and, although frequent inquiries were made, no writing was found on the paper.

"Mrs. Simmons happening to call at the house one day, accompanied by her husband, made the usual inquiry and received the usual answer. The drawer had been opened not two hours before, and nothing was seen in it but the pencil lying on the blank paper. At the suggestion of Mrs. Simmons, however, another investigation was made; and on the paper were found a few pencil lines, resembling the handwriting of the deceased, but not so closely as to satisfy the mother's doubts. Mrs. Simmons handed the paper to her husband: he thought there was a slight resemblance, but would probably not have remarked it had the writing been casually presented to him. Had the *signature* been given him, he should at once have decided on the resemblance. He proposed, if the spirit of his son were indeed present, as alphabetical communications received through the sounds affirmed him to be, that he should, *then* and *there*, affix his signature to the suspicious document.

"In order to facilitate the operation, Mrs. Simmons placed the closed points of a pair of scissors in the hand of the medium and dropped her pencil through one of the rings or bows, the paper being placed beneath. The hand presently began to tremble; and it was with difficulty it could retain its hold of the scissors. Mr. Simmons then took the scissors into his own hand, and dropped the pencil through the ring. It could not readily be sustained in this position. After a few moments, however, it stood as if firmly poised and perfectly still. *It then began slowly to move. Mr. Simmons saw the letters traced beneath his eyes. The words, James D. Simmons, were distinctly and deliberately written; and the handwriting was a fac-simile of his son's signature.*

"But what Mr. Simmons regards as the most astonishing part of this seeming miracle is yet to be told. Bending down to scrutinize the writing more closely, he observed, just as the last word was finished, that the top of the pencil leaned to the right. He thought it was about to slide through the ring; but, to his infinite surprise, *he saw the point slide slowly back along the word 'Simmons,' till it rested over the letter i, when it imprinted a dot.* This was a punctilio utterly unthought of by him. He had not noticed the omission, and was therefore entirely unprepared for the amendment. He suggested the experiment, and he thinks it had kept pace only with his will or desire. But how will those who deny the agency of disembodied spirits in these marvels, ascribing all to the unassisted powers of the human will, or to the blind action of electricity, — how will they dispose of this last significant and curious fact?

"The only peculiarity observable in the writing was that the lines seemed sometimes slightly broken, as if the pencil had been lifted, then set down again.

"One other circumstance I am permitted to note, which is not readily to be accounted for on any other than spiritual agency. Mr. Simmons, who received no particulars of his son's death until several months after his decease, proposing to send for his remains, questioned the spirit as to the manner in which the body had been disposed of, and received a very minute and circumstantial account of the means which had been resorted to for its preservation, it being at the time unburied. Improbable as some of these statements seemed, they were, after an interval of four months, confirmed as literally true by a gentleman then recently returned from California, who was with young Simmons at the period of his death. Intending soon to return to California, he called on Mr. Simmons to learn his wishes in relation to the final disposition of his son's remains. The above particulars I took down in writing, by the permission of Mr. Simmons, during his relation of the facts."

In the "British Standard," of Aug. 14, 1863, Dr. Campbell remarks of these and similar phenomena, "The conclusion of the whole matter is this: we believe in the existence of angels and of devils, in the existence of the spirits of men both good and bad; we believe that all are capable of acting in their disembodied state on the minds of men still in the flesh; we believe in the possibility of intercourse between man and these disembodied intelligences, whether good or bad; we believe, on the authority of Scripture, that spirits are capable of entering human bodies, of speaking through them and acting in them; and hence we believe in the possibility of spirits operating on matter in the way of rapping out the letters of the alphabet, or in the way of writing with the pencil. We see nothing in Scripture or in the nature of the case that militates against these conclusions. All that we require is proof, indubitable, sensible proof, from our own eyes and ears. On that condition, we at once give full credence."

To the question often put by the inconsiderate, in regard to the phenomena, "What good have they all done? — What's the use of them all?" Dr. Campbell replies, "We are sometimes met with the question *cui bono?* We deny our obligation, as a condition of rational faith, to prove the *cui bono*. It may exist where we see it not, and have important ends to accomplish with which we are unacquainted."

Dr. Campbell relates some singular occurrences in his own experience, and concludes, "Explanation of such phenomena we have none to offer; but we stand by the facts as here stated."

It is astonishing how often this *cui bono* interrogatory is put by persons who ought to see how a little reflection would silence them. Once when Dr. Franklin was asked in regard to some discovery, "What's the use of it?" he retorted by saying, "What's the use of a new-born baby?" And as for that matter, it might be asked, "What's the use of any thing?"

"I do not see that people have been made better men and women by these things," says a popular editor, in reference to the spiritual phenomena, the genuineness of which he admits. And by a superficial thinker, the remark will be taken as sound common sense, and as settling the whole question of their importance.

But you will observe that precisely the same objection might be brought against the discoveries of Copernicus, of Newton, and even of Morse and Fulton. Have people been made better men and women by the theory of gravitation, by the steamboat, the railroad, and the electric telegraph? Indeed have the printing-press and the photographic art been exclusively servants in the cause of morality? Such questions, if not always put in the spirit of "the loud laugh that speaks the vacant mind," certainly indicate rather a narrow view of the great facts of existence.

CHAPTER III.

MANIFESTATIONS THROUGH MISS KATE FOX.

"The spiritual world
Lies all about us, and its avenues
Are open to the unseen feet of phantoms
That come and go, and we perceive them not,
Save by their influence, or when at times
A most mysterious Providence permits them
To manifest themselves to mortal eyes." — *Longfellow.*

WE come now to a narrative of phenomena so remarkable that they will probably excite many an exclamation of incredulity, although the authority on which they rest is above suspicion.

We have already had occasion to quote the testimony of Dr. John F. Gray, of New York. He was one of the earliest and most persevering investigators of the Hydesville phenomena. To us he has been personally known for more than a quarter of a century; and he is well known to a large circle of intelligent patients in the great city where he has had a lucrative professional practice until, a few years ago, he retired from active occupation.

Dr. Gray accepts the spiritual hypothesis as the only one covering all the phenomena he has witnessed. His reasons for believing that spirits communicate with men in the body are thus stated in a succinct summary of the results that have come to his knowledge during the last twenty years: —

"I. *Phenomena of a physical nature* not referable to the laws of physical relation; such as the moving of ponderable bodies, independent of earthly mechanics; the production of a great variety of sounds, also independent of any known or conceivable mechanical apparatus; the production of lights of various colors, sizes, shapes, degrees of brilliancy, and duration of incandescence, in every case without the presence of any chemical agents or apparatus known to or usable by man; and, lastly, *the reproduction of living material bodies, through which extemporaneous, but real and tangible physical organizations, the spirits have re-appeared to their friends on earth,* expressing their peculiarities of physical form and movement, and likewise their peculiar and distinctive modes of apprehension, feeling, and intellection. Through these temporarily organized effigies of their former earth-bodies, they have (as I know from several instances of recent date) spoken to and sung with their relatives here, and have given many other equally palpable proofs of their ability to reconstruct and inhabit a physical form.

"II. *Phenomena of a mental nature* not referable to earthly volition and intelligence; such as the contrivance and production of the physical phenomena above cited; the production of writings in various ancient and modern languages, wholly unknown to those in whose presence they have been executed; the utterance of prophecy; the narration of events, and the recital of mental facts that are transpiring in distant places, often across broad oceans; the improvisation and incredibly rapid production of symbolic drawings and elaborate pictures by persons not versed in the pictorial art, and unable to explain the symbols they have executed and combined in such a way as to convey a good lesson of life, or renew a long-buried personal reminiscence; lastly, the felicitous and accurate impersonation of persons long departed this life, and who were wholly unknown to and unheard of by the personators.

"The philosophy of spirit-intercourse sheds a mellow light over human history and human science. It founds a positive psychology, and teaches where to look for wellsprings of invention and progress; and it reconciles us to the hard ministry of sin and sorrow, of ignorance and suffering."

In 1860, Mr. C. F. Livermore, an opulent and well-known banker of New York (formerly of the firm of Livermore &

Clewes, but now retired from business), lost his wife, to whom he had been much attached, and who had been attended during her last illness by Dr. John F. Gray, an old friend of the husband. Mr. Livermore, an inveterate skeptic, was now induced by Dr. G. to call on Miss Kate Fox, the young woman through whose quick-wittedness these rapping phenomena were originally interrogated and developed at Hydesville.

In February, 1861, Mr. L. accordingly had a sitting with Miss Fox; and the result was an entire change in his views concerning life and death.

At a small gathering of inquirers at which our friend, Mr. Benjamin Coleman, of London, was present, in 1861, Dr. Gray read the following extraordinary account by Mr. Livermore of the manifestations which Mr. L. obtained through Miss Fox. After describing the precautions he took to prevent the possibility of deception, Mr. L. proceeds as follows : —

"The lights being extinguished, footsteps were heard as of persons walking in their stocking-feet, accompanied by the rustling sound of a silk dress. It was then rapped out by the alphabet, 'My dear, I am here in form; do not speak.' A globular light rose up from the floor behind me; and, as it became brighter, a face, surmounted by a crown, was distinctly seen by the medium and myself. Next, the head appeared, as if covered with a white veil: this was withdrawn after the figure had risen some feet higher; *and I recognized unmistakably the full head and face of my wife*, surrounded by a semi-circle of light about eighteen inches in diameter. The recognition was complete, derived alike from the features and her natural expression. The globe of light was then raised, and a female hand held before it was distinctly visible. Each of these manifestations was repeated several times, as if to leave no doubt in our minds. Now the figure, coming lower down and turning its head, displayed, falling over the globe of light, *long flowing hair*, which, even in its shade of color, appeared like the natural tresses of my wife, and like hers was unusually luxuriant. This whole mass of hair was whisked in our faces many times, conveying the same sensations as if it had been *actually human natural hair*. This also was frequently repeated, and the hair shown to us in a variety of ways. The light and the rustling sound then passed round the table and approached me, and what seemed to the touch a skirt of muslin was thrown over my head, and a hand was felt as if holding it there. A whisper was now heard; and the words, 'Sing, sing,' were audibly pronounced. I hummed an air, and asked, 'Do you like that?' 'Yes, yes,' was plainly spoken in a whisper; and in both cases I recognized distinctly the voice of my wife, to which I had become sensitively familiarized during her last illness, when she had become too weak to talk aloud."

At another sitting, a few days after, the same precautions and conditions being observed, the following phenomena were witnessed: —

"The table was lifted from the floor, the door violently shaken, the window-sash raised and shut several times; and, in fact, every thing movable in the room seemed in motion.

"Questions were replied to by loud knocks on the door, on the window, ceiling, table, everywhere; all being the work of several powerful spirits, who were present, and whose presence was necessary, as it was afterwards explained, to support or induce the manifestations of a more beautiful and interesting character.

"An illuminated substance, like gauze, rose from the floor behind us, accompanied by a rustling sound, like that of a silk dress. The previously described electrical rattle became very loud and vigorous. The figure of a female passed round the table, and, approaching us, touched me. The gauzy substance was shaped as though covering a human head, and seemed as if drawn down tight at the neck. Upon close examination, as it approached near me a second time, it changed its form, and now seemed in folds over a melon-shaped oblong, concave on one side; and in this cavity there appeared an intensified brilliant light. By raps, I was requested to look beyond the light. I looked as directed, and saw the appearance of a human eye. Again receding with the rattle, the light became still brighter; and then, re-approaching, the gauze, which had changed in form, was grasped by a naturally-formed female hand; and unfolding, revealed to me, with a thrill of indescribable happiness, *the upper half of the face of my wife*, the eyes, forehead, and expression in perfection. The moment the emotion of recognition had passed into my mind, it was acknowledged by a succession of quick raps.

"The figure disappeared and re-appeared several times, the recognition becoming each time more nearly perfect, with an expression of calm and beautiful serenity. I asked her to kiss me if she could; and, to my great astonishment and delight, an arm was placed around my neck, and a real palpable kiss was implanted on my lips, through something like fine muslin. A head was laid upon mine, the hair falling luxuriantly down my face. The kiss was frequently repeated, and was audible in every part of the room. The light then moved to a point about midway between us and the wall, which was distant about ten feet. The rattling increased in vigor; and the light, gradually illuminating that side of the room, brought out in perfection an entire female figure facing the wall, and holding the light in her outstretched hand, shaking it at intervals, as the light grew dim. My name and her name were repeated in a loud whisper; and among other things which occurred during this remarkable sitting, the figure at the close stood before *the mirror, and was reflected therein*."

The incidents of another evening were thus described: "The lights and electrical rattle were as strong as on the previous occasions. Hands were placed upon my forehead, a head placed upon mine, the hair, as before, falling down my face into my hand. I grasped it, and found it positively and unmistakably human hair. It was afterwards whisked playfully at me, creating as much wind as an ordinary fan. The spiritual robe was then dropped over my head and face, as real and material in substance as cotton or muslin of a very fine texture. At one time, the globe of light extended to about two feet in diameter. At last, it was shaken with another sharp rattle; and, shining brightly, revealed again the full head and face of my wife, every feature in perfection, but spiritualized in shadowy beauty such as no imagination can conceive, or pen describe. In her hair, just above the left temple, was a single white rose, the hair being arranged with great care. The next appearance, after a brief interval, revealed the same face, with a pink rose instead of a white one. The whole head and face were shown to us, at least twenty times during the sitting, and each time was recognized by me, the perfection of the recognition being in proportion to the brilliancy of the light. During the whole of these manifestations, cards of a large size, provided by myself, were placed on the floor, with a pencil; and long messages were found to have been written upon them," &c.

Dr. Gray, in conclusion, said, "These manifestations could not have been produced by human means; and if you admit the competency of the witness, of which, from my knowledge of him, I have no doubt, they are, in my opinion, conclusive evidence of spirit identity."

Several persons in the assembly rose to ask questions of Dr. Gray, respecting this very startling narrative; and onc gentleman said, he really could not, though a believer in Spiritualism, receive such statements without great misgivings of delusion being mixed up with them. "Now," he said, "I put it to you, Dr. Gray, Do you believe that such things can and did occur?" Dr. Gray replied very calmly, "Yes, my friend; I believe as implicitly every word of those narratives as I do in my own existence."

Previous to leaving New York, Mr. Coleman made a special visit to Miss Kate Fox, the medium for these wonders; and she fully corroborated all that Mr. Livermore had told him.

Of Miss Kate Fox, Dr. Gray writes: "She has been intimately known to my wife and me from the time she was a very young girl; that is to say, from 1850 to this date [1861]. At that early day in the history of the manifestations, she was frequently a visitor in my family; and then, through that child alone, without the possibility of trick from collusion with others, or, I may truly add, of imposture of any kind, all the various phenomena recorded by friend L., except the reproduction of visible human forms, were witnessed by Mrs. Gray and myself, and many other relatives and friends of our family. Among these I may mention, as frequent, attentive, and very able observers, the late Dr. Gerald Hull,* my brother-in-law; and Dr. Warner, my son-in-law. Miss Fox is a young lady of good education, and of an entirely blameless life and character."

Of Mr. Livermore, Dr. Gray says, "Besides his general character for veracity and probity, Mr. L. is a competent witness to the important facts he narrates, because he is not in any degree subject to the illusions and hallucinations which may be supposed to attach to the trance or ecstatic condition. I have known him from his very early manhood, and am his medical adviser. He is less liable to be misled by errors of his organs of sense than almost any man of my large circle of patients and acquaintance."

Mr. Livermore is of opinion that the electrical conditions, both of the atmosphere and of the persons receiving manifestations, are even more important and subtle than mental conditions. He says of himself, "My condition has always been highly electrical. I find no difficulty in lighting gas by applying the end of my finger to the burner, after having excited the electricity of my system, by friction of my feet on the carpet. This, however, is not an uncommon occurrence here; though I have repeatedly tried it in England without success."

"You ask if I believe all the manifestations are from one spirit. Most certainly not; for it has been repeatedly explained, and I think proved, that the spirit made itself visible to me through the powerful aid of other spirits."

Cards were written on, in a very neat small hand, exactly like the natural handwriting of "Estelle," the wife, when in the flesh. *Fac-similes* of two of these cards, the one purporting to

* Dr. Hull, who was universally respected and beloved, both as a physician and a friend, has often corroborated to us, personally, the most remarkable of the facts to which Dr. G. bears witness.

be written by the spirit of Mr. L.'s wife, and the other by the "spirit of Benjamin Franklin," are published in the "London Spiritual Magazine," of November, 1861.

A spirit, assuming to be Franklin, was afterwards repeatedly visible. In a letter, dated Nov. 23, 1861, Mr. Livermore writes: "I now aver, that no doubt of the identity* of the spirit longer remains upon my mind. His appearance [the same on several occasions] corresponds with the original portrait of the philosopher; the difference being simply that which one would expect to find between a painting and a face replete with life and expression. His presence was a wonderful and startling reality, seated in the chair opposite me at the table, vividly visible, and even to each article of dress. There could be no mistake."

This *eidolon* of Franklin, as well as that of Estelle, was afterwards seen by the brother-in-law of Mr. Livermore, and by Dr. Gray. The following are extracts, taken somewhat at random from Mr. Livermore's spiritual diary, of 1861–1863:—

"*Aug.* 18, 1861, 8 P.M.—Present, the medium and myself. Atmosphere heavy and warm. Carefully examined the room, locked the door, took the key, and made all secure. Sat in quiet half an hour, when a spherical oblong light, enveloped in folds, rose from the floor to our foreheads, and rested upon the table in front. By raps, 'Notice how noiselessly we come.' Heretofore the light had generally appeared after a succession of startling sounds and movements of movable objects; but in the present instance all was quiet. From this time, 8.30, till 11.30, the light was constantly visible, but in different forms. It remained upon the table a full half-hour, the size and shape

* If spirits have the power, attributed to them by many seers, of assuming any appearance at will, it is obvious that some high spiritual sense must be developed in us before we can reasonably be sure of the *identity* of any spirit, even though it come bearing the exact resemblance of the person it may claim to be. We think, therefore, that the fact that the spirit, described by Mr. L., bore the aspect of Franklin, and called itself Franklin, is no sufficient reason for dismissing all doubts as to its identity. It may be, that we must be in a spiritual state before we can really be wisely confident of the identity of any spirit.

of a large melon. As during this time it was passive, I asked if it could rise, whereupon it immediately brightened, flashed out, and rising, seemed a living, breathing substance. By raps, 'This is our most important meeting; for it brings to our circle two powerful spirits great and good.' The light became gradually more powerful, and so brilliant upon the side opposite us as to illuminate that part of the room. It now rose from the table, resting upon my head and shoulder; the drapery in the mean time touching and falling upon our faces, with a peculiar scent of violets. After resting upon, and pressing my head and shoulder *with the weight of a living head*, it descended to the floor. I was now satisfied that the purpose of this meeting was some other than the appearance of the spirit of my wife. The light now rose with increased brilliancy, showing a head upon which was a white cap surrounded by a frill. Seeing no face, I asked what this meant. The reply was by raps, '*As when I was ill.*' This was correct; for it was to all appearances the peculiar cap worn by my wife during her last illness. This having passed away, the light appeared again very brilliantly, showing a crown composed apparently of oak-leaves and flowers, a very, *very* beautiful manifestation. I had brought with me on this occasion some new cards of a larger size, different from any before used, and had placed upon two of them private marks. These I put upon a book on the table. In a few minutes they were taken from the book, and one of them appeared near the floor, suspended three or four inches from the carpet, — I could not judge accurately; but the light brightly showed the centre card and radiated from each side to a distance of some three or four inches; or, in other words, the card was the centre of a circle of spirit-light of a foot in diameter; while an imperfectly-shaped hand, holding my small silver pencil, was placed upon the card and moved quietly across from left to right, as though writing, and when finishing a line, it moved quickly back to recommence another. We were not permitted to look at this very long at a time, as our steady gaze disturbed the operating forces; but it remained more or less visible for nearly an hour. The full formed hand was seen only a portion of the time; but, during all this time, a dark substance, rather smaller than the natural hand, held the pencil, and continued to write. One side of the card being finished, *we saw it reversed and the other page commenced.* This is satisfactory evidence of the reality of spirit-writing, if any evidence can be satisfactory. There could have been no possible deception here. I held the medium's hand: the door was locked, and every precaution was taken by me as in previous instances. The identical cards were returned subsequently, covered with the finest writing. . . .

"*Sept.* 26, 1861.— . . . After five or six appearances of my wife, the light rested upon the floor some ten feet distant from me; then, rising, it suddenly darted across the room backwards and forwards, until, having gained sufficient power, it flashed brightly upon the wall, and brought into relief the entire figure of a large, heavy man, who stood before us. He was rather below the medium height; but broad-shouldered, heavy, and dressed in black, his back towards us, and his face not visible. He appeared thus three times very perfectly, remaining in view each time for about a minute. The moment his entire form was discerned by us, rappings commenced simultaneously in all parts of the room, which continued during the time he was in sight, as if to express delight at the achievement of a new success. On asking if the spirit we saw was that of Dr. Franklin, we were answered in the affirmative by three heavy dull knocks upon the floor, as though made by a heavy foot, which were several times repeated. During this sitting, the spirit of my wife approached, tapping me upon the shoulder, smoothing my hair, and caressing me; *while her long tresses, as natural as in life, dropped over my face, with the peculiar scent of delicate, freshly gathered violets.* A new and very curious manifestation now took place, showing us how the echoes were produced; and there was spelled out, '*Darling, have you not been rewarded?*' The light in producing these echoes or explosions assumed a lily shape, nearly the size of my head, and so brilliant as to light the entire surface of a table and the centre of the room, so that

Miss Fox and I could see each other distinctly, as well as various objects in the room. Then bounding up and down from the surface of the table some twelve or eighteen inches, it struck the table, and, descending on my arm, produced the raps or echoes.

"*Friday Evening, Oct.* 4, 1861. — A bouquet of flowers was placed upon the mantel in a vase with water. As soon as the gas was turned down, a movement was heard; and we were requested to 'get a light.' Upon doing so, we found the flowers, with the vase and other articles, had been removed from the mantel to the table, which stood in the centre of the room. We again extinguished the light, when immediately the heavy curtains of the window were drawn aside, and raised and lowered repeatedly, admitting the light from the street. Rustlings were heard after an interval of quiet, with sounds as of persons walking in stocking-feet. A peculiar sound was produced by striking against the wall, as though with a bag of keys or broken earthenware. This same bag of keys, or whatever it might have been, also seemed to be dropped from a height of several feet, and to fall heavily upon the floor, while we were told to listen. Tremendous concussions were then made upon the floor, jarring the whole house. The spirits of my wife and Dr. Franklin came to me in form at the same time, — he slapping me heavily upon the back, while she gently patted me upon the head and shoulder. The electrical rattle was now heard; and the light increasing in brilliancy disclosed to our view the full figure of a heavy man. At my request, the figure 'walked' across the floor, and appeared many times in different positions with entire distinctness. My wife now appeared in great vividness and beauty. Her figure floated gracefully through the room, her white robes falling back as she glided through the air, *brushing away pencils, cards, &c., as she passed over and swept across the table.* This spirit-robe was shown us in a variety of ways; and the manifestation of texture was exquisitely beautiful. We saw her plainly withdraw her face behind it, pushing the robe forward while it swung in the air. It was brought over the table, the light being placed behind, so that it became transparent and gossamer like, as though a breath of air would dissolve it. This was frequently repeated, and the robe drawn across my head, as palpably as though of material substance. Whenever it approached closely, we discovered a peculiar scent of purity, like a very delicate perfume of newly gathered grass or violets.

"*Oct.* 20, 1861. — This manifestation was a powerful one, showing the whole figure of my wife, but not her face. She stood before us enveloped in gossamer, her arm and hand as perfect as in life, the arm bare from the shoulder, with the exception of the gossamer, which was so transparent that it was more beautiful for being thus dressed. I asked to be touched; when she advanced, laid her arm across my forehead, and permitted me to kiss it. I found it as large and as real in weight as a living arm. At first it felt cold, then grew gradually warm. She held up the little finger, and moved it characteristically; and while we were looking at that, she let her hair fall loosely down her back. The manifestation was concluded by her writing a card, *resting it upon my shoulder*, caressing me upon the head and temple, and kissing me for good-night.

"*Nov.* 3, 1861. — This evening, according to promise, my wife came in full form, placing her arms completely around my neck; but the most remarkable and novel manifestation was the production of perfume from spirit-flowers. Something, resembling a veil in its contact, was thrown over my head; and, while it was resting there, spirit-flowers were placed at my nose, exhaling the most exquisite perfume I have ever smelt. I asked what this was; and was told 'My wreath of spirit-flowers.' At my request the same was brought to the medium, who experienced similar sensations. This was repeated probably a dozen times, the perfume being as strong as that of tuberose, but entirely different and far more exquisite.

"*Sunday Evening, Nov.* 10, 1861. — Immediately upon sitting down, there was communicated by raps, '*No failure.*' . . . My wife tapped upon my shoulder, informing me that she should give all her aid to Dr. Franklin, who now became visible, *his face* for the first time being seen. The light was apparently held by another figure enveloped in dark covering, from behind which the light approached, shining full upon the face of Dr. Franklin, about whose identity there can be no longer any doubt or mistake. I should have recognized it anywhere as Dr. Franklin's face, as I have learned to know it from the original paintings I have seen of him; but the strong points of his character were manifest as no painting could exhibit them. He was apparently dressed in a white cravat, and a brown coat of the olden style; his head was very large, with gray hair behind his ears; his face was radiant with benignity, intelligence, and spirituality: while my wife's was an angel face of shining beauty, spiritualized in its expression of serenity and happiness. His appearance was that of a man full of years, of dignity, and of fatherly kindness, in whom one could find counsel, affection, and wisdom. He came, perhaps, a dozen times, and once or twice so near that *his eyes were seen full and clear.* My wife appeared three times in white robes and enveloped in flowers.

"*Monday Evening, Nov.* 12, 1861. — Electric rattlings were heard; and the light becoming very vivid discovered to us *Dr. Franklin seated, his whole figure and dress complete.* Indeed, so vivid was the light, and so real was the man sitting there, that his shadow was thrown upon the wall as perfectly as though a living human being were there, in his earth-form. His position was one of ease and dignity, leaning back in the chair, with one arm upon the table, occasionally bending forward in recognition of us, his gray locks swinging in correspondence with the movement. We closed our eyes by request. Upon opening them, he was standing on the chair, his form towering above us like a statue. Again he resumed his seat, the act being accompanied by loud rustlings, which attend each movement of the spirit. A message from my wife informed me that a card would be visibly handed to Dr. Franklin. During all these appearances, there seemed to be two other forms or spirits assisting, one of whom held the light. One of these enveloped figures approached Dr. Franklin, and, extending an arm, held a card directly before his face, so that the card was distinctly visible, and then placed it on his knee, and afterwards handed it to me. The power was great, remaining vigorous during the evening; and Dr. Franklin, my silent companion, sat in his chair, my *vis-à-vis*, for an hour and a quarter.

"*Wednesday Evening*, *Nov.* 21, 1861. — . . . Something like a handkerchief of transparent gossamer was brought; and we were told to look at the hand which now appeared under the gossamer, as perfect a female hand as was ever created. I advanced my own hand, when the spirit-hand was placed in it, grasping mine; and we again grasped hands with all the fervor of long-parted friends, my wife in the spirit-land and myself here. The expression of love and tenderness thus given cannot be described; for it was a reality which lasted through nearly half an hour. I examined carefully that spirit-hand, squeezed it, felt the knuckles, joints, and nails, and kissed it, while it was constantly visible to my sight. I took each finger separately in my hand, and could discern no difference between it and a human hand, except in temperature; the spirit-hand being cold at first, and growing warm. I wore a glove, however, and could not perhaps judge accurately in all respects. At last 'good-night' was spelled out, by the spirit-hand tapping upon mine, and then for a parting benediction, giving it a hearty shake. Nothing in all these manifestations has been more real to me, or given me greater pleasure, than thus receiving the kindly grasp of a hand dearer to me than life, but which, according to the world's theory, has long since with all its tenderness and life mouldered into the dust of the earth.

"*Friday Evening*, *Nov.* 29, 1861. — My brother and I and the medium present. Conditions unfavorable. Heavy rain-storm. Darkened the room, and immediately a spirit-light rose from the floor. I put on my glove, and my brother did the same. The light soon came in my hand, when I felt that it contained a female hand. It was frequently placed in mine, and by me grasped tightly, so that I felt every part of it, *both the medium's hands being at the time held by me*. The spirit of my brother's deceased child also placed his hand in mine; and a large man's hand, purporting to be that of Dr. Franklin, was placed in mine, seizing and shaking it so violently, that it shook my whole frame, and also the table. My brother, also, had each of these hands placed in his. Thus, three distinct and different-sized hands were within a few minutes placed in each of ours, and recognized unmistakably as, first, a female hand; second, a child's; third, that of a full-sized man, each with its characteristic weakness or strength. At my request, the folding-doors of the room were opened and shut with great force repeatedly.

"*Saturday Evening, Nov.* 30, 1861. — At home in my own house; carefully locked the door. Conditions favorable; weather clear and cold. Soon after darkening the room, heavy knocks came upon the table with the electric rattle, but without any light. By raps, the encouraging 'No failure to-night' was communicated. My cane and hat and a glass of water were called for. A vacant chair by the table moved and got into position without being touched by us. A request was made 'to close eyes,' when a sound, like drawing a match, was heard several times repeated upon the table, with no result. Matches were then asked for. I procured a number of wax vestas; and holding one over the table, it was instantly taken by a spirit-hand, drawn across the table, and ignited at the third attempt. We opened our eyes: ***the room was illuminated by the burning match; and Dr. Franklin was before us, kneeling***, the top of his head about a foot above the table. We looked at him as long as the match burned; and he became invisible as it expired. . . . Soon after the male figure first appeared, the following was communicated by raps: 'Now, dear son, can the world ever doubt? This is what we have so long labored to accomplish. — B. F.' Also, 'My dear, now I am satisfied. — ESTELLE.' Upon cards there was subsequently written by the spirit, as follows: 'This meeting is the most important we have ever had. Long have we tried to accomplish this manifestation, and success has crowned our efforts. You saw that I had only to light the match to show you that I was as naturally in form as you are. I have long tried to come in an earthly light, and have at last succeeded.'

"*Dec.* 15, 1861.—The figure of Dr. Franklin appeared perfectly delineated, seated in the window, and permitted me to examine his hair with my hand. The hair was to sight and touch as real as human hair.

"*Saturday Evening, Dec.* 28, 1861.—In my own house and room, which was carefully examined, and door locked by myself. Soon after extinguishing the gaslight, the spirit-light rose, and requested us, by raps, to follow it across the room to the window, which was heavily curtained, to exclude the light from the street. By raps, the following was communicated: '*I come; I come in a cloud.*' Immediately the light became very vivid: the 'cloud' appeared against the curtain, a portion of it overhanging from the top; while the face and figure of my wife, from the waist, was projected upon it with stereoscopic effect. White gossamer, intertwined with violets and roses, encircled her head; while she held in her hand a natural flower, which was placed at my nose, and subsequently found upon the bureau, having been carried by the spirit from a basket of flowers on the table, standing in the centre of the room. We were told to notice her dress, which seemed tight-fitting, of a substance like delicate white flannel. She was leaning upon her right hand; the cuff of her sleeve was plain and neatly turned back. In answer to my inquiry, whether this appearance was not like a *bas-relief*, I was answered, '*No; but you see the fine spirit-form. You notice I come in health, and not as one year ago to-night.*' This appearance is new, and quite different from those originally seen, and is effected without noise or demonstrations of any kind.

"*Thursday Evening, Jan.* 23.—My wife made her appearance standing against the door. She was exquisitely robed in white, and enveloped in blue gossamer. A white ribbon, tied or knotted in the centre, passed across her waist; and a large and perfect *bow-knot* of white silk ribbon was attached to her breast diagonally. In her hand, near her face, she held a small oval mirror, about two inches in diameter. We had seen the mirror before, but at a distance. On this occasion I determined to examine it closely, and approached to within six or eight inches. The mirror was apparently glass, and reflected objects perfectly, — not only the light itself, but I saw my own face in it. The spirit-finger held opposite was reflected with all its motions. We asked for certain movements of the finger, which were made as requested, and simultaneously reflected in the mysterious glass. The flowers in her hair and on her person were real in appearance. Over her forehead was a crown of flowers. In the centre was a button or flower of black and gold upon a background of white. A card taken from me, and upon which I had written a private question, was held by the spirit in front of her face, and behind the oval mirror, which thus hung suspended and swinging against the white card, rendering it a real, palpable object. The light shone vividly upon her face and figure; and while we stood looking intently, she instantly, as quick as thought, disappeared, with a rushing sound. Then, by raps, was communicated, 'The electricity is very strong; and we did this to show you how quickly we can disappear.' Very soon she returned, as real as before. The light was subsequently placed upon the floor, near the door; while we receded to the middle of the room, remaining thus, at a distance of some ten feet from the medium, for twenty minutes. We were then requested to open the window to admit air, to enable them to dissipate the electricity. Immediately upon the fresh air being admitted, the light grew dim and disappeared.

"*Jan.* 24. — A stormy night with hail and sleet, ending in a severe gale. Conditions favorable. My wife appeared dressed precisely as last night, except having white gossamer around the top of her head. The 'bow,' which was in the same place upon her breast, was the same as then; and on this occasion was taken in our fingers for examination, being to sight and touch as real as silk. A low, murmuring sound was heard, something like the buzzing of a bee. I listened carefully, and noticed that it came from the lips of the spirit. This was an unsuccessful attempt to speak, or rather the preparatory process, eventually to result, doubtless, in success. The light approached her face.

We were told to look in her mouth. Upon doing so, we discovered what seemed a piece of dried grass projecting from her lips about three inches. This was then placed in my hand and in my mouth. I closed my teeth upon it, finding it a real substance. By raps, I was told it was a spiritual substance, when it was withdrawn, and disappeared. A large musical box was standing upon the table, which required considerable force to start it or to stop it by means of springs. At my request, the spirit-light rose, resting upon the keys, and started the music, then stopped it, changing or repeating the tunes, and finally *wound it up.*

"*Jan.* 30, 1862. — A manifestation of great power and 'solid form.' A veiled figure robed in white stood by us; and, opening the drapery which enveloped the head, we distinctly saw the eyes, forehead, and hair of Estelle, life-like, '*like flesh and blood.*' The lower part of the face was covered with the gossamer. This figure walked and floated through the room; kissed me, rested its arm, while fully visible, upon my head and shoulders, repeating the same to the medium. The arm was round, full, and flesh-like. I examined it both with my eyes and hands.

"*Jan.* 31, 1862. — Estelle and Dr. Franklin appeared alternately. Dr. Franklin's shirt-bosom and collar were as real to appearance as though made of linen. We handled them, and examined in the same manner his tunic, which was black and felt like cloth: his face and features were perfect and distinctly visible. This manifestation differs from that of last night, this having been spoken of by them as 'the fine spiritual form,' which seems like the projection of form, color, and expression, with stereoscopic effect. We now see that the rustling is produced by movements of the envelope or robe, and is doubtless electrical.

"*Sunday Evening, Feb.* 9, 1862. — My wife appeared leaning upon the bureau, with white lace hanging in front of and around her head. This lace or open work (like embroidery) was so real, that the figures were plainly discernible, and could have been sketched. As she stood in front of the bureau, the top of the mirror was plainly visible over her head, reflecting her form and surroundings. There were flowers in her hair; and in other respects her appearance was similar to those previously described. The body of her dress or robe was of spotted white gossamer, while the lace-work was in diamonds and flowers.

"*Wednesday Evening, Feb.* 12, 1862.—I found the power strong; and soon after entering the room messages were rapped out upon the door across the entire width of the room, fifteen feet distant from the medium and myself. About fifteen minutes after extinguishing the light, my wife came to us in exquisite beauty; if possible, more vividly than ever, and directly over the table. In her bosom was a white rose, green leaves and other smaller flowers. A card which she had written upon was visibly given to me, handed back, and returned to me repeatedly by her, while she was in full view. Her hand, real in form and color, was affectionately extended to me, and caressed me with a touch so full of tenderness and love that I could not restrain my tears; for to me it was really her hand, her native gentleness was expressed through it. The card was as follows: 'Dear C.,—Beautiful spring is approaching; flowery spring. Over you lightly fall its shadows; and may no sorrow, no clouds, touch the brightness of your future. Have you not noticed, dear C., that all your life you have been prospered, guided, and directed by the guardians of your happiness? You have always been followed by an invisible protecting power, which will ever be near when danger threatens, to step between you and difficulty, to lead you into paths of happiness and peace. We are now more closely linked, from our constant intercourse. There is not a day closes without a lasting blessing from us. As life is short, live well and live purely. . . . Fear not the world: there will be a day when this great truth will be seen in its true light and prized as it should be. . . . Be happy: all is well. Good-night.—ESTELLE.'

"*Saturday Evening, Feb.* 15.—Atmosphere unfavorable and damp. This meeting was held especially for Mr. G——, my brother-in-law. There were present, the medium, Mr. G——, and myself. I asked for a manifestation of power; and we at once received the following message: '*Listen, and hear it come through the air; hands off the table.*' Immediately a terrific metallic shock was produced, as though a heavy chain in a bag swung by a strong man had been struck with his whole power upon the table, *jarring the whole house.* This was repeated three times, with decreasing force. A heavy marble-topped table moved across the room; and a large box did the same, no person touching or being near either of them. An umbrella which had been lying upon the table floated through the room, touching each of us upon the head, and was finally placed in G——'s hand. These physical manifestations were given doubtless to convince an additional witness of the reality of spirit or invisible power. If such was the object, the purpose was well served; for every possible precaution had been taken by him, *even to the sealing of the doors and windows.*

"*Sunday Evening, Feb.* 16, 1862.—Appearance of my wife and of natural flowers. I had been promised a new manifestation, '*something natural as life.*' We sat longer than usual in quiet, and received the infallible message, '*No failure.*' The spirit announced her presence by gentle taps upon my shoulder, accompanied by rustlings, kissed me, and asked for a card and a pin, then another pin; all of which I handed over my shoulder, together with a small strand of my hair, which latter was particularly requested. The taking of each of these articles was accompanied by rustlings; and, as the spirit-hand was extended over my shoulder visibly, the drapery fell upon my hand and arm. Some ten minutes were now occupied by the spirit in arranging the card, pins, &c., when the following message was received: '*I will give you a spirit-flower.*' Immediately afterwards an apparently *freshly gathered flower* was placed at my nose, and that of the medium. My wife now appeared in white, holding the card in one hand, and the spirit-light in the other; while we discovered, fastened to the card, a leaf and flower. I asked if I could have the flower, and was answered in the affirmative. My hand was then taken by the spirit, opened, and the card placed thereon; while I was particularly and repeatedly enjoined to '*be very careful,*' and '*do not not drop or disturb it.*'

With the other hand I now lighted the gas, and found, to my surprise and astonishment, a leaf of laurel, about two and a half inches in length, pinned upon the card, and a pale pink flower pinned to the centre of the leaf, with the strand of hair passed through and tied in the leaf. We examined it carefully, smelled it, touched it, and found it fragrant and fresh. The card had not been during all this time within reach of the medium, who sat on my right, while the spirit stood at my left, and the doors were as usual carefully and securely locked. After a careful examination of five or ten minutes, we were requested to darken the room. Before doing so, wishing to preserve the leaf and flower, I placed them and the card upon a book in a remote part of the room, and returning to the medium, turned out the gas. The following message was then communicated: 'I gave you the sacred privilege of seeing this flower from our spirit-home: it has vanished.' I immediately relighted the gas, and directed my steps across the room, when I found the card and the pins precisely as I had left them; but the leaf and flower were gone. By raps, 'Next time you shall see the flowers dissolve in the light.' The following was also written upon another card by the spirit of Benjamin Franklin: 'My son, we are achieving a great victory at this moment. — B. F.'*

"*Saturday Evening, Feb.* 22, 1862. — Appearance of flowers. Cloudy. Atmosphere damp. Conditions unfavorable. At the expiration of half an hour, a bright light rose to the surface of the table, of the usual cylindrical form, covered with gossamer. Held directly over this was a sprig of roses, about six inches in length, containing two half-blown white roses, and a bud with leaves. The flowers, leaves, and stem were perfect. They were placed at my nose, and smelled as though freshly gathered; but the perfume in this instance was weak and delicate. We took them in our fingers, and I carefully examined the stem and flowers. The request was made as before to 'be very careful.'

* Fort Donelson, on the Tennessee River, was taken on this day by the Federal forces, February 16th.

I noticed an adhesive, viscous feeling which was explained as being the result of a damp, impure atmosphere. These flowers were held near and over the light, which seemed to feed and give them substance in the same manner as the hand. I have noticed that all these spiritual creations are nourished and fed or materialized by means of the electrical reservoir or cylinder, and that when they begin to diminish or pass off, incrassation or increase takes place the moment they are brought in contact with, or in proximity to, the electrical light. By raps, we were told to '*Notice and see them dissolve.*' The sprig was placed over the light, the flowers drooped, and, in less than one minute, melted as though made of wax, their substance seeming to spread as they disappeared. By raps, '*See them come again.*' A faint line immediately shot across the cylinder, grew into a stem; and, in about the same time required for its dissolution, the stem, bud, and roses had grown into created perfection. This was several times repeated, and was truly wonderful. We were promised the phenomenon of their probable disappearance in the gaslight when the atmosphere became pure and clear.

"*Sunday Evening, Feb.* 23, 1862. — Flowers. Atmosphere very damp. Conditions unfavorable. The flowers were reproduced in the same manner as last evening. I felt them carefully; and a rose was placed in my mouth, so that I took its leaves between my lips. They were delicate as natural rose-leaves, and cold; and there was a peculiar freshness about them, but very little fragrance. The following message was written upon a card: 'My dear C——, — Again we have to contend with the atmosphere; but how much we have been able to do, owing to the many powerful aids who have been so kind to us! Do you realize the great blessings we are giving you? Do you realize what a great proof you have received in being permitted to see the flowers which decorate our sacred walks? . . . The time is coming, has come, when this subject will be honored. Good-night. — ESTELLE.'

"*Tuesday Evening, Feb.* 25, 1862. — Appearance in presence of a third witness, Mr. G——, the medium, and myself. The room in which we sat was connected with another smaller room by sliding-doors; but the doors and windows leading into these two were carefully sealed. After sitting about half an hour, we were directed to open these sliding-doors; while the medium and myself proceeded to a window against which was hung a dark curtain to exclude the light as usual. Meanwhile Mr. G—— remained by the table. Upon reaching the window, a vivid light rose from the floor, discovering to us the form of a male spirit standing against the white wall adjoining the window. At first his face was not visible, or rather was concealed by the unusual quantity of dark drapery by which he was enveloped; but after two or three efforts the face of Dr. Franklin was recognized. During this time Mr. G—— was not permitted to leave the table. At last the conditions having become stronger, or rather the effect of his presence having been partially overcome, the following message was received: '*Dear friend, approach.*' Mr. G—— now came to us, when the spirit of Dr. Franklin immediately became visible to him. He saw the hair was real; for while we stood before him it was frequently placed over and on the light to show its substantiality. He did not, however, see the spirit in the same degree of perfection that we do, but sufficiently well to recognize the face of Dr. Franklin as represented in his portraits. The eyes, hair, features, and expression, together with a portion of the drapery, were all visibly perfect; but the power of the electrical light was considerably weakened from the effects of Mr. G——'s presence. These effects were very curious. With Mr. G—— in the other room, the light was bright and vivid, decreasing as he approached in proportion to the distance; again brightening as he receded, and *vice-versâ*, showing that the sphere of a person in the earth-form has a direct influence upon these creations of the invisible world; and that this influence may be a disturbing one, from no other cause except surprise, fear, or any violent emotion resulting from inexperience in the phenomena."

In a letter to Mr. Coleman, dated June 10th, 1862, Mr. Livermore writes, "I have the pleasure of announcing to you the initiation of Dr. Gray as a witness of the visible presence of Dr. Franklin on Friday night last. He saw the spirit less distinctly than has generally been my experience, but sufficiently well to recognize him. This being, however, the first time of seeing him, he may expect to attain by progressive steps the same vividness that has been manifested to us, after the first emotions of surprise have been overcome by familiarity with the phenomenon. The doctor actually saw and took the gray hair of Franklin's spirit, as well as a portion of the clothing in his hand, and examined them. To me this is now a very common occurrence; but the additional corroborative testimony of Dr. Gray is very important."

Dr. Gray, on his part, fully confirms all this. He writes (January, 1867), "I can only reply to your latest request, that I would write out my testimony in this case for publication, that Mr. Livermore's statements are each, one and all of them, fully reliable. His recitals of the *séances* in which I participated are faithfully and most accurately stated, leaving not a shade of doubt in my mind as to the truth and accuracy of his accounts of those at which I was not a witness. I saw with him the philosopher Franklin, in a living, tangible, physical form, several times and on as many different occasions. I also witnessed the production of lights, odors, and sounds; and also the formation of flowers, cloth-textures, &c., and their disintegration and dispersion.

"These phenomena, including the apparition of Dr. Franklin and also many other phenomena of like significance, have all been shown to me when Mr. Livermore was not present and not in the country even.

"Mr. L. is a good observer of spirit phenomena; brave, clear and quick sighted, void of what is called superstition, in good health of body and mind, and remarkably unsusceptible to human magnetism. Moreover, he knows that all forms of spirit communication are subject to interpolation from earth-minds, and are of no other or greater weight than the truths they contain confer upon them.

"Miss Fox, the medium, deported herself with patient integrity of conduct, evidently doing all in her power, at all times, to promote a fair trial and just decision of each phenomenon as it occurred. — JOHN F. GRAY."

The narrative of Mr. Livermore includes nearly all the most important phenomena which have been experienced in connection with these modern manifestations. His observations in respect to the costume of the supposed spirits appear to have been careful and minute. This question of the dress of spirits has been often discussed. When Joan of Arc was in mockery asked by her judges about the clothing of the spirits who visited her, she replied, "Is it possible to conceive that a God who is served by ministering spirits cannot also clothe them?"

Swedenborg affirms that in the spirit-world all clothing is representative, and is outwrought from the affections and states of its several inhabitants.

Some seers have asserted that the spiritual body is composed of a subtle ether, and that spirits make themselves visible by means of its vibration, and can give what forms they please, by a mere effort of the will, to their coverings; that the human body itself, and the garments we wear, are composed of the same ultimate particles of matter; and that the spiritual fabric is nothing but those ultimate particles in their most attenuated state. Of the power of spirits to use the elements of our own atmosphere, in giving concretion, visibility, and tangibility, odor and color, to forms, the experiences of Mr. Livermore and others offer strong testimony. The subject is one which a more advanced science may some day be able to explore.

CHAPTER IV.

MANIFESTATIONS THROUGH MR. HOME.

"We all are at once mortal and immortal; inhabitants of time and dwellers in eternity." — *H. J. Slack.*

DANIEL DUNGLASS HOME was born near Edinburgh, March, 1833. When about a year old, he was adopted by an aunt. Some eight years afterwards he accompanied her and her husband to America. At the age of seventeen, he was residing at Norwich, Conn. Soon after the developments at Hydesville, through the Fox family, he began to manifest extraordinary powers as a medium, and in 1851 had acquired considerable reputation among those interested in the phenomena in the United States.

He went to Europe early in the spring of 1855; and his career there, in the exercise of his wonderful gifts, has been of a character to bring him repeatedly before the public.

Not long since he was a party to a lawsuit, at the trial of which he was the subject of a good deal of abuse and misrepresentation by the English press. It was the celebrated case of Lyon *versus* Home. The plaintiff, Mrs. Lyon, was a widow lady, seventy years old or more, possessed of a considerable fortune, and without any child or near relative. Having read Mr. Home's "Incidents of My Life," she called on him, introduced herself (Oct. 30, 1866), and asked him to visit her. He did so; and, after two or three interviews, she proposed to make him her adopted son. In November, she executed a will in his favor; and the next month he took the name of Lyon, advertising the fact. She executed a deed, confirming a gift of £24,000, and adding £6,000; and, finally, in January, 1867, she conveyed to him, after the reservation of a life-interest, a further snm of £30,000. All this was done in legal form, and after deliberation and consultation.

Whether it was the part of good taste and manly independence in Mr. Home to accept these large sums, we decline to discuss; but we will venture the remark, that, among the self-righteous ones who have made him the subject of their denunciations, there is probably not an individual who, under similar circumstances, would not have consented to be enriched in the same way.

From the facts in Mr. Home's affidavit, we are led to infer that it was not till after he had been thus formally adopted by the old lady as a son, that he discovered she had been calculating on his marrying her. "Do you know," said she, "that nothing would be greater fun than that I should marry you? How the world would talk!" Mr. Home does not appear to have been agreeably impressed by the intimation.

In her bill of complaint, Mrs. Lyon asserted that she was made to believe by Home, "that the spirit of her deceased husband required her to adopt the said defendant." It very soon appeared on the trial, by her own displays of wilfulness and headstrong unveracity, that the old lady was one whom neither spirits out of the flesh nor in the flesh would be likely to influence to do what was contrary to her own caprice. She contradicted her own testimony so grossly, that even the presiding Vice-Chancellor — bitterly prejudiced as he was against Mr. Home and against Spiritualism — could not avoid speaking of her testimony as "clearly untrustworthy, and such as no man ought to have his case decided upon against him."

And yet there was no evidence whatever, except her own assertion, that Mr. Home had tried to get her to adopt him, by representing that her departed husband recommended it. Mrs. Lyon seems to have been dazzled by the social position which she fancied that Home occupied, by his presents from kings and emperors, and to have aspired to mix in the aristocratic world, and to assume in her old age a rank from which she had been all her life excluded.

She soon found she had miscalculated in regard to Mr. Home. Instead of taking her matrimonial hints, he was so unaccommodating as to fall ill, and threaten to die. He had a little boy for whom Mrs. Lyon conceived a deadly dislike; and she now saw before her the prospect of the large sums she had parted with going to enrich this youth. One fine day, as Mr. Home was about starting for Paris, he was arrested and thrown into prison under a writ of *ne exeat regno.*

The trial came on in the spring of 1868, before Vice-Chancellor Giffard, who decided the case adversely to Mr. Home, ordering him to restore all the money he had received from Mrs. Lyon. From this decision Mr. Home appealed; but lately there has been a compromise between the parties, which ends the affair.

The fable of the wolf and the lamb is recalled by Mrs. Lyon's attempt to show that she was under the "undue influence, ascendency, and power" of Mr. Home. Hers appears to have been the stronger will in the case; and she had every thing her own way.

The affidavit of Mr. Home sets forth, that from his childhood he has been subject to the occasional happening of singular physical phenomena in his presence, which are most certainly not produced by him or by any other person in connection with him. "I have," he affirms, "no control over them whatever: they occur irregularly, and even when I am asleep. Sometimes I am many months, and once I have been a year, without them. I cannot account for them further than by supposing them to be effected by intelligent beings or spirits. Similar phenomena occur to many other persons. . . . These phenomena, occurring in my presence, have been witnessed by thousands of intelligent and respectable persons, including men of business, science, and literature, under circumstances which would have rendered, even if I had desired it, all trickery impossible."

Mr. Home proceeds to affirm that they have also been witnessed in their own private apartments, when any contrivance of his must have been detected, by the emperor and empress of the French, the emperor of Russia and his family, the king of

Prussia, and other royal personages, who have had ample opportunities, which they have used, of investigating the phenomena and inquiring into the character of the medium.

"I have resided," continues Mr. Home, "in America, England, France, Italy, Germany, and Russia; and in every country I have been received as a guest and friend by persons in the highest position, who were quite competent to discover and expose, as they ought to have done, any thing like contrivance on my part to produce these phenomena. I do not seek, and never have sought, the acquaintance of any of these exalted personages. They have sought me; and I have thus had a certain notoriety thrust upon me. I do not take money, and never have taken it; although it has been repeatedly offered me for or in respect of these phenomena. . . . Some of the phenomena in question are noble and elevated, others appear to be grotesque and undignified. For this I am not reponsible, any more than I am for the many grotesque and undignified things which are undoubtedly permitted to exist in the material world. I solemnly swear that I do not produce the phenomena aforesaid, or, in any way whatever, aid in producing them."

In the course of the cross-examination, Mr. Home said, "I have seen spirits; have conversed with them orally. They have called to me in sounds audible to my ear; and I have talked to them. Strange sounds are heard, like a rapping. It does not indicate who the spirit is. We take it for granted, the same as in the call of the telegraph wire, that there is an intelligence there at the end of it. The language used by the spirits is exceedingly beautiful and elevated.

"I have been bodily displaced in violation of the ordinary rules of gravity. (I must protest against its being supposed that I am the only person to whom this has occurred.) Chairs and tables have been moved in the same way. I have found a useful result of Spiritualism in convincing those who did not believe in it of the immortality of the soul."

Mr. Home is a person of very delicate constitution and extreme nervous sensibility. He is tall, slender, and fair-haired, and does not convey the idea of robustness, physical or mental. His acquaintances generally appear to have mingled in their regard for him a sort of tenderness, as if he were one to be shielded from the rougher experiences of life. Those who have known him best, testify to his character as "a man of honor and proper moral feeling."

Our first call on Mr. Home was made without signifying our intention to any one. We had never seen him or corresponded with him, and did not suppose that he even knew us by name. But as we rang the bell, he, without having seen us, said to Mrs. R., at whose house he was stopping, "That is Mr. —— who rings. He has come to call on me."

Dr. Winslow Lewis, long known as one of the most eminent surgeons of Boston, informed us, in Home's presence (Feb. 21, 1865), that he (Dr. L.) took up the "Boston Directory" the day before to look for a name which he had not mentioned to any human being. "Here, I'll find it for you," said Home, taking the book out of his hand, and instantly pointing to the name.

Dr. Lewis also told us that he handed to Home a photograph-album, full of likenesses, the originals of which were unknown to him; and Home pointed to those persons who had deceased, and in every instance he was right.

"Second sight," said Home, joining in the conversation, "is my strong point." (His mother had been a seeress. From her he had probably derived his gift.) "Being at a party once in London, I heard one man say to another, 'Do you know that fellow?'—'Oh, yes! that's that humbug, Home.' At once I turned to the last speaker, and said, 'Excuse me, sir; but I am at this moment vividly impressed with the particulars of an affair in which you were an actor*—let me see—when you were twenty-two years of age. But I cannot help wondering why you took the course you did, when you might have'—here

* Instances of a similar faculty in the lives of Zschokke, the late Forceythe Willson, and others, are well authenticated.

I whispered the rest in his ear. The man looked aghast, and, drawing me aside, said, 'There should be no human being but myself who knows a word of that affair. Say no more. You have said enough.' This man subsequently became one of my best friends."

As these are comparatively very slight manifestations of power, we will not pause to anticipate the obvious objections which skepticism might raise to the uncorroborated form in which they are here put.

From the numerous published accounts, amounting now to several hundred, by many different witnesses, of the phenomena produced through the mediumship of Home, we select the account, which we slightly abridge, by the late Robert Bell, contributed to the "Cornhill Magazine" (London, August, 1860), when the late Mr. Thackeray — so justly celebrated for his writings — was the editor.

In introducing the account, Mr. Thackeray says, "I can vouch for the good faith and honorable character of our correspondent, a friend of twenty-five years' standing."

Of Mr. Thackeray's own convictions on the subject we have the following record, which we extract from Weld's "Last Winter in Rome" (1865): —

"I remember well meeting the late Mr. Thackeray, at a large dinner-party, shortly after the publication in the 'Cornhill Magazine,' then edited by him, of the paper entitled 'Stranger than Fiction.' In this paper, as will be remembered by many readers, a detailed account was given of a spiritual *séance*, at which Mr. Home performed, or caused to be performed, many surprising things, the most astounding being his floating in the air above the heads of persons in the room. There were several scientific men at the dinner-party, all of whom availed themselves of the earliest opportunity to reproach Mr. Thackeray with having permitted the paper in question to appear in a periodical of which he was editor, holding, as he did, the highest rank in the world of letters. Mr. Thackeray, with that imperturbable calmness which he could so well assume, heard all that was said against him, and the paper in question, and thus replied: 'It is all very well for you, who have probably never seen spiritual manifestations, to talk as you do; but, had you seen what I have witnessed, you would hold a different opinion.' He then proceeded to inform us that, when in New York, at a dinner-party, he saw the large and heavy dinner-table, covered with decanters, glasses, dishes, plates — in short, every thing appertaining to dessert — rise fully two feet from the ground, the *modus operandi* being, as he alleged, spiritual force. No possible jugglery, he declared, was or could have been employed on the occasion; and he felt so convinced that the motive force was supernatural, that he then and there gave in his adhesion to the truth of Spiritualism, and consequently accepted the article on Mr. Home's *séance*. Whether Mr. Thackeray thought differently before he died, I cannot say; but this I know, that every possible argument was used by those present to endeavor to shake his faith in Mr. Home's spiritual manifestations, which were, as they declared, after all but sorry performances compared with the surprising tricks of Houdin or Frikell."

We will not longer detain the reader from that part of Mr. Bell's paper relating to Mr. Home:—

"'I have seen what I would not have believed on your testimony, and what I cannot, therefore, expect you to believe upon mine,' was the reply of Dr. Treviranus to inquiries put to him by Coleridge as to the reality of certain magnetic phenomena, which that distinguished *savant* was reported to have witnessed. It appears to me that I cannot do better than adopt this answer as an introduction to the narrative of facts I am about to relate. It represents very clearly the condition of the mind before and after it has passed through experiences of things that are irreconcilable with known laws. I refuse to believe such things upon the evidence of other people's eyes; and I may possibly go so far as to protest that I would not believe them even on the evidence of my own. When I have seen them, however, I am compelled to regard the subject from an entirely different point of view. It is no longer a question of mere credence or authorty, but a question of fact. Whatever conclusions, if any, I may have arrived at on this question of fact, I see distinctly that I have been projected into a better position for judging of it than I occupied before; and that what then appeared an imposition, or a delusion, now assumes a shape which demands investigation.

"But I cannot expect persons who have not witnessed these things, to take my word for them; because, under similar circumstances, I certainly should not have taken theirs. What I do expect is, that they will admit as reasonable, and as being in strict accordance with the philosophical method of procedure, the mental progress I have indicated, from the total rejection of extraordinary phenomena upon the evidence of others, to the recognition of such phenomena as matter of fact, upon our own direct observation. This recognition points the way to inquiry, which is precisely what I desire to promote. . . .

"Our party of eight or nine assembled in the evening; and the *séance* commenced about nine o'clock, in a spacious drawing-room, of which it is necessary to give some account, in order to render perfectly intelligible what is to follow. In different parts of the room were sofas and ottomans, and in the centre a round table, at which it was arranged that the *séance* should be held. Between this table and three windows, which filled up one side of the room, there was a large sofa. The windows were draped with thick curtains, and protected by spring-blinds. The space in front of the centre-window was unoccupied; but the windows on the right and left were filled by geranium-stands.

"The company at the table consisted partly of ladies and partly of gentlemen; and amongst the gentlemen was the celebrated Mr. Home. . . . He looks like a man whose life has been passed in a mental conflict. The expression of his face in repose is that of physical suffering; but it quickly lights up when you address him, and his natural cheerfulness colors his whole manner. There is more kindliness and gentleness than vigor in the character of his features; and the same easy-natured disposition may be traced in his unrestrained intercourse. He is yet so young, that the playfulness of boyhood has not passed away and he never seems so thoroughly at ease with himself and others as when he is enjoying some light and temperate amusement. . . .

"The *séance* commenced in the centre of the room. I pass over the preliminary vibrations to come at once to the more remarkable features of the evening. From unmistakable indications, conveyed in different forms, the table was finally removed to the centre-window, displacing the sofa, which was wheeled away. The deep space between the table and the window was unoccupied, but the rest of the circle was closely packed. Some sheets of white paper, and two or three lead-pencils, an accordion, a small hand-bell, and a few flowers were placed on the table. Sundry communications now took place, which I will not stop to describe; and at length an intimation was received, through the usual channel of correspondence, that the lights must be extinguished. As this direction is understood to be given only when unusual manifestations are about to be made, it was followed by an interval of anxious suspense. There were lights on the walls, mantel-piece, and console-table; and the process of putting them out seemed tedious. When the last was extinguished, a dead silence ensued, in which the tick of a watch could be heard.

"We must now have been in utter darkness, but for the pale light that came in through the window, and the flickering glare thrown fitfully over a distant part of the room by a fire which was rapidly sinking in the grate. We could see, but could scarcely distinguish, our hands upon the table. A festoon of dull gleaming forms round the circle represented what we knew to be our hands. An occasional ray from the window now and then revealed the hazy surface of the white sheets, and the misty bulk of the accordion. We knew where these were placed; and could discover them with the slightest assistance from the gray, cold light of a watery sky. The stillness of expectation that ensued during the first few minutes of that visible darkness was so profound that, for all the sounds of life that were heard, it might have been an empty chamber.

"The table and the window, and the space between the table and the window, engrossed all eyes. It was in that direction everybody instinctively looked for a revelation. Presently, the tassel of the cord of the spring-blind began to tremble. We could see it plainly against the sky; and, attention being drawn to the circumstance, every eye was upon the tassel. Slowly, and apparently with caution, or difficulty, the blind began to descend: the cord was evidently being drawn; but the force applied to pull down the blind seemed feeble and uncertain. It succeeded, however, at last; and the room was thrown into deeper darkness than before. But our vision was becoming accustomed to it; and masses of things were growing palpable to us, although we could see nothing distinctly. Several times, at intervals, the blind was raised and pulled down; but, capricious as the movement appeared, the ultimate object seemed to be to diminish the light.

"A whisper passed round the table about hands having been seen or felt. Unable to answer for what happened to others, I will speak only of what I observed myself. The table-cover was drawn over my knees, as it was with the others. I felt distinctly a twitch, several times repeated, at my knee. It was the sensation of a boy's hand, partly scratching, partly striking, and pulling me in play. It went away. Others described the same sensation; and the celerity with which it frolicked, like Puck, under the table, now at one side and now at another, was surprising. Soon after, what seemed to be a large hand came under the table-cover, and with the fingers clustered to a point, raised it between me and the table. Somewhat too eager to satisfy my curiosity, I seized it, felt it very sensibly; but it went out, like air, in my grasp. I know of no analogy in connection with the sense of touch by which I could make the nature of that feeling intelligible. It was as palpable as any soft substance, velvet, or pulp; and at the touch it seemed as solid; but pressure reduced it to air.

"It was now suggested that one of the party should hold the hand-bell under the table; which was no sooner done than it was taken away, and after being rung at different points was finally returned, still under the table, into the hand of another person.

"While this was going forward, the white sheets were seen moving, and gradually disappeared over the edge of the table. Long afterwards we heard them creasing and crumpling on the floor, and saw them returned again to the table; but there was no writing upon them. In the same way, the flowers which lay near the edge were removed. The semblance of what seemed a hand, with white, long, and delicate fingers, rose up slowly in the darkness, and, bending over a flower, suddenly vanished with it. This occurred two or three times; and although each appearance was not equally palpable to every person, there was no person who did not see some of them. The flowers were distributed in the manner in which they had been removed; a hand, of which the lambent gleam was visible, slowly ascending from beneath the cover, and placing the flower in the hand for which it was intended. In the flower-stands in the adjoining window, we could hear geranium-blossoms snapped off, which were afterwards thrown to different persons.

"Still more extraordinary was that which followed, or rather which took place, while we were watching this transfer of the flowers. Those who had keen eyes, and who were in the best position for catching the light upon the instrument, declared that they saw the accordion in motion. I could not. It was as black as pitch to me. But, concentrating my attention on the spot where I supposed it to be, I soon perceived a dark mass rise awkwardly above the edge of the table, and then go down, the instrument emitting a single sound, produced by its being struck against the table as it went over. It descended to the floor in silence; and a quarter of an hour afterwards, when we were engaged in observing some fresh phenomena, we heard the accordion beginning to play where it lay on the ground.

"Apart from the wonderful consideration of its being played without hands, no less wonderful was the fact of its being played in a narrow space, which would not admit of its being rawn out with the requisite freedom to its full extent. We listened with suspended breath. The air was wild, and full of strange transitions, with a wail of the most pathetic sweetness running through it. The execution was no less remarkable for its delicacy than its power. When the notes swelled in some of the bold passages, the sound rolled through the room with an astounding reverberation; then, gently subsiding, sank into a strain of divine tenderness. But it was the close that touched the hearts, and drew the tears of the listeners. Milton dreamt of this wondrous termination when he wrote of 'linkèd sweetness long drawn out.' By what art the accordion was made to yield that dying note, let practical musicians determine. Our ears, that heard it, had never before been visited by 'a sound so fine.' It continued diminishing and diminishing, and stretching far away into distance and darkness, until the attenuated thread of sound became so exquisite that it was impossible at last to fix the moment when it ceased.

"That an instrument should be played without hands, is a proposition which nobody can be expected to accept. The whole story will be referred to one of the two categories under which the whole of these phenomena are consigned by 'common sense.' It will be discarded as a delusion or a fraud. Either we imagined we heard it, and really did not hear it; or there was some one under the table, or some mechanism was set in motion to produce the result. Having made the statement, I feel that I am bound, as far as I can, to answer these objections, which I admit to be perfectly reasonable. Upon the likelihood of delusion, my testimony is obviously worth nothing. With respect to fraud, I may speak more confidently. It is scarcely necessary to say, that in so small a circle, occupied by so many persons, who were inconveniently packed together, there was not room for a child of the size of a doll, or for the smallest piece of machinery, to operate. But we need not speculate on what might be done by skilful contrivances in confines so narrow, since the question is removed out of the region of conjecture by the fact, that, upon holding up the instrument myself in one hand, in the open room, with the full light upon it, similar strains were emitted, th regular action of the accordion going on without any visible agency. And I should add that, during the loud and vehement passages, it became so difficult to hold, in consequence of the extraordinary power with which it was played from below, that I was obliged to grasp the top with both hands. This experience was not a solitary one. I witnessed the same result on different occasions, when the instrument was held by others.

"It is not my purpose to chronicle all the phenomena of the evening, but merely to touch upon some of the most prominent; and that which follows, and which brought us to the conclusion of the *séance*, is distinguished from the rest by this peculiarity, — that it takes us entirely out of that domain of the marvellous in which the media are inanimate objects.

"Mr. Home was seated next to the window. Through the semi-darkness his head was dimly visible against the curtains, and his hands might be seen in a faint white heap before him. Presently, he said, in a quiet voice, 'My chair is moving; I am off the ground: don't notice me; talk of something else,' or words to that effect. It was very difficult to restrain the curiosity, not unmixed with a more serious feeling, which these few words awakened; but we talked, incoherently enough, upon some indifferent topic. I was sitting nearly opposite to Mr. Home; and I saw his hands disappear from the table, and his head vanish into the deep shadow beyond. In a moment or two more he spoke again. This time his voice was in the air above our heads. He had risen from his chair to a height of four or five feet from the ground. As he ascended higher, he described his position, which at first was perpendicular, and afterwards became horizontal. He said he felt as if he had been turned in the gentlest manner, as a child is turned in the arms of a nurse. In a moment or two more, he told us that he was going to pass across the window, against the gray, silvery light of which he would be visible. We watched in profound stillness, and saw his figure pass from one side of the window to the other, feet foremost, lying horizontally in the air. He spoke to us as he passed, and told us that he would turn the reverse way, and recross the window; which he did. His own tranquil confidence in the safety of what seemed from below a situation of the most novel peril, gave confidence to everybody else; but, with the strongest nerves, it was impossible not to be conscious of a certain sensation of fear or awe. He hovered round the circle for several minutes, and passed, this time perpendicularly, over our heads. I heard his voice behind me in the air, and felt something lightly brush my chair. It was his foot, which he gave me leave to touch. Turning to the spot where it was on the top of the chair, I placed my hand gently upon it, when he uttered a cry of pain; and the foot was withdrawn quickly, with a palpable shudder. It was evidently not resting on the chair, but floating; and it sprang from the touch as a bird would. He now passed over to the farthest extremity of the room; and we could judge by his voice of the altitude and distance he had attained. He had reached the ceiling, upon which he made a slight mark, and soon afterwards descended, and resumed his place at the table. An incident which occurred during this aërial passage, and imparted a strange solemnity to it, was that the accordion, which we supposed to be on the ground under the window, close to us, played a strain of wild pathos in the air from the most distant corner of the room.

"I give the driest and most literal account of these scenes, rather than run the risk of being carried away into descriptions which, however true, might look like exaggerations. But the reader can understand, without much assistance in the way of suggestion, that at such moments, when the room is in deep twilight, and strange things are taking place, the imagination is ready to surrender itself to the belief that the surrounding space is inhabited by supernatural presences. Then is heard the tread of spirits, with velvet steps, across the floor; then the ear catches the plaintive murmur of the departed child, whispering a tender cry of 'Mother!' through the darkness; and then it is that forms of dusky vapor are seen in motion, and colored atmospheres rise round the figures that form that circle of listeners and watchers.

I exclude all such sights and sounds because they do not admit of direct and satisfactory evidence, and because no sufficient answer can be made to the objection, that they *may* be the unconscious work of the imagination.

"Palpable facts, witnessed by many people, stand on a widely different ground. If the proofs of their occurrence be perfectly legitimate, the nature of the facts themselves cannot be admitted as a valid reason for refusing to accept them as facts. Evidence, if it be otherwise trustworthy, is not invalidated by the unlikelihood of that which it attests. What is wanted here, then, is to treat facts as facts, and not to decide the question over the head of the evidence.

"To say that certain phenomena are incredible, is merely to say that they are inconsistent with the present state of our knowledge; but, knowing how imperfect our knowledge is, we are not, therefore, justified in asserting that they are impossible. The 'failures' which have occurred at *séances* are urged as proofs that the whole thing is a cheat. If such an argument be worth noticing, it is sufficient to say that ten thousand failures do not disprove a single fact. But it must be evident that, as we do not know the conditions of 'success,' we cannot draw any argument from 'failures.' We often hear people say that they might believe such a thing, if such another thing were to happen; making assent to a particular fact, by an odd sort of logic, depend upon the occurrence of something else. 'I will believe,' for example, says a philosopher of this stamp, 'that a table has risen from the ground, when I see the lamp-posts dancing quadrilles. Then, tables? Why do these things happen to tables?' Why, that is one of the very matters which it is desirable to investigate, but which we shall never know any thing about so long as we ignore inquiry.

"And, above all, of what use are these wonderful manifestations? What do they prove? What benefit have they conferred on the world? Sir John Herschel has answered these questions with a weight of authority which is final. 'The question, *Cui bono?* — to what practical end and advantage do your researches tend? — is one which the speculative philosopher, who loves knowledge for its own sake, and enjoys, as a rational being should enjoy, the mere contemplation of harmonious and mutually dependent truths, can seldom hear without a sense of humiliation. He feels that there is a lofty and disinterested pleasure in his speculations, which ought to exempt them from such questioning. But,' adds Sir John, 'if he can bring himself to descend from this high but fair ground, and justify himself, his pursuits, and his pleasures, in the eyes of those around him, he has only to point to the history of all science, where speculations, apparently the most unprofitable, have almost invariably been those from which the greatest practicable applications have emanated.'

"The first thing to be done is to collect and verify facts. But this can never be done if we insist upon refusing to receive any facts, except such as shall appear to us likely to be true, according to the measure of our intelligence and knowledge. My object is to apply this truism to the case of the phenomena of which we have been speaking; an object which I hope will not be overlooked by any persons who may do me the honor to quote this narrative."

Of this account, Dr. J. M. Gully (known to many American as well as English patients), who was present at the *séance*, and the neighbor of Mr. Bell, says, "I can state with the greatest positiveness that the record is, in every particular, correct; and that no trick, machinery, sleight-of-hand, or other artistic contrivance, produced what we heard and beheld. . . . I may add that the writer omits to mention several curious phenomena. A distinguished *littérateur*, who was present, asked the supposed spirit of his father whether he would play his favorite ballad for us. Almost immediately the flute-notes of the accordion (which was on the floor) played through, 'Ye banks and braes of Bonnie Doon,' which the gentleman assured us was his father's favorite air, whilst the flute was his father's favorite instrument. He then asked for another favorite air, not Scotch, of his father's, and 'The Last Rose of Summer' was played in the same note. This, the gentleman told us, was the air to which he had alluded."

Mr. C. F. Varley, the electrician, in a letter dated May 7th, 1868, gives an account of a sitting at his own house, with Mr. Home; when a large ottoman, capable of seating eight persons, was moved all over the room, and a side-table was driven up to him by invisible means; Mr. V. having hold of both Mr. Home's hands and legs at the time. "Imposture," says Mr. V., "was impossible."

In the third chapter of the Book of Daniel, we read that King Nebuchadnezzar caused three men, Shadrach, Meshach, and Abednego, to be bound and cast into a burning, fiery furnace. But a fourth form, like unto "the Son of God," was seen walking with the three, loose from their bonds, in the fire. "And the princes, governors, and captains, and the king's counsellors, being gathered together, saw these men upon whose bodies the fire had no power, nor was a hair of their head singed, neither were their coats changed, nor the smell of fire had passed on them."

Investigators into modern spiritual phenomena will not question the literal truth of this narrative. The facts have been paralleled repeatedly during the last twenty years.

The ordeal by fire is of great antiquity. It was known to the Greeks. In one of the plays of Sophocles, a suspected person declares himself ready "to handle hot iron, and to walk over fire" in proof of his innocence.

Blackstone, the great legal authority, writes, "Fire-ordeal was performed either by taking up in the hand, unhurt, a piece of red-hot iron, of one, two, or three pounds' weight; or else by walking, barefoot and blindfold, over nine red-hot ploughshares, laid lengthwise at unequal distances; and if the party escaped being hurt, he was adjudged innocent; but if it happened otherwise, he was then condemned as guilty. By this method, Queen Emma, the mother of Edward the Confessor, is mentioned to have cleared her character when suspected of familiarity with Alwyn, Bishop of Winchester."

The ordeal was accompanied with religious service, within consecrated walls; and the solemnity with which the Church superintended the appeal to Heaven invested it with a sacred character. A form of ritual was appointed by ecclesiastical authority. It will be familiar to many readers, from its being given by Sir Walter Scott in the historical Notes to his "Fair Maid of Perth."

The theory that the exemption, in these cases, from harm by fire was the result of trick, or fraud, or the contrivance of priestcraft; that chemical agencies were applied to protect the body from the natural effects of fire; that some liniment was used to anoint the soles of the feet; that asbestos was mixed with a composition to cover the skin; that the hands were protected by asbestos gloves, so made as to imitate the skin, — is all pure supposition. There is no evidence to support it: it is simple conjecture as to how it is supposed these things *might* have been done, not evidence as to how they really *were* done. To prevent the defendant from preparing his hands by art, and in order to ascertain the result of the ordeal, his hands were covered up and sealed during the three days which preceded and followed the fiery application; and it is an entirely gratuitous conjecture that those in whose care the accused was placed made use of these opportunities to apply preventives to those whom they wished to acquit, and to bring back the hands to their natural condition. "Even were the clergy, generally, base enough, and impious enough," says Mr. Shorter, "to resort to these juggling tricks, and blasphemously appeal to Heaven with a lie in their mouths, and with the consciousness of so monstrous a fraud, this could scarcely have been done without the connivance of magistrates and civil rulers, who were not always well disposed to the Church, but not unfrequently looked upon the ecclesiastical authorities with a jealous eye."

The instances are quite numerous in which American mediums have thrust their hands into the flames of hot fire, and held them there for a minute or more.

At a *séance* in London, in 1860, in the presence of several persons (whose names are at the service of the curious), Mr. Home, being entranced, did, in the presence of all, lay his head on the burning coals, where it remained several moments, he sustaining no injury: not a hair of his head was singed.

A writer, to whose intelligence and veracity Mr. Shorter* bears testimony, has witnessed this fire-test several times; and, to bring up our chain of evidence to this year of grace 1868, we quote from his letter, of March of that year, to Mr. S.: —

"'The evening on which the phenomena I am about to relate occurred, had been full of interest, several very remarkable manifestations having taken place: such as the absorption of water by an unseen agency, and the retention of water in an open-necked bottle, though the same was *inverted* and violently moved and swung about. Mr. Home, who was all the time in a deep trance, now poured several drops of water upon his finger-points; and I noticed a slight jet of steam rise, hissingly, from the ends of his fingers, and accompanied by flames of electric light, or odic, of a violet, bluish color, half an inch to an inch in length, much resembling the drawings given in Reichenbach's works. Still continuing in a trance, Mr. Home now approached the fire, and, kneeling down before the hearth, proceeded to explain how great the power of spiritual beings was over matter, not because they worked miracles, but from their superior chemical knowledge; adding, 'We gladly have shown you our power over fluids: our power over solids is as great. Now see how I handle burning coal;' then laying hold of the burning back of coal in the hearth with his hands, he deliberately broke it asunder; and, *taking a large lump of incandescent coal into the palm of his hand* (the size of an orange), Mr. Home arose and walked up to Mrs. ——, whose alarm at what she was witnessing had quite unbalanced her. *I examined his hand and wrist;* the heat was so intense that it struck through the back of his hand, all but scorching his wristband; and Mr. Home then, addressing Mrs. ——, said, 'That is a burning

* Under the Latinized name of Thomas Brevior, Mr. Thomas Shorter, of London, a man of most unimpeachable integrity, and of rare ability, has contributed to the literature of Spiritualism some of the most valuable writings with which it has been enriched.

coal, A——; it is a burning coal; feel the heat of his hand. A burning coal will not hurt Daniel!—have faith!' *I closely examined his hand, and by the light of the glowing coal I could trace every line in the palm of the hand.* The skin was not, as will be surmised, covered by a glove, or steeped in a solution of alum: it was as clean as soap and water could make it. Mr. Home now explained that spiritual beings had the power of abstracting heat as a distinctive element; and to prove this he said, now mark:—

"'We will cool it now,—draw out the heat.' My doubts were by this time thoroughly aroused: I closely watched the process. On laying hold of the coal, which had become black, I found it to be comparatively cooled; and, taking it from his hand, I examined it carefully; so also the skin of his hand. At his request, I returned the coal into the palm of his hand; almost instantaneously, the heat returned; not to incandescence, only the caloric. On applying my hand to the coal, I burnt myself, and took conviction at the cost of a slight injury. I cannot say I doubted any more. The scrutiny I had submitted the hand of Mr. Home to precluded this; but, desirous of making certain of the fact of an unprotected surface of the hand of the medium being 'fire-proof,' I took Mr. Home's hand, rubbed it, moistened it; not a trace of any foreign matter, and, strange enough, no smell of smoke, or the burnt smell of fire observable. Mr. Home, who was still in a trance, smiled good-temperedly at my persevering efforts to undo my own conviction. . . .

"On another evening, Mr. Home, after he had shown us some truly remarkable phenomena, all whilst in a trance, knelt down before the hearth: deliberately arranging the bed of burning coal with his hands, he commenced fanning away the flames; then, to our horror and amazement, *placed his face and head in the flames*, which appeared to form a bed, upon which his face rested. I narrowly watched the phenomenon, and could see the flames touch his hair. On withdrawing his face from the flames, *I at once examined his hair; not a fibre burnt or scorched*,—unscathed he came out from the fire-test, a true medium.

"I am aware that great incredulity will reward my narrative. I give what I have seen, as a fact, refraining from explanation.

"That the fire-test has played its part in the records of every race of people, the veriest tyro in history knows. Fire-test was the crucial test of religious fanatics, whose unreasoning orthodoxy sought strength by imitating the wondrous phenomenon I have just been recording."

Thus, then, the credibility of the narratives from the Hebrew Scriptures is confirmed by, and they in turn confirm, the similar narratives which we find in various countries and centuries, even to our own. Their range is too extensive, many of them are too circumstantial and well attested, the testimony to the facts is too clear, too independent and concurrent, to permit us to assign them wholly to imposture. Make what large and liberal abatement you will for fraud on the one hand, and credulity on the other, you cannot altogether dispose of the question in that way; and any attempt to do so can only be fitly characterized as itself an experiment on the credulity of mankind.

Another extraordinary experience, of which Mr. Home has been the subject, is the elongation and shortening of his body. This was a phenomenon not unknown to the ancients, and to inquirers into the facts of witchcraft. Jamblichus, who flourished in the fourth century after Christ, writes, "The signs of those that are inspired are multiform. Sometimes there are pleasing harmonies, &c. . . . Again, the body is seen to be taller or larger, or is elevated, or borne aloft through the air; or the contraries of these are seen to take place about it."

Mr. H. D. Jencken, of Norwood, England, communicates, under date of December, 1867, his experiences at four *séances*, at which the body of Mr. D. D. Home was elongated and shortened; and on all these occasions Mr. J. used his utmost endeavor to make certain of the fact. On two of them, he had the amplest opportunity of examining Mr. Home, and measuring the actual elongation and shortening. At one, the extension appeared to take place from the waist; and the clothing separated eight to ten inches. Mr. J., who is six feet, hardly reached up to Mr. Home's shoulder. Walking to and fro, Mr. Home especially called attention to the fact of his feet being firmly planted on the ground.

"He then," says Mr. J., "grew shorter and shorter, until he only reached my shoulder, his waistcoat overlapping to his hip. . . . Encouraging every mode of testing the truth of this marvellous phenomenon, Mr. Home made me hold his feet, whilst the Hon. Mr. —— placed his hands on his head and shoulders. The elongation was repeated three times. Twice, whilst he was standing, the extension, measured on the wall by the Hon. Mr. ——, showed eight inches; the extension at the waist, as measured by Mr. ——, was six inches; and the third time the elongation occurred, Mr. Home was seated next to Mrs. ——, who, placing her hand on his head, and her feet on *his* feet, had the utmost difficulty in keeping her position, as Mr. Home's body grew higher and higher; the extreme extension reached being six inches."

"I could name many," writes the well-known Mrs. S. C. Hall, "who have been lifted out of the slough of materialism by, in the first instance, seeing the marvellous manifestations that arise from Mr. Home's mediumship." And she adds, "Mediumship is a mystery we cannot fathom, nor understand why the power should be delegated to one more than to another."

But it is equally perplexing why other gifts should be delegated to one person and not to another; why Mozart should be a consummate musician at five years of age, and another person should not, at fifty, be able to tell one tune from another; why an idiot boy should possess an astonishing power of computation, and another person, well-endowed in most respects, should not be able to do in a week what the other will do in a few seconds.

A certain "secularist" denies all authority to *instinct* in supplying hopes of a future life, inasmuch as *he* does not happen to be conscious of the existence in *himself* of that instinct which others undoubtedly have in a strong degree. But it is just as irrational for a man to deny immortality to others, because he himself may be unconscious of those transcendent faculties which are developed in mediums, as it would be for him to deny, because of his own deficiencies as a mathematician or a musician, the possibility of the existence of such mortals as Newton and Beethoven.

"Why has not Providence made the possession of all good things *universal and unexceptional?*" it may be asked. In other words, Why has not God made all intelligences perfect like himself? Why does he permit any existence but his own? The advocate of the theory of pre-existence says we bring our faculties from our anterior states; so that what we make our own we keep.

It is inscrutable, and seems unjust, that Providence should give my neighbor a faculty, and deny it to *me;* especially when I greatly desire and covet it. We cannot explain why Providence should be so partial; but *let us not, on that account, deny the fact.* Because Swedenborg, or the Seeress of Prevorst, or Andrew Jackson Davis, or Daniel Home, or Emma Hardinge may see a spirit, and we may not, let us not jump to the conclusion that they are either dupes or liars; especially when they prove to us, as they do, that they possess powers of prevision, or clairvoyance, which we do not possess (at least in our normal state), and which are such as we ascribe only to spirits.

There may be a faculty for apprehending spiritual truths, and for communicating with spiritual beings, just as there is for grasping the fundamental principles of mathematical or musical science. Where the faculty is deficient, let us beware how we deny the rightfulness of its existence in others, — pronouncing it a mere excrescence upon the human soul, to be removed by the surgery of those "secular" doctors, who think to cure the great heart of humanity of the hope of rejoining the loved ones gone before.

CHAPTER V.

THE SALEM PHENOMENA OF 1692 AND 1868.

"When a man is so fugitive and unsettled, that he will not stand to the verdict of his own faculties, one can no more fasten any thing upon him than he can write in the water, or tie knots of the wind." —*Henry More.*

AN elaborate work on Salem Witchcraft, from the pen of the Rev. Charles W. Upham, an esteemed Unitarian clergyman, was published in Boston, U.S., in 1867. Of it the "Edinburgh Review" (July, 1868) remarks, "No more accurate piece of history has ever been written."

Accurate in its facts it may be, and yet of questionable accuracy in the construction it puts on them.

If there is any thing in human history that is established by human testimony, it is the occurrence, in all the ages of which we have any authentic record, of phenomena, still familiar to multitudes, but which are now denied by a large class of minds; not because the phenomena are not vouched for by abundant testimony, but because they do not happen to accord with individual notions of the possible or the actual.

Sir William Blackstone did not depart from this world till four years after the declaration of American Independence. He was the contemporary of our immediate ancestors. His "Commentaries on Law and Testimony" are still so highly esteemed, that they have not to this day been superseded as the first work proper to be placed in the hands of the law student. Few men better qualified to weigh and scrutinize testimony, at once in a practical and philosophical spirit, have ever lived; and, on the subject of Witchcraft, Blackstone remarks, in the fourth book of his Commentaries, "To deny the possibility, nay, actual existence of witchcraft and sorcery, is at once flatly to contradict the revealed Word of God in various passages of both the Old and New Testament; and the thing itself is a truth to which every nation in the world hath borne testimony, *either by examples seemingly well attested*, or by prohibitory laws, which, at least, suppose the possibility of commerce with evil spirits."

Mr. Lecky, in his "History of Rationalism" (1864), shows that the testimony establishing the facts of witchcraft is of the most irresistible character. The accumulations of evidence are such as to amaze the skeptical student. "The wisest men in Europe shared in the belief of these facts; the ablest defended it; the best were zealous foes of all who assailed it. For hundreds of years no man of any account rejected it. Lord Bacon could not divest himself of it. Shakespeare accepted it, as did the most enlightened of his contemporaries. Sir Thomas Browne declared that those who denied the existence of witchcraft were not only infidels, but also, by implication, atheists."

The phenomena of witchcraft were real enough for the authorities in England and Scotland to burn the supposed witches by thousands; for Geneva (1515) to burn five hundred in three months; for the diocese of Como in Italy to slaughter one thousand; for a single diocese in France to destroy more than could be reckoned; for the little town of Salem in Massachusetts to put to death some of its best men and women.

All at once a re-action in public opinion took place, and the belief in witchcraft declined. From one extreme men went to the other. The re-action was at first not so much against the facts as against the fanatical construction put upon them; but the general discredit soon involved both. An unfavorable public opinion undoubtedly checked the development of *mediums* for the phenomena. Grief and indignation succeeded the wild credulity that had made innocent parties responsible for acts, the interpretation of which was to be found in a purely scientific study of the matter, free from all religious prepossession.

The marvels of witchcraft, as they were developed in Salem,

and are recorded by Upham, were of the same class with those phenomena which the present writer and thousands of other persons have witnessed, during the last thirty years, in cases of somnambulism, whether induced by mesmerism or independent of that influence, and, in the more recent manifestations, through persons called mediums. In the Salem phenomena there were violent convulsions of the bodies of those afflicted, especially when the supposed witch was near. There were surprising and apparently superhuman exhibitions of muscular strength. Violent motions in objects around, as if attracted and impelled by some mysterious force, were witnessed. A staff, an iron hook, shoes, keys, and even a chest, were seen to move, as if tossed by an invisible hand. A bed, on which a sufferer lay, shook most violently, even when several persons were seated on it. Stones were hurled against houses and persons: articles of iron, pewter, and brass were tossed about, a candlestick being thrown down, a spit flying up chimney, and a pressing-iron, a stirrup, and even a small anchor, being moved; of which facts many persons were eye-witnesses.

Mysterious *rappings* were also heard. Audible scratchings on the bedstead of a person affected were made. A drumming on the boards was heard; and a voice seemed to say, "We knock no more! we knock no more!" A frying-pan rang so loud that the people at a hundred yards' distance heard it. Sounds as of steps on the chamber-floor were heard. Divers noises as of the clattering of chairs and stools were heard in an adjoining room. Very varied are these instances.

Wonderful *powers of thought* and grace of expression were exhibited by the most ignorant and uneducated, and by persons of ordinary, and even of small, mental capacity. Of one person it is recorded, "He had a speech incessant and voluble, and, as was judged, in various languages." Of a little girl it is mentioned, "She argued concerning death, with paraphrases on the thirty-first Psalm, in strains that quite amazed us."

Cases of *mysterious knowledge*, like those now called clairvoyance, are reported, even by the coolest witnesses. Brattle mentions that "several persons were accused by the afflicted whom the afflicted never had known." Little girls thus affected described, as their tormentors, persons they had never seen; and by these descriptions the parents or friends of the girls sought out the accused, even in remote places.

Perhaps the most consistent explanation of this implication of innocent persons, by the children, and others who were the *mediums* on the occasion, is, that they were under the control of mischievous and malignant spirits, who found their pleasure in fixing suspicion on the wrong parties.

If we may believe Swedenborg, spirits are very human in their weaknesses. In his spiritual diary, he says, "When spirits begin to speak with man, he must beware, lest he believe them in any thing; for they say almost any thing. Things are fabricated by them, and they lie. . . . If man then listens and believes, they press on, and deceive and seduce in divers ways."

Of one of the little daughters of John Goodwin, of Boston, Mather says, "Perceiving that her troublers understood Latin, some trials were thereupon made whether they understood Greek and Hebrew, which, it seems, they also did; but the Indian languages they did not seem so well to understand."

We have repeatedly known a medium to be lifted in her chair from the floor on to a table, where there was no means of its being done by any known human agency, or mechanical contrivance. How like is this to the testimony of respectable citizens of Boston in 1693, in the case of Margaret Rule! "I do testify," says Samuel Aves, "that I have seen Margaret Rule lifted up from her bed, wholly by an invisible force, a great way towards the top of the room where she lay." "We can also testify to the substance of what is above written," say Robert Earle, John Wilkins, and Daniel Williams. "We do testify" — to a precisely similar occurrence, say Thomas Thornton and William Hudson.*

"We have in history," says Calmet, "several instances of persons full of religion and piety, who, in the fervor of their

* See Calef's "More Wonders of the Invisible World," p. 75.

isons, have been taken up into the air, and have remained here for some time. We have known a good monk, who rises sometimes from the ground, and remains suspended, without wishing it. I know a nun to whom it has happened, in spite of herself, to be thus raised up."

He mentions the same thing as occurring to St. Philip of Neri, St. Catherine Colembina, and to Loyola, who was "raised up from the ground to the height of two feet, while his body shone like light."

Savonarola, before his tragical death at the stake, and while absorbed in devotion, was seen to remain suspended at a considerable height from the floor of his dungeon. "The historical evidence of this fact," says Elihu Rich, in the "Encyclopædia Metropolitana," "is admitted by his recent biographer."

Indeed, the authentic instances of this phenomenon are far too numerous to mention.

Of certain children supposed to be bewitched, Mr. Upham writes, "The convulsions and paroxysms of these girls; their eyes remaining fixed, bereft of all light and expression; their screams, the sounds of the motions and voices of the invisible beings they heard; their becoming pallid before apparitions, of course seen only by themselves, &c., — were the result of trickery, was nothing but acting, but such perfect acting as to make all who witnessed their doings to believe it to be real. They would address and hold colloquy with spectres and ghosts; and the responses of the unseen beings would be audible to the fancy of the bewildered crowd. . . . But none could discover any imposture in the girls, . . . who had by long practice become wonderful adepts in the art of jugglery and probably of ventriloquism."

According to Mr. Upham, the witchcraft which manifested itself in Salem, in 1692, was attributable "to childish sportiveness; to the mischievous proceedings of the children in the Rev. Mr. Parris's family"!

There is an incredulity which it requires a good deal of credulity to arrive at in the face of notorious facts. Even the

"Edinburgh Review" — eulogistic as it is, and, for the most pa. justly, of Mr. Upham — rebukes him for confining his vie almost exclusively, to the theory of fraud and falsehood, a. affording the true key in dealing with these phenomena.

"Mr. Upham," says the reviewer, "is evidently *very far indeed from understanding or suspecting* how much light is thrown on the darkest part of his subject, by physiological researches carried on to the hour when he laid down his pen. . . . In another generation, the science of the human frame may have advanced far enough to elucidate some of the Salem mysteries, together with some obscure facts in all countries, which cannot be denied, while as yet they cannot be understood."

So far so good. But the reviewer, while reluctantly admitting facts that Spiritualism has forced upon the attention of the world, cannot avoid going out of his way to speak an ill word of those who have adopted the spiritual hypothesis, and to bring against them the charge of making several thousand lunatics for our asylums; a charge which the statistics of those asylums have repeatedly disproved, and which, if it were true, would be no argument against the prosecution of truth, any more than the fact that many thousands become insane from religious excitement would be an argument against religion.

Dr. Maudsley, a writer quoted with approval by the reviewer, shows the absurdity of thus charging a morbid tendency of the brain, ending in insanity, upon the mere *topic*, toward which the mind may have directed itself at a certain time. This "topic" may be denounced by shallow observers as the exciting cause; but a deeper diagnosis will prove that the true cause lay in the cerebral cells.

The reviewer calls the Spiritualists "a company of fanatics, . . . who can form no conception of the modesty and patience requisite for the sincere search for truth, . . . who wander in a fool's paradise," . . . and who are "partly answerable for the backwardness" of conservative men of science.

"Who excuses accuses," says the proverb; and the reviewer's apology for the men of science will be accepted only by simple-is. Here for twenty years have the Spiritualists been proiming certain facts and phenomena, which they have called upon the *savans* to investigate. The hypothesis as to the origin of these facts, whether mundane or ultra-mundane, had nothing to do with the facts themselves. A man who sees Mr. Home lifted to the ceiling may believe it was done by a spirit, or by a latent force in the individual himself, or in the surrounding spectators. All that Spiritualists have said has been, "Come and see the fact, and explain it then as you please. But do not denounce us as dupes and fanatics for believing the testimony of our senses. Do not expect us to be laughed out of the verdict of our own faculties, as poor Sir David Brewster was, after seeing the table move."

This, it is well known, has been the position of all intelligent Spiritualists; there being many, so-called, who believe simply in the facts, without attempting to explain them. And now the "Edinburgh Review," seeing that the time is coming when it must prepare for *a change of base* in regard to these facts (as it has done in regard to mesmerism), charges it upon the so-called Spiritualists, that by their *hypothesis* they have frightened off investigation! Bold investigators they must be who can be terrified by an hypothesis.

The late Dr. William T. G. Morton, when he was told that sulphuric ether would produce insensibility to pain, went on fearlessly and tested the fact, and became a great discoverer. As the "Edinburgh Review" would have said of him, "He could form no conception of the modesty and patience requisite for the sincere search for truth."

When Kate Fox heard the raps, she said, "Do as I do," and found that they were regulated by intelligence. She, too, could form no conception of this vaunted "modesty and patience." She imagined an hypothesis: she tried it; and the result was the production of the phenomenon.

Subsequent investigators into the phenomena have followed her example. They have interrogated the invisible power, whatever it may be, producing the manifestations; and, by adopting the hypothesis that it was intelligent, and could answer que tions, they have found that it could do so; and they have arriv at great results, just as other discoverers have, by simply leanin on an hypothesis.

And so when this learned reviewer charges Spiritualism with "deluding and disporting itself with a false hypothesis about certain mysteries of the human mind," he merely utters words of resentment that have no philosophical significance. He might as well abuse Columbus for finding America through the false hypothesis of its being the eastern end of Asia. If an hypothesis is adequate to the desired result, what absurdity to denounce a man for using it as a temporary scaffolding on which to mount!

"Hypotheses," says Novalis, "are nets: only he who throws them out will catch any thing."

But for the earnestness of investigators, a large class of facts, discovered, or, rather, rediscovered, by Spiritualism, would have been relegated to the oblivion where they have lain for ages. To this day, it has been a constant warfare on the part of Spiritualists, to establish these facts. Men of science and of learning, with here and there an exception, have done all they could to discredit and crush them out.

And now, when the facts number their believers by millions, and it begins to be impossible to ignore them longer, the "Edinburgh Review" — while it timidly admits some of the least remarkable of them — would blacken with its harmless ink the fair fame of the men through whose intrepidity, fidelity to truth, and impenetrability to precisely such sneers as the reviewer's, those pregnant facts have become the property of science once more.

And he stigmatizes these men as "fanatics"! Is he aware in what company the fanatics now find themselves? Not to mention those eminent men of the last generation, — such as Lavater, the physiognomist; Schubert, the philosopher; Goethe, Zschokke, Görres, Oberlin, Von Meyer, Ennemoser, Kerner, and many others, who were Spiritualists before the phenomena of 1848, —

we need but refer to the late Archbishop Whately, the late Lord Lyndhurst, the late Mr. Senior, the late Mr. Thackeray, the late Mrs. Browning, and other distinguished persons, by whom these phenomena were accepted as spiritual. Cardinal Wiseman admits them. So do Professor De Morgan, Mr. Robert Chambers; Bishop Clark, of Rhode Island; Mr. Varley, the electrician; the eloquent Jules Favre, a member of the French Academy; Garibaldi, Mazzini, and hundreds of eminent men, towards whom for this reviewer to affect contempt, would be simply ridiculous.

But it is only a short distance in the admission of facts that he has as yet gone. When, by and by, he is compelled to go further, and to accept the most surprising of the phenomena recorded in this volume, and to abandon his complacent theory that the marvels which Spiritualists proclaim are merely the chimera of "an objective world of their own subjective experience," he will, we hope, be a little more cautious in his sneers at men who, if they had heeded such ridicule as his, would long since have been checked in the investigation of facts, so repugnant to the preconceived notions of quarterly reviewers.

We refer to Mr. Upham's book, simply to call attention to the fact, that in his own town of Salem, at the very time he was writing a history of witchcraft, in which he sets down as delusions and tricks certain phenomena that were established as true in the minds of judges, juries, clergymen, and magistrates, by the overwhelming amount of evidence that was adduced, there lived (1865–68), hardly a stone's throw from his own house, a young man of the name of Charles H. Foster, born in Salem, 1838, through whom similar phenomena, quite as remarkable as any in the annals of witchcraft, might have been witnessed and fairly tested. Mr. Upham must have known of him by reputation; for Mr. Foster was and is widely celebrated, both in America and England, for his marvellous displays of a knowledge such as we call *spiritual*, inasmuch as it far transcends all that we can conceive of in our normal state of consciousness, as accessible to our bodily senses.

When Justinus Kerner was investigating (1826), through

Madame Hauffé (the Seeress of Prevorst), phenomena belonging to the same group with those of modern Spiritualism, the critics and reviewers, he tells us, instead of coming to see the facts for themselves, as they were invited, all rushed home, mounted their high stools, and began to write against the phenomena and everybody connected with them.

So has it been with Spiritualism; and now it looks as if the reviewers could never forgive the despised "fanatics" for getting hold of facts in advance of them, and making them commit themselves against them.

Of our own experiences with Mr. Foster, we will record only one class; but with this we have repeatedly been made familiar, both at his rooms and our own house. We have reason to believe that there are several thousand persons at this time, in America and England, who could confirm our experience by their own with the same medium.

Some time in 1861, seeing Mr. Foster's advertisement in the newspapers, we called on him at his temporary boarding-place, near the United-States Hotel in Boston. We had intimated our purpose to no one, either at the moment or previously. We had been asked by no one to attend. We had never seen Mr. Foster. He had never seen us, as he said and as we believe. We sought him simply in his capacity of a professional medium to test his powers.

He was alone in a small room, and we two remained alone during the sitting. The room was about 15 by 15, with two windows looking on the area back of the house. The curtains were up. It was noonday. There was no possibility of deception.

At his request, we wrote twelve names of departed friends on twelve scraps of paper, and rolled the scraps into pellets. We were at liberty to use our own paper, or to tear from what was lying on the table. Mr. Foster walked away from us while we wrote; and we were careful that he should not see even the motion of our hand.

The paper we used was fine as tissue paper. We folded, and then rolled up each piece separately, and pressed it till it was hardly larger than a common grape-stone. We placed the pellets on the uncovered mahogany of the table, and mixed them up. Mr. Foster ran his fingers rapidly over them, without taking up any one of them. Then, almost instantly, he pushed one after the other towards us, and, as he did so, gave us, without pause or hesitation, name after name, until he came to one which was a name so unusual, that we know of but two persons alive at this moment who bear it. "The name of this person will appear on my arm," said Mr. Foster; and, rolling up his sleeve, he showed us the name *Arria*, in conspicuous red letters, on the skin of his left arm.

He had given the names on eight of the pellets correctly in their order.

Having had enough to astonish us for one sitting, we did not ask him to do more that day. On many subsequent occasions, similar tests of power were given to us by Mr. Foster.

In this experiment it was impossible that he could get his knowledge from our mind. This is a favorite theory of the scoffers; but it will not apply here. We knew, it is true, the names that were on the pellets; but the pellets were so mixed up that we could not have told which was which, had our life depended on it. We might have guessed right once; but to do it eight times in succession was hardly in the range of possibility.

Where did Mr. Foster get the faculty of telling us what was written on each of those pellets?

In a pellet on which we had written the name of *George Bush*, we had added, as a further test of Mr. Foster's clairvoyance, these words: "Are these things truly from human spirits?" Seizing a pencil, Mr. F., with a nervous rapidity, wrote off the following reply before the pellet had been opened, and before we knew the name that was on it: "These communications are truly from the spirit-world. And is it not a glorious thought thus to be able to communicate with the beloved ones who have gone to the far-off spirit-land?" This, though not an exact reply to the inquiry, was near enough to excite astonishment. To Rufus Dawes, we wrote, "Old friend, what shall I think of it?" The reply was, "Think it is all right now. It is a boon given to man to prove to him the immortality of the soul. [Signed] Rufus."

The replies do not afford any satisfactory evidence of spirit identity, nor were the questions framed with that view; but what explanation can we give of the faculty that could read in every particular pellet, rolled into illegibility as it was, the name and the question?

The venerable John Ashburner, of London, editor of Reichenbach's "Dynamics of Magnetism," and long one of the most successful practising physicians in England, has given a narrative of his experiences in the presence of Mr. Foster. As many of these accord with our own, we give them in preference to extracts from our own notes.

"I have myself," says Dr. Ashburner, "so often witnessed spiritual manifestations that I could not, if I were inclined, put aside the evidences which have come before me. When Mr. Charles Foster was in London, in 1863, he was often in my house; and numerous friends had opportunities of witnessing the phenomena which occurred in his presence. . . . The second morning that he called on me was about two weeks after his arrival in England. Accidentally, at the same time arrived at my door Lady C. H. and her aunt, wife of the Rev. A. E. I urged them to come in, and placed them on chairs at the sides of my dining-table. Their names had not been mentioned; Mr. Foster having retired to the further extremity of the room, so as not to be able to see what the ladies wrote. I induced them each to write, upon separate slips of paper, six names of friends who had departed this world. These they folded into pellets, which were placed together.

"Mr. Foster, coming back to the table, immediately picked up a pellet, and addressing himself to Mrs. A. E., 'Alice,' he said, which made the lady start, and ask how he knew her name. He replied, 'Your cousin, John Whitney, whose name you wrote on that little piece of paper, stands by your side, and desires me to say, that he often watches over you, and reads your thoughts, which are always pure and good. He is delighted at the tenderness and care which you exhibit in the education of your children.' Then he turned towards me, and said, 'Alice's uncle is smiling benignantly, as he is looking towards you. He says, you and he were very intimate friends.' I said, 'I should like to know the name of my friend;' and Mr. Foster instantly replied 'Gaven. His Christian name will appear on my right arm.'

"The arm was bared; and there appeared, in red letters, fully one inch and a quarter long, the name *William* raised on the skin of his arm. Certainly, William Gaven was my dear old friend, and the uncle of the lady whose name is Alice. How, without yielding to the truth of the assertion of Mr. Foster, that he was a discerner of spirits, the fact could be known to a complete stranger, who had all his life resided in America, and could know nothing, even of the names of the ladies whom I had brought into my dining-room from the street-door, where I had accosted them, their names not having been known to my servants, is a phenomenon well calculated to puzzle the intellect of any one not having faith in Spiritualism. Mr. Foster's arm retained, on the surface of his skin, the raised, red letters for fully five minutes. I applied a powerful magnifying lens over them, and my two young friends and I watched them until they subsided and disappeared. It has been said that the skin was scratched by a pointed lead-pencil, and I knew some persons who wrote on their arms, and succeeded in raising red letters; but the letters did not so quickly subside, and in some instances left sore scratches, marks or tokens of the want of common sense.

"Mr. Foster next addressed himself to Lady C., whom he had never seen before in his life, until he met her in my dining-room. 'Your mother,' said he, 'the Marchioness of ——, stands by your side, and desires to give you her fond blessing and very affectionate love.' He added, 'Lady C., you wrote on a piece of paper I hold here the name of Miss Stuart. She stands by the side of your mother, and is beaming with delight at the sight of her pupil. She was your governess, and was much attached to you.' He added, 'That charming person, the Marchioness, was a great friend of the doctor's. She is so pleased to find you all here! Her Christian name is to appear on my arm.' Mr. Foster drew up his sleeve, and there appeared in raised, red letters, on the skin, the name *Barbara*, which subsided and disappeared gradually, as the former name *William* had done. Here were cases in which it was quite impossible that the medium could have known any single fact relating to the families or to the intimacies of any of the persons present. I had myself formed his acquaintance only two days, and the ladies had arrived from a part of the country with which he could not possibly be acquainted. It may be inquired very fairly, how it is proposed to connect such a narrative with any philosophical view of our mental functions? One need be at no loss for a reply; but it is more advisable at present to multiply our facts.

"My father was, in his youth, addicted to the pursuit of knowledge, and besides physics and chemistry, although he never proposed to become a professional physician, he studied anatomy at the Borough Hospitals, and had the late Mr. Cline for his teacher, and Sir Astley Cooper for his fellow-student. Mr. Foster had passed his life of twenty-four years in America. The son of a captain in a merchant ship, sailing from and to the port of Salem, in Massachusetts, he had never heard of Sir Astley Cooper. One evening, in my drawing-room, a hand, as palpable as my own hand, appeared a little above the table, and soon rested gently upon the thumb and four fingers on the surface of it. Several persons were seated round the table. Mr. Foster, addressing me, said, 'The person to whom that hand belongs is a friend of yours. He is a handsome man, with a portly presence, and is very much gratified to see you, and to renew his acquaintance with you. Before he mentions his name, he would like to know if you remember his calling your father his old friend, and yourself his young friend.' I had forgotten it; but I remembered it the moment the name was mentioned. 'He calls himself Sir Astley Cooper,' said Mr. Foster, 'and wishes me to tell you, that certain spirits have the power, by the force of will, of creating, from elements of organic matter in the atmosphere, fac-similes of the hands they possessed on earth.' Shortly, the hand melted into air. Then Mr. Foster said, 'Two friends of yours desire to be remembered to you. They accompany Sir Astley Cooper: one was a military surgeon, and went to Canada. He was at Edinburgh your fellow-student. He calls himself Bransby Cooper. The other was your intimate friend, George Young, who has communicated with you once before, since he left your sphere.'

"It would not be difficult to multiply facts relating to the spiritual manifestations of this very extraordinary medium. My friend, Sir William Topham, well known among all who have investigated mesmeric phenomena, as the person who induced on Wombwell, at the Wellow Hospital, that profound unconscious sleep, which enabled Mr. Squire Wood to amputate a most excruciatingly painful limb, above the knee, without the patient's knowledge, asked me to give him the opportunity of inquiring minutely into the phenomena, respecting which our friend Elliotson* and I were so completely divided in opinion. Sir William, with the concurrence of Foster, fixed an early day for dinner. There were only the three of us at the dinner-table. The servant placed the soup-tureen on the table. No sooner had I helped my friends to soup, than Sir William, who had preferred the seat with his back to the fire, requested permission to alter his mind, as the fire was too much for him. He went to the opposite side of the table, forgetting to take his napkin with him. Immediately, a hand, apparently as real as the hand of any one of us, appeared, and lifted the napkin into the air gently and gracefully, and then dropped it carefully on the table.

* Elliotson, as we have already seen, after having been a materialist up to his seventieth year, came round finally to Ashburner's views of the phenomena, and died a happy believer in them.

Almost simultaneously, while we were still engaged over our soup, one side of the dining-table was lifted up, as our philosophic friend Mr. Faraday would conclude, by unseen and *unconscious muscular energy;* and the moderator lamp did not fall from its place on the centre of the table. The decanters, saltcellars, wineglasses, knives and forks, water-carafes, tumblers, all remained as they were in their places: no soup was spilled; and Faraday's unconscious muscular force, or some correlative or conserved agency, prevented the slightest change among the correlative ratios of the table furniture, although the top sloped to very nearly an angle of forty-five degrees. There was a wonderful conservation of my glass, china, and lamp. The servant who was waiting upon us stared, lifting up both arms, exclaimed, 'Law! well, I never!' and the next minute he cried out, 'Do, do look at the pictures!' which, with their ten heavy frames, had appeared to strive how far they could quit the wall, and endeavor to reach the dinner-table.

"The appearance of hands was by no means an unusual phenomenon. One evening, I witnessed the presence of nine hands floating over the dining-table.

"On one occasion, the Hon. Mrs. W. C. and her sister-in-law desired to try some experiments in my *dunker kamer*, a room the Baron von Reichenbach had taught me how to darken properly for experiments on the od force and the odic light emanating from living organized bodies. This room afforded opportunities for marvellous manifestations. When the light was excluded, the two ladies were seated on one side of a heavy rosewood occasional table with drawers, weighing at least seventy or eighty pounds; Mr. Foster and I were on chairs opposite to them. Suddenly a great alarm seized Mr. Foster: he grasped my right hand, and beseeched me not to quit my hold of him; for he said there was no knowing where the spirits might convey him. I held his hand, and he was floated in the air towards the ceiling. At one time, Mrs. W. C. felt a substance at her head, and, putting up her hands, discovered a pair of boots above her head. At last Mr. Foster's aërial voyage ceased, and a new phenomenon presented itself. Some busts, as large as life, resting upon book cupboards seven feet high, were taken from their places. One was suddenly put upon Mrs. W. C.'s lap; others, on my obtaining a light, were found on the table. I removed these to a corner of the room, and put out the light. Then the table was lifted into the air, and there remained for some seconds. Then it gently descended into the place it had before occupied, with the difference that the top was turned downwards, and rested on the carpet. The ladies were the first to perceive that the brass casters were upwards.

"One of these ladies had missed, on another occasion, her pocket-handkerchief. Mr. Foster told her she would find it in the conservatory behind the back drawing-room. It was behind a flower-pot. Mrs. W. C. went up-stairs, and found the handkerchief in the spot indicated. A similar event happened a second time. The question was, How the pocket-handkerchief could travel from the dining-room, all doors being shut, to the floor above, where it was deposited on a shelf in the conservatory. Mr. Faraday would aver that my facts were corroborative of his conservation of force.

"In that back drawing-room stands a heavy Broadwood's semi-grand pianoforte. Mr. Foster, who is possessed of a fine voice, was accompanying himself while he sang. Both feet were on the pedals, when the pianoforte rose into the air, and was gracefully swung in the air from side to side, for at least five or six minutes. During this time, the casters were about at the height of a foot from the carpet."

Mr. Foster's first indications of mediumship took place when he was about fourteen years old, at the Phillips school in Salem, where his attention was arrested by raps near him on his desk during school-hours. The next change was to violent noises near his bed at night, which at once awakened him, and brought his parents into his room, where the furniture was found tossed about in all directions. At first this happened only in the dark; but soon it came in the light, and furniture would be heard moving about in rooms where no person in the flesh was present.

At his manifestations on one occasion, when letters were coming on his skin, two men seized him rudely by the arm to discover the trick, as they called it. "We know," said they, "that no letters will come on the arm while we hold it." "What will you have?" asked Foster. "Something that will be a test," cried they; "something that will fit our case." Immediately, while they were holding the arm, as in a vice, and glaring upon it with all their eyes, appeared in large round characters the words, *two fools!*

Of this phenomenon of *stigmata* on the flesh, the instances are numerous and thoroughly authenticated. Ennemoser, Passavent, Schubert, and other eminent German physiologists, admit the fact as not only established as regards many of the so-called saints of the Catholic Church, but in undoubted modern instances, as in the case of the *ecstaticas* of the Tyrol, Catherine Emmerich, Maria Dorl, and Domenica Lazzari, all of whom exhibited the stigmata. The signatures of the fœtus are analogous facts; and if the mind of the mother can act on another organism, why, it is asked, should not the minds of mediums act on their own? The fact of the phenomenon has been placed by testimony beyond the dispute of any but the ignorant. We have witnessed repeatedly under circumstances where to doubt would have been to reject all rational proof as worthless.

We have spoken of the Salem phenomena of 1692, as analogous with those of 1868. While we write, additional proof of this is offered. Indeed, the candid chronicler will find himself embarrassed by the number of confirmatory narratives, old and new.

In July, 1868, occurrences of an inexplicable character took place in the house of a Mr. Travis at Thorney, a small hamlet near Muchelney, and about two miles from the town of Langport, in England. The following account is from the "Bristol [Eng.] Daily Post," of August, 1868: "It is said that even the walls shake at times; while the doors and windows are opened and closed again very frequently in a most forcible manner.

Pillows and bolsters are taken from beneath the occupants of beds. Noises, ranging from the reports of many muskets to the distant boom of a field-piece, are heard in different parts of the house. Scores of persons attest the accuracy of these statements. Most of them avow that no human agency could do what they have seen done and escape detection. If there is any thing true in the doctrines which the Spiritualists preach, they may make converts by the hundred in this neighborhood."

Another English journal, the "Western Gazette," of July 31, 1868, describing these occurrences, remarks, "Of course, a great philosopher cannot be expected to investigate a 'trumpery ghost story,' or a 'silly tale of a haunted house.' He *knows* that it is impossible for a table to move without hands; and it would, therefore, be only a waste of his valuable time to inquire whether a table has ever done so or not. This, we fear, is the view which too many of our all-knowing *savans* will take of the Muchelney business. But is such a view truly philosophical? Do we know every thing yet? Are there no natural laws or forces yet to be discovered? — no exceptions, or apparent exceptions, to the operation of known laws to be determined?

"We may safely assert that it is impossible that one and one can ever make three, or that the three angles of a triangle can ever make more or less than two right angles; but, once clear of mathematics, we can never be safe in using the word 'impossible.'

"A generation that sees two men on opposite sides of the globe, conversing with each other by means of an ubiquitous agent, that is known only by its effects, can surely believe in almost any thing, except the incorrectness of the multiplication-table.

"We have no well-defined theory on the subject of these phenomena; but we are convinced that there is no trickery in *this* case; that the phenomena are due to causes of which science has, as yet, taught us nothing; and that we should act in an unphilosophical spirit if we rejected the evidence of our own and others' senses because of its apparent inconsistency with the littl which we happen to know of nature's laws."

How much longer will a false conservatism think to put out of existence verified facts, like these, by uttering its shrill negative cry, shutting its eyes, and burying its ostrich head in the sand!

CHAPTER VI.

VARIOUS MEDIUMS AND MANIFESTATIONS.

"Now there are diversities of gifts, but the same spirit. . . . Have all the gifts of healing? do all speak with tongues? do all interpret?" — *St. Paul.*

THE number of persons in the United States through whom phenomena similar to those in the case of Mr. Home and of Mr. Foster have taken place, is now so large, that to mention them all would almost require a volume. Charles Colchester, young and of a fine personal appearance, but wayward and infirm of purpose, like many similar *sensitives*, gave us several sittings, at which he manifested remarkable powers, not unlike those of Mr. Foster.

We know not how true it is, but Mr. Colchester told us, that on his meeting Hermann, the celebrated *prestidigitateur*, the latter said to him, "If you can give me the name of my father, I will believe that your intelligence is preternatural; for no person in America, I am convinced, knows that name." Colchester at once wrote out the name, *Samuel Hermann Radesky;* and Hermann said it was right.

Mr. William Ambisy Colby relates (July 6, 1861) that he called on Colchester in New York. "I first asked him," says Mr. Colby, "if he could tell me what I had lost. He told me I had lost a pocket-book with papers in it of no value; that it was picked from my pocket in a Broadway stage. I then told him that he was mistaken; for there was a paper amongst them of value. 'Oh, no!' said Colchester, 'I am not mistaken; but it is *you* who are mistaken. The paper you have reference to is a check for $315, which, instead of putting in your wallet, you put in your hat, inside the lining.' I immediately looked in my l and, sure enough, the check was there, just where Colchester t(me it was."

Here, surely, was information, on the medium's part, quite independent of any *conscious* knowledge in the mind of the inquirer. Yet how often we are told, that the medium's knowledge is always got from the mind of the inquirer!

In a letter to our friend, Mr. Coleman, of London, Professor W. D. Gunning writes (1868) from Boston, as follows in regard to Mrs. Cushman, a medium who resides in Charlestown, Mass.: "I visited her house in company with a Boston clergyman. A guitar was laid on my knees, and after a few minutes lifted up, held in the air, and played upon by unseen hands. This was in full daylight. The concert lasted an hour. It was utterly impossible for the lady to touch the strings. No mortal, under the circumstances, could have made the music. Of this we were both satisfied. We did not decide hastily, but only after the fullest investigation. Now, the agent that played the guitar, whatever it was, acted wonderfully like a human being. We requested a particular tune: it was played; then another: that was also played; and so on for an hour. How could we resist the conviction that here, unseen by us, was a spiritual being, a man or woman, knowing the music that we knew, hearing our words or reading our thoughts, and able, under conditions we may not understand, to move material things? 'We are compassed about with a cloud of witnesses.' We need to return to the early faith, the faith of the founders of Christianity, the faith of all great poets of all ages. This age is steeped in materialism; but re-action has begun. Men are crying out for the knowledge of Eternal Life. With the eloquent Bishop of Rhode Island, I hail this influx from the spirit-world as a gift of the Father, sent in his own good time to his children, to wean them from doubt, to confirm them in faith, to take away the sting of death by the knowledge that immortality means no gauzy abstraction, but *real human* life."

In the winter of 1860–61 we tested, on some forty occasions, e mediumship of Miss Jennie Lord (now Mrs. Webb), through whom physical manifestations, of quite a startling character, were produced. These took place several times at our own house, where the possibility of trick or collusion was carefully excluded. In two or three instances a friend (the same who wrote the incredulous letter from Rochester in Chapter II. of this volume) was present, at our invitation; and once we were present at his, when the phenomena took place in his own house in Boston. While the occurrences of the evening were fresh in our minds, we each wrote an account of them; and his narrative was afterwards put in the form of a letter and sent to our mutual friend, William M. Wilkinson, in England. We subjoin it:—

"DEAR SIR,—I wish to send you an account of some spiritual manifestations which I have lately witnessed; and which, indeed, have been the only experiences of the kind, which I have had since I saw you in London last June. As you know, I have been long absent from this city, sojourning in France and Italy for four years.

"On my return here, I found that, among my immediate friends, Spiritualism was regarded as a something dead. But the only reason which my informants could give for their belief, was that they had not heard the subject mentioned for a year or two. However, I asked them, and they smiled when I did so, whether the northern lights would become incredible by not being talked about.

"Through a friend, whose name and judgment are a sufficient guarantee for whatever he may choose to vouch, I heard lately of a medium whom I had never known before. That medium is a young fragile woman. Last Tuesday evening she came to my house. I had some friends to meet her. Altogether for the *séance*, we were eight in number. It was explained to us, that the medium would pass into a state of trance, and that the room would have to be darkened. 'Oh!' says some skeptic, 'a dark room! That is enough for me.' Perhaps so; and perhaps also it would be enough for him, equally, if it were insisted that mediumship was impossible in the dark, and possible only in room all ablaze with light. But, before we advance furthe I will ask this skeptic, why is it that an iron ball will retain hea in the dark longer than in the light? And perhaps in ascertaining that, he may learn something which may help in the inquiry, why spiritual mediumship is sometimes stronger or more effective when the light is excluded?

"There were ranged on a table, about two feet behind the medium, the articles which it was understood would be in requisition during the evening. About the placing of the articles, there was no mystery made, nor was any jugglery possible, in connection with the manner in which they were disposed. We sat round a table; and, after a little singing, the medium passed into a state, apparently, of trance. The expression of her face was much changed, was much refined and beautified. The last light was extinguished. All round the table, we held one another's hands, except the medium; and she, instead of holding my hand, laid her hand upon mine, drawing her hand along it, as though for some mesmeric purpose. Her other hand was placed similarly on the hand of one of my friends, who sat on the other side of her.

"For persons hard of belief, I would remark, that if darkness be unfavorable, in some respects, for detecting imposture, it is also very unfavorable, in a strange place, for the operations of one who would cheat. I wish it too, to be fully understood, that, throughout all the wonders which happened, we had full knowledge of each other's hands every moment. Several times when the phenomena were most remarkable, I said to my friends, 'Now are we all sure, that we, every one, have charge of the hands which we ought to be holding?' And the answer was, 'Yes: we are all satisfied.'

"A bell was carried round the room, ringing; was rung over our heads, and was placed against my cheek. A guitar was played upon, as it was carried about the room. It was laid on our heads and played upon. It was whirled over our heads so rapidly, that we felt the wind of it, as it went round and round. It was rapped on the heads of five or six persons; it was rattled among the glasses of the chandelier; it was struck on the floor, and thrown on to the table, — and all this, as it seemed, in a moment. The quick, versatile movement of the instrument I can liken to nothing so much as to the darting of a fly to and fro.

"A glass of water was placed to my lips, in the neatest manner possible; and I drank from it. And it was carried round to the lips of other persons at the table. A tambourine was beaten as it was borne about the room. It was struck on our heads; and it was shaken above us with great force. A horn was blown, and made a noise almost terrific. With several of us a sheet of paper was spread over the face, and through it we felt distinctly the pressure of a hand. A hand, without any thing intervening, was placed on my head. It was a large hand; and it grasped my head firmly, and shook it. It took hold of a lock of hair over my forehead, and pulled it. That these things were not done by persons of flesh and blood, I know thoroughly well.

"I have an acquaintance, who was wont to be a very fierce and bitter opponent of Spiritualism. He used to account mediumship as an imposture, a transparent and a gross imposture; a most cunning imposture; and also a most simple kind of imposture. Now, lately he said to me, 'Blowing a horn, playing a guitar! What is the good of that?'

"I answered him, 'My friend, I did not say there was any good in it. I merely said there was a fact in it, and that fact, the operation of a spirit. And if you think that to be nothing, why, then you must think very differently now from what you did, when the mere supposition that a spirit might rap on the table used to make you foam with excitement, as you remember.'

"'Ah, well!' he said, 'but what now do you think is the use of it? And why cannot it be done anywhere by anybody? And if spirits can do such things as you say, why can they not tell us something useful, whether there is going to be a war'——

"'And perhaps you would add,' I replied, 'how to square the circle, how to be infallible as to latitude and longitude at sea, and how to find the philosopher's stone. But, my friend, it may be that many a spirit is less intelligent than you yourself are. For, when you think of it, what a way to wisdom that would be, for a spirit to become omniscient with merely slipping off his overcoat of flesh!'

"'But — but — but why do they not teach us something — some of them? And is it not true, that they often tell lies? And, in fact, somehow I can make nothing out of it.'

"To this I answered, 'That is very probable! and no great wonder. And by the way of mediumship, as to spirits telling falsehoods, as you suppose they do sometimes, why that would show at least that there are lying spirits. And that thing made certain to you as a fact, would be a matter of more importance, infinitely, than the discovery of twenty new comets. And now as to a spirit blowing a horn or beating a tambourine, you think it is nothing. But, for myself, I think that it implies a spirit present who is the actor; that it proves that, under certain circumstances, spirits have power over matter; and that it suggests many subjects for the most serious consideration of the theologian, the moralist, and the man of science.

"I am, yours truly, W. M."

On several occasions we have known Miss Lord to be lifted bodily with her chair, while she was seated in it, from the floor on to the table, by some unknown force. In a like mysterious manner, at our own house, a large bass-viol was played on vigorously and with fair skill, while the medium's hands were held by us, and deception was impossible. "Coronation" and other sacred tunes were thus played. The *power*, whatever it was, would, before playing, spend a minute or two in tuning the instrument, and would then indicate its readiness by tapping the heads of certain persons in the audience with the bow. A large, flexible hand, full of life and guided by intelligence, and which was nearly twice the size of Miss Lord's hand, touched us and others repeatedly on the head, pulled our hair, took down the hair of our sister, and then put it up as before, placed a tumbler filled with water at our lips, and this at the right angle, and with the nicest adjustment, so that not a drop was spilt. These manifestations, though in utter darkness, were of such a character, and produced under such conditions, as to render imposture impracticable.

A writer in "Once a Week," a London journal, recently undertook to account for the phenomenon of the "spirit-hand" by the theory that the effect was accomplished by the aid of an instrument he calls the *lazy-tongs*. It is perhaps superfluous to say that his explanation is now obsolete, like the toe-joint theory to explain the rappings: it does not begin to cover the facts.

Another medium, through whom we have witnessed some astonishing phenomena, though we have not had opportunities of testing them as thoroughly as those through Miss Lord, is Miss Laura V. Ellis, of Springfield, Mass. This young lady was only fourteen at the time we first saw her in the summer of 1866. She entered a small movable cabinet or closet, and while she was tied there in the most thorough manner, the door was closed, whereupon, in an incredibly short space of time, various manifestations requiring the free use of hands or feet took place. The following account by Mr. L. J. Fuller of a sitting at Willimantic, Conn., February, 1867, corresponds with our own experience: —

"After Miss Ellis was tied in the usual way with strips of cloth, the knots were sewed through and through, and then the ends of the cloth sewed strongly to the sleeves of her dress; after which she was firmly secured in the cabinet, when the following manifestations were given: A string was tied around her neck in a square knot in six seconds; this was repeated twice, with the same results. A string was tied around the waist in four seconds; repeated twice, once in four seconds, and once in three; tied around the back of her neck in eight seconds; front of her neck, fifteen seconds; repeated in fourteen seconds; untied from her neck in fifteen seconds; untied from front of neck in three seconds; bell rung in two seconds; repeated in four seconds; loud raps with stick in two seconds; repeated in one second; stick thrust through the aperture of the cabinet fourteen inches, and afterward thrown ten feet from the cabinet; playing on the tambourine in one second; playing on the trombone in one second; also singing, and keeping time with the trombone; drumming, whistling, and keeping time with the jews-harp, and other instruments; besides many other and varied manifestations. Her hands were then untied and extended horizontally, and tied to staples, so that by turning the hands toward the head, which was fastened back to the cabinet, the nearest they could come to the ends of the knot was twelve inches from them. The knot was untied the first time in thirty seconds, and the second time in twenty seconds.

"The whole was done under the closest scrutiny of a committee of three, no one of whom could detect the slightest evidence of collusion during the whole entertainment. The medium's hands were repeatedly examined during the whole time of the entertainment, and found in the same condition as when first tied. No show of any effort on her part could by the closest scrutiny be detected; and all unprejudiced minds were satisfied that the manifestations were produced by some power outside of Miss Ellis."

On another occasion, at Keene, N.H., according to the report of Mr. Henry Woods, "a trombone, harmonicon, tambourine, and drum were played, and other feats performed; all these feats being done while Miss Ellis's wrists were securely tied at her back, and to the cabinet, her ankles tied, and neck also fastened to the cabinet. Last, but not least, a knife with the blade shut, having been laid in her lap, was taken and used to cut her loose from the cabinet, and to disengage her wrists, the knife being then left half-way open in her lap. Let none say that these things are accomplished by trickery, until they have been personal witnesses of the wonderful phenomena presented."

Various attempts have been made to prove that these phenomena are mere tricks; and several imitators, some perhaps with the partial aid of forces similar to those operating through Miss Ellis, have undertaken to show that the manifestations could all be accomplished by manual dexterity; but, thus far, no one has succeeded in indicating this to the satisfaction of candid committees. It is not uncommon for partially gifted mediums to try to excite attention by denouncing the manifestations through their more successful brethren or sisters as fraudulent; but, when it comes to the proof, they always fail of proving *in the light*, that all the phenomena can be produced by trick or skill.

Under date of Nov. 24, 1867, Mr. W. A. Danskin, of Baltimore, Md., gives an account of a youth, about nineteen years old, and whose head measured twenty-two inches round, from whose neck a solid iron ring weighing fourteen ounces, and measuring but fifteen inches on its inner circle, was taken and replaced. The ring was submitted to the closest inspection, both before the experiment and while on the neck.

On one occasion another ring, precisely similar in appearance to the one ordinarily used at the exhibition, was made, marked by four indentations while the metal was soft, and brought to the hall, at one of the public exhibitions, without the knowledge of the medium or his friends. The parties having it in charge watched their opportunity, and substituted the marked ring for the original. The manifestation was successfully given, though the time of it was somewhat extended, and the medium was much exhausted.

"Once," says Mr. Danskin, "when only three persons were present, — the medium, a friend, and myself, — we sat together in a dark room. I held the left hand of the medium, my friend held his right hand, our other hands being joined; and, while thus sitting, the ring, which I had thrown some distance from us on the floor, suddenly came around my arm. I had never loosened my hold upon the medium, yet that solid iron ring, by an invisible power, was made to clasp my arm, thus demonstrating the power of our unseen friends to separate and *re*-unite, as well as to expand, the particles of which the ring was composed."

The following testimonial is signed by thirty-one gentlemen of Baltimore, whose names may be found published in the "Banner of Light," of Jan. 11, 1868:—

"We, the undersigned, hereby testify that we have attended the social meetings referred to; and that a solid iron ring, seven inches less in size than the young man's head, was actually and unmistakably placed around his neck. There was, as the advertisement claims, no possibility of fraud or deception, because the ring was freely submitted to the examination of the audience, both before and while on the neck of the young man."

This extraordinary medium died of consumption of the lungs, July 2, 1868. Since his death, Mr. Danskin writes, "The ring manifestation was entirely free from deception or fraud; and, under the conditions established, fraud was absolutely impossible." He is confident that in no single instance did this medium attempt to impose on any one.

Some surprising manifestations, through Mr. Charles H. Read, of Buffalo, have been witnessed during the summer and autumn of 1868. The "Daily Times," of Brooklyn, N.Y., in its issue of April 3, 1868, has a clear and accurate account of these phenomena, of which ours is an abridgment.

Precautions were taken against the possible intrusion of any confederate. Mr. Read was securely tied. The wrists were made fast until the cord settled well into the flesh; it was then drawn between the knees, the ends being carried down with two well-jammed turns on the front rung of the chair, and then back to the rear rung, where the end was made fast with several half-hitches. The arms were secured and tied to the back of the chair, and the legs fastened at the ankles to the rear legs of the same. Being seated in position, and at a distance from the table, the gas was turned off; and in *about one-half of a minute*, on being re-lighted, *one of the rings encircled his arm.*

The fastenings were instantly examined, and found undisturbed. During the dark interval, some singing was indulged in. Supposing a confederate to have been able to pass the twine barrier without ringing the bell, he could not, in half a minute, have untied the ropes so as to slip the ring on the arm, and re-tie them again; for it required more than five minutes to adjust them in the first instance, and the same knots could not have been even simulated. The ring still remaining on the right arm, the gas was again extinguished; and in less than a minute the light revealed *the stool on his arm;* or, in other words, the ring was on the floor, where it had been heard to fall, and the stool had taken its place. There was no movable ring which could have been removed so as to slip the stool-leg down between the arm of the medium and his body. The ropes and knots were still intact.

Once more was darkness; and the next revelation was the medium's *coat off and on the floor*, against the wall, at some distance from his position. The fastenings were again examined: not the least slackening was found. A further test was made, and the stool appeared on the other arm.

At the request of the demonstrator, the writer placed his own coat on the table; and, in less time than this sentence may be written, he beheld one of Mr. Read's arms in the sleeve of the garment, which could not be removed without cutting or untying the ropes. A moment or two of darkness, however, sufficed to find it thrown to one side of the apartment. During these demonstrations the medium seemed to become gradually weak and exhausted, as if he had been rudely handled. Finally, there was more darkness; and in a little more than a minute, counted by a healthy and regular pulse, there came a sound as of something thrown aside, which the gas revealed as *the ropes on the chandelier*.

The man was entirely free, and before him dangled the fastenings. His wrists showed deep indentations; and his hands were swollen, from partial suspension of the circulation of the blood. The reader may be assured that in all this there was not, and could not be, the slightest collusion. Mr. Read could not untie himself, nor could he be approached by a confederate.

Similar phenomena through Mr. Read, accompanied with

touches from spirit-hands,* on the persons of several among the audience, were witnessed on the evening of Sept. 8, 1868, at the residence of Mr. Z. A. Willard, of Boston.

Vocal manifestations have been not unfrequent in the history of supposed spiritual disturbances. Some very singular occurrences took place in the family of John Richardson, in Hartford, Trumbull County, Ohio, the latter part of the year 1854. The affidavits of himself, his wife, and Mr. James H. More, bearing date Jan. 8, 1855, were duly made before Mr. William J. Bright, a justice of the peace, who, in communicating them to the public, says, "The facts are of public notoriety here, and can no doubt be sustained by any amount of evidence."

The wildest doings of the days of witchcraft are paralleled in the following narrative, which we quote for its explicit testimony in regard to the vocal manifestations: —

"About five weeks ago," deposes Mr. Richardson, "my attention was arrested by a very sharp and loud whistle, seemingly in a small closet in one corner of my house. This was followed by loud and distinct raps, as loud as a person could conveniently rap with the knuckles. The closet-door is secured or fastened by a wood button that turns over the edge of the door. This button would frequently turn, and the door open, without any visible agency. This was followed by a loud and distinct (apparently) human voice, which could be heard, perhaps, fifty rods.

"After repeating a very loud and shrill scream several times, the voice fell to a lower key, and, in a tone about as loud as ordinary conversation, commenced speaking in a plain and distinct manner, assuring the family that we would not be burned, and requesting us to have no fear of any injury, as we were in no danger. Those manifestations being altogether

* Prof. Denton, the accomplished geologist, author of a remarkable work, entitled "The Soul of Things," says, "I have seen spirit-hands over and over again, — have taken impressions of them in flour and putty and clay." We have a letter from Dr. J. F. Gray, describing his examination of a spirit-hand in the light.

unaccountable to myself and family, we searched the entire house, to find, if possible, the cause of this new and startling phenomenon, but found no one in or about the premises but the family. Again we were startled by a repetition of the screams, which were repeated perhaps a dozen times, when the voice proceeded to inform us that the conversation came from the spirit of two brothers, calling themselves Henry and George Force, who claim to have been murdered some eleven years since; and then gave us what they represented as a history of the tragedy, and insisted that we should call on some of the neighbors, to hear the disclosure. John Ranney, Henry Moore, and some dozen others, were then called in, to whom the history was detailed at length. We could readily discover a difference in the voice professing to come from the two spirits.

"About the third day after these manifestations commenced, my wife brought a ham of meat into the house, and laid it on the table, and stepped to the other side of the room, when the ham was carried by some invisible agency from four to six feet from the table, and thrown upon the floor. At another time, a bucket of water was, without human hands, taken from the table, carried some six feet, and poured upon the floor. This was followed by a large dining-table turning round from its position at the side of the room, and being carried forward to the stove, a distance of more than six feet. This was done while there was no person near it. The same table has, since that time, been thrown on its side without human agency, and often been made to dance about while the family were eating around it. At one time, dishes, knives, and forks were thrown from the table to the opposite side of the room, breaking the dishes to pieces.

"On another occasion, the voice requested Mrs. Richardson to remove the dishes from the table, which was done immediately, when the table commenced rocking violently back and forward, and continued the motion, so that the dishes could not be washed upon it, but were placed in a vessel and set upon the floor, from which a number of them flew from the tub to the chamber-floor overhead, and were thus broken to pieces. What crockery remained we attempted to secure by placing it in a cupboard, and shut the doors, which were violently thrown open; and the dishes flew, like lightning, one after another, against the opposite side, and broke to pieces. At another time, a drawer in the table was, while there was no person near it, drawn out; and a plate that had been placed there carried across the room and broken against the opposite wall. And this kind of demonstration has continued until nearly all the crockery about the house has been broken and destroyed.

"At different times, the drawers of a stand in a bedroom have been taken out, and at one time carefully placed on a bed. A large stove-boiler has been, while on the stove, filled with water, tipped up, and caused to stand on one end, and the water was turned out upon the floor, and at this time taken off from the stove, and carried some six feet, and set down upon the floor, and this while untouched by any person. A teakettle has often been taken from the stove in the same manner, and thrown upon the floor. At one time, a spider, containing some coffee for the purpose of browning, was taken from the stove, carried near the chamber-floor, and then thrown upon the floor. And frequently, while Mrs. Richardson has been baking buckwheat cakes on the stove, the griddle has, in the same unaccountable manner, been taken from the stove, and thrown across the house; and often cakes have been taken from the griddle while baking, and have disappeared entirely.

"At one time, the voice, speaking to my wife, said it (the spirit) could bake cakes for George, a boy eating at the table. Mrs. Richardson stepped away from the stove, when the batter (already prepared for baking cakes) was by some unseen agency taken from a crock sitting near the stove, and placed upon the griddle, and turned at the proper time, and when done taken from the griddle, and placed upon the boy's plate at the table. The voice then proposed to bake a cake for Jane, my daughter, who was at work about the house. The cake was accordingly baked in the same manner as before stated, and carried across the room and placed in the girl's hand.

"During all these occurrences, the talking from the two voices and others has continued, and still continues daily, together with such manifestations as I have detailed, with many others not named. The conversation, as well as the other demonstrations, have been witnessed almost daily by myself and family, as well as by scores of persons, who have visited my house to witness these strange phenomena.

"I will only add, that the spirit (the voice) gave as a reason for breaking crockery and destroying property, that it is done to convince the world of the existence of spirit presence."

Several instances are related in which photographs of supposed spirit-forms have been taken. In the autumn of 1862, the "spirit photographs," said to be got through Mr. Mumler, a Boston photographer, were a subject of much controversy. Of course, the many believed Mr. Mumler an impostor; but no evidence whatever of this was ever adduced, and now, after a lapse of years, we cannot learn that any one who knows him personally can harbor a doubt of his honesty. On the contrary, the testimony from friends and acquaintances, familiar with his character and social position, is strongly in his favor. What gave rise to distrust was the fact that many persons who sat for photographs of some spirit-friend, to accompany their own, would find, instead of a likeness they could recognize, a figure that would be a puzzle to them and all their acquaintances. Of course, the immediate cry, in these cases, would be "Humbug!" But the failure was no evidence whatever that the questionable figure was not photographed on the plate in some way wholly inexplicable to the operator.

Mr. William Guay, an experienced photographer from New Orleans, and who, though a stranger, was admitted to scrutinize the whole process, reported favorably to Mr. Mumler's entire sincerity in the matter, and satisfied himself that *a second form did appear on the negative without a visible object to produce it.* Mr. Mumler allowed him to superintend all the operations. "He desired me," says Mr. Guay, "to select my glass out of a large number that stood on the table, or work-bench. I picked up one from the lot, examined it, rubbed it, threw my breath on it, and found it to be nothing more than a piece of ordinary glass, to all appearance new.

"He insisted on my going through the operation of coating, silvering, &c., which I did, as follows: Having the glass in my hand, I coated it with the collodion and placed it in the bath, previously examining the inside, or at least looking into the bath by the light of the lamp. While the plate was in the bath, I examined the plate-holder, which stood beside the bath. When the plate was done, I took it out, placed it in the plate-holder, and carried it to the camera-stand, under the skylight. Keeping my eye constantly on it while it was at my feet, resting on the camera-stand, I examined the tube and camera-box. Finding every thing all right, I took my seat in such a position as to see every thing going on. Being seated profile, I could, by turning my eyes around, see pretty well the background and also the camera-box, Mr. Mumler by it, and the young man (his assistant) off in the corner, I having previously made sure that there was nobody else about beside us three."

The focus was adjusted, the cloth removed; and Mr. Guay sat, hoping that the picture of his departed wife "might come on the negative, standing in front and by him." The sitting over, he immediately passed to the camera-box, took out the plate-holder, and went to the dark room, followed by Mr. Mumler. Having thrown on the developing solution, Mr. Guay closely watched what was coming; and, to his astonishment, he saw *two pictures come out*, one of which was a faint picture of his wife. The negative having dried, he varnished it and placed it on the shelf.

On a subsequent day, he went through the same process, with similar precautions, and this time silently wished that the "spirit likeness" might be one of his father. His wish was answered. "It is impossible," says Mr. Guay, "for Mr. Mumler to have procured any pictures of my wife and father. The likeness of my father is clear and perfect; that of my wife is not." Mr. Guay says he had much conversation with Mr.

Mumler, and could not detect a single syllable suggestive of fraud.

Professor W. D. Gunning (1867) relates an instance in which a spirit-hand appeared on the photograph of a young girl. He says, "While sitting before the camera, she was smitten with partial blindness. She described it to me as 'a kind of blur coming suddenly over her eyes.' She spoke of it to the artist, who told her 'to wink and sit still.' In developing the plate, he noticed an imperfection, but did not observe it closely. He sat the girl again, and took a sheet of eight tintypes. She felt no blur over her eyes, and there was no blur on the pictures. The artist *now examined the first sheet, and found hands on the face and neck of every tintype, eight in all!* I have examined four of these, and find the hands in precisely the same position on each picture. Now the artist affirms that no human being but himself and the girl was in the room when these pictures were taken. He has no theory: he only knows that these hands came on the picture through no agency of his. What, then, shall we say?"

Professor Gunning shows that the theory that the plate was an old one, and the hands had been photographed there before, is absurd. "As well talk of making an Iliad by throwing down a ton of types at random!" Other explanations he rejects as equally unsatisfactory, and says, "The best part of my life has been spent in the study and interpretation of science; and, in all humility, I should be competent to weigh and interpret facts so simple as these. And, to my mind, this picture is a fact quite as important to science as an Amazonian fish. I will not cross an ocean for a new bug, and cry 'Humbug!' to a fact like this at my very door. . . .

"In paintings of the creation, done in the Middle Ages, you will see the hand of Deity moving over chaos; only the hand, for clouds and darkness veil His form. Belief in the Infinite Being and the life eternal was nourished and sustained in our fathers, by art. And now art comes to us even more divine; for she is Nature's own, painting with sunbeams. And our loved ones now and then lift the veil, and reach forth a hand from out that world of light and beauty, — from that world a hand clothed upon with elements from this; and art, in her new era, ministers again to our hope of immortality."

Of the numerous speaking mediums, the writing, the drawing, the musical (such as Blind Tom, the colored boy), the healing, the letter-reading, &c., we have left ourselves little room to speak in this place. Many of the phenomena through these various mediums are quite as wonderful and significant as those we have described; but as they are more open to partial explanation by causes not outside of the individual, we shall not do more than to refer to them at present.

CHAPTER VII.

THE SEERESS OF PREVORST.—KERNER.—STILLING.

"I gaze aloof
On the tissued roof,
Where time and space are the warp and woof,
Which the King of kings
As a curtain flings
O'er the dreadfulness of eternal things."—*Rev. Thos. Whitehead.*

THE most remarkable of the phenomena we have recorded had their counterpart in those known in the little village of Prevorst, amid the mountains of Northern Wurtemberg, twenty-two years before the Fox family first heard the rappings at Hydesville.

Frederica Hauffé, the seeress, was born in Prevorst, in the year 1801. She died in 1829. "She lived," writes the late Margaret Fuller, "but nine-and-twenty years; yet in that time had traversed a larger portion of the field of thought than all her race before in their many and long lives."

The biography of the seeress, published in 1829, was from the pen of Justinus Kerner, chief physician at Weinsberg, a man of unquestionable ability and stainless integrity. His proclamation of the phenomena, and the spiritual facts developed in the life of his subject, brought upon him a storm of ridicule and denunciation, from which there are few men who would not have shrunk. He met it bravely, and maintained his ground with a steadiness which no sneers from the *savans* and wits among his contemporaries could impair; and at last his veracity as a biographer, his philosophical sagacity, and his skill as a cool observer of facts have been completely vindicated by the events of the last twenty years.

After her marriage, in her 19th year, to Herr Hauffé, a worthy man, the seeress, who was of a remarkably delicate organization, became subject to spasmodic attacks, and would often pass into a somnambulic state. She at last became so sensitive to magnetic influences that even the nails in the walls had to be removed. Articles, the near neighborhood of which to her person was found injurious, would be taken away by an unseen hand. Such objects as a silver spoon would be perceptibly conveyed from her hand to a more convenient distance, and laid on a plate; not thrown, for the things would pass slowly through the air as if borne by invisible agency.

In 1826, Dr. Kerner took charge of Mrs. Hauffé. He soon found that drugs had no effect upon her. Even the homœopathic pharmacopœia was discarded. The seeress, in her clairvoyant state, prescribed for herself better than any physician could have done.

The rapping phenomena were common in her presence. Kerner says, "As I had been told by her parents, a year before her father's death, that, at the period of her early magnetic state, she was able to make herself heard by her friends, as they lay in bed at night, in the same village, but in other houses, by a knocking, — as is said of the dead, — I asked her, in her somnambulic state, whether she was able to do so now, and at what distance? She answered, that she would sometime do it; that to the spirit space was nothing. Sometime after this, as we were going to bed, — my children and servants being already asleep, — we heard a knocking, as if in the air, over our heads. There were six knocks, at intervals of half a minute. It was a hollow, yet clear sound, soft, but distinct. We were certain there was no one near us, nor over us, from whom it could proceed; and our house stands by itself. On the following evening, when she was asleep, when we had mentioned the knocking to nobody whatever, she asked me whether she should soon knock to us again; which, as she said it was hurtful to her, I declined."

And again he tells us, "In my own house, I can bear witness ;ot only to the sounds of throwing, knocking, &c.; but a small table was flung into a room without any visible means; the pewter plates in the kitchen were hurled about, in the hearing of the whole house, — circumstances laughable to others, and which would be so to me, had I not witnessed them in my sound mind; but which become doubly significant, when I compare them to many accounts I have heard of the like nature, where there was no somnambule in question."

Here we have phenomena precisely like those with which the records of witchcraft, and the accounts of haunted houses, are filled.

Speaking of a spirit who frequently came to her, Kerner says, "His appearance was always preceded by knockings; however suddenly a person flew to the place to try and detect whence the noise proceeded, they could see nothing. If they went outside, the knocking was immediately heard inside, and *vice-versâ*. However securely they closed the kitchen-door, nay, if they tied it with cords, it was found open in the morning; and though they frequently rushed to the spot on hearing it open or shut, they never could find anybody. Sounds as of breaking wood, of pewter plates being knocked together, and the crackling of a fire in the oven, were also commonly heard; but the cause of them could not be discovered. A sound resembling that of a triangle was also frequently heard; and not only Mrs. Hauffé, but others of her family, often saw a spectral female form. The noises in the house became at length so remarkable, that her father declared he could stay in it no longer; and they were not only audible to everybody in it, but to the passengers in the street, who stopped to listen to them, as they passed."

The Rev. Mr. Hermann wrote several questions for a spirit who visited Mrs. Hauffé to answer. From the time these were shown to the spirit, Mr. Hermann "found himself awakened at a particular hour every night, and felt immediately an earnest disposition to prayer. There was always, at the same time, a knocking in his room, sometimes n the floor, and sometimes on the walls, which his wife heard as well as himself; but they saw nothing."

Several experiments were made to test the reality of the seeress's spirit-vision. Kerner relates that "An acquaintance of Mrs. Hauffé's who sometimes visited her, one day informed us that a friend of hers was dead. This person had promised her that he would appear to her after death, and we consequently hourly expected to learn that she had seen his ghost; but days, weeks and months passed without any such event happening. Then the acquaintance owned, that, not believing in the reality of these apparitions, he had said it for an experiment; the person was not dead. Another experiment was made as follows: Mrs. Hauffé was frequently visited by the spectre of a deceased person, of whom she had never seen or heard any thing whatever. A friend bade her learn of this ghost the period of his birth, which neither she nor I knew. This was done; but when our friend made inquiry of his relations whether the time mentioned was correct, they said, 'No.' This our friend wrote to us; and I read the letter to Mrs. Hauffé, advancing it as a strong argument against the reality of the apparitions. She answered, unmoved, that she would inquire again. She did so, and the answer was the same. I wrote again to my friend, saying so, and begging him to ascertain more particularly the period of the birth in question; and, on doing this, he found that the relations had been in error; the time had been correctly named."

He adds, "I could relate many other equally remarkable facts, but that I should be encroaching too much on the privacy of the parties concerned." He details twenty-two facts that occurred at Weinsberg in evidence of the presence and operations of spirits. Concerning these he says, "Of the greatest number, I was myself a witness; and what I took upon the credit of others, I most curiously investigated, and anxiously sought, if by any possibility a natural explanation of them could be found; but in vain." These facts are further corroborated by councillors, professors, and other official persons.

Mrs. Hauffé's statement concerning the spirits who appeared

to her is interesting. Her words are, "I see many with whom I come into no approximation, and others who come to me, with whom I converse, and who remain near me for months: I see them at various times by day and night, whether I am alone or in company. I am perfectly awake at the time, and am not sensible of any circumstance or sensation that calls them up. I see them alike whether I am strong or weak, plethoric or in a state of inanition, glad or sorrowful, amused or otherwise; and I cannot dismiss them. Not that they are always with me; but they come at their own pleasure, like mortal visitors, and equally whether I am in a spiritual or corporeal state at the time. When I am in my calmest and most healthy sleep, they awaken me: I know not how; but I feel that I am awakened by them, and that I should have slept on had they not come to my bedside. I observe frequently that, when a ghost visits me by night, those who sleep in the same room with me are, by their dreams, made aware of its presence: they speak afterwards of the apparition they saw in their dream, although I have not breathed a syllable on the subject to them. Whilst the ghosts are with me, I see and hear every thing around me as usual, and can think of other subjects; and though I can avert my eyes from them, it is difficult for me to do it: I feel in a sort of magnetic *rapport* with them. They appear to me like a thin cloud, that one could see through; which, however, I cannot do. I never observed that they threw any shadow. I see them more clearly by sun or moonlight than in the dark; but whether I could see them in absolute darkness, I do not know. If any object comes between me and them, they are hidden from me. I cannot see them with closed eyes, nor when I turn my face from them."

"Here then," says Mr. Shorter, in his review of these occurrences, "nearly forty years ago, in the life of this poor, untaught peasant woman, we have brought together those modes of spirit manifestation which call forth so much denial when their occurrence at the present day is affirmed; manifestations in dream, vision, voice, touch, writing, drawing, presentiment, prediction, apparitions, second-sight, clairvoyance, crystal-seeing, movements of objects, rappings, trance-speaking, thought-reading, and the spirit-language."

According to Kerner, Eschenmayer, Schubert, Görres, and others, who observed Madame Hauffé long and carefully, she seemed to be more in the spiritual world than in the physical. "She was," says Kerner, "more than half a spirit, and belonged to a world of spirits: she belonged to a world after death, and was more than half dead. In her sleep only was she truly awake. Nay, so loose was the connection between soul and body that, like Swedenborg, she often went out of the body, and could contemplate it separately."

Like many other clairvoyants she could, in her somnambulic state, read any thing laid on the pit of her stomach, and inclosed between other sheets of blank paper. Her perception of different sensations from plants, precious stones, and other minerals, was repeatedly tried by placing them in her hands, when she would always ascribe the same property to the same thing. She was at times lifted into the air, as has been the case with Mr. Home, Miss Lord, and other modern mediums, as well as with many saints and devotees of all countries and times.

Science in its progress is daily supplying, in connection with these and kindred facts, many new analogies. "However incomprehensible," says Friedrich von Meyer, "a world of spirits may be to the natural reason, the progress of our knowledge of the physical world and of the extraordinary nature of man is every day rendering it more comprehensible."

Kerner, who died in 1859, full of years and honors, was a writer of no ordinary force and culture. In the spirit with which he handles his assailants, he often reminds us of that matchless master of controversial weapons, Lessing.

In his "Leaves from Prevorst," published subsequently to the seeress's death, Kerner, after relating some striking cases of spirit-agency, of recent occurrence, through others than the seeress, says that any person wishing to convince himself of one of them "has only to make the little journey from Stuttgart to Oberstenfeld."

"But," adds Kerner, with a fine irony, "it is much more conenient to sit at your writing-table by the fireside, and decide on such things without seeing them."

His picture of the class of critics who pronounce judgment on facts in this way is one for all time. Some of these philosophers, indeed not a few may be found in our own country, mounted on reviewers' stools, and sending forth their oracular criticisms, weekly or monthly, on matters they know nothing about, in any practical or experimental sense.

"None of those gentlemen," writes Kerner, "who call themselves the friends of truth, set so much value upon it, as to move a single foot over the Resenbach: no one takes the least trouble to prove these things at the time, and on the spot. For many years the extraordinary manifestations of the Seeress of Prevorst were made public; but none of the gentlemen who now, all at once, pretend that they would have liked so very much to have seen her, and who sit and write whole blue-books about her, ever took a moment's trouble, whilst she lived, to see, to hear, and to test her.

"At their writing-tables they continued sitting, but professed to have seen, heard, and proved every thing, — much more than the quiet, earnest, and deeply thinking psychologist, Eschenmeyer, who *did* take the trouble to examine and prove every thing at the time and on the spot, for the truth's sake, shunning no journey, when necessary, in the severest cold of winter. Only by such a method can such things be probed to the truth: the learned way of knowing and speculating by the pounce-box proves nothing.

"These gentlemen *who construct their heaven and their hell according to their own wishes*, and push the love and grace of God before them in any direction that is convenient to them, rather than give themselves up to believe what, from their pride and sensual indulgences, is most unpleasant and repugnant to them, labor hard, by all the arts of intellectual acuteness and of dialectics, to persuade themselves, though it be but for the brief moment of this life, that the future inevitably awaiting them, will correspond with the wishes and feelings which exis' in this body.

"Probably it is very difficult for the pride of man to believe that he shall, one day, come into a condition where the nothingness of his inner being shall issue to the light; when the mask shall fall, under which he has endeavored here to conceal himself, and to parade himself complacently in the public eye. It is difficult, too, for the so-called intellectual* *to believe in spirits that do not show themselves spiritual.* According to them, every man after his death should at once arrive at the intellectual knowledge and eminence of a Hegel. But now come spirits, trifling and foolish, and spirits like those who came to the Seeress of Prevorst; who longed after Scripture texts and hymns; at the name of Jesus became clearer, and asserted that only in the name of Jesus can rest and joy be found. In such spirits it is impossible for the learned and intellectual to believe; and such apparitions are to them only the product of a sick fancy.

"And spirits now come, who are much poorer and more destitute than spirits in this life ever showed themselves, so that to our critics such a spirit-world must appear unworthy of God; and if they could convince themselves that such a spirit-world did exist, they would doubt the wisdom of the Creator: since spirits, they think, should either not show themselves at all, or in a manner to do honor to their Maker. This signifies nothing, however; for God and Nature will have the mastery! †

"Let us suppose, for a moment, that those creatures on our

* Witness the silly remarks of the "London Saturday Review" of Dec. 17, 1862, which says, "If this is the spirit-world, and if this is spiritual intelligence, and *if all that spirits can do is to whisk about in dark rooms, and pinch people's legs under the table, and play 'Home, Sweet Home,' on the accordion, and kiss folks in the dark, and paint baby pictures, and write such sentimental namby pamby as Mr. Coleman copies out from their dictation, it is much better to be a respectable pig and accept annihilation than to be cursed with such an immortality as this.*" Kerner anticipates and answers the sneers of witlings like this.

† Bacon says, "The voice of nature will consent, whether the voice of man do or not."

earth, which constitute a transition class, and find themselves, as it were, in an intermediate state, as seals, bats, megatherians, were so formed that they could only be seen by men of a peculiar condition of nerves, and by others not at all, the latter would protest that no such creatures existed, or could possibly exist. They would exclaim excitedly, 'A creature half-mouse, half-bird, a creature half-calf, half-fish, would be unworthy of the Creator, who never brings forth helpless, crippled, half-existences. Such things, they would say, are the mere births of a sick fancy; and, were they really existent, which, however, it would be the height of folly to believe, would make one doubt the wisdom of the Creator.' That is precisely what the critics say of what they call low and undignified spirits.

"But these creatures, now mentioned, do exist at this very time, my beloved! spite of thy belief and thy critical judgment; and thou shalt not, therefore, doubt the wisdom of their Creator, but shalt fall down, and, with all humility, shalt worship and say, 'What I here in the dust, with the eye of a mole, regard as so great a disharmony, will hereafter, when the scales fall from my mole's-eye, appear as harmony.'

"And so is it also with those wretched spirits! Beloved! they are there! However thou mayest, in thy notions of the Creator, consider them so unworthy; however, in thy intellectual wealth, mayest struggle against them in thy spirit, — there they are, contrary to all the systems of such learned, acute, and intellectual men! There they are in truth, as real as the helpless caterpillars, out of which slowly the butterflies shall unfold themselves. There they are, and you cannot hinder them; cannot do otherwise than disbelieve in them, and, disbelieving, fight against them with all your dialectic arts, ready-writings, wit, and acuteness, *but which, in fact, does not at all annihilate this spirit-world;* but it goes on its way, troubling itself not in the least about all your intellectual skirmishing.

"On this point an able writer has said already, 'Suppose a critic to write an article that turned out and was decided by the public to be a poor affair, are we to consider it unworthy of the Creator to have made such a "wretched stick"? And suppose this critic to have suddenly departed into the other world, without having got any more sense, are we to doubt the wisdom of the Creator, if the man should manifest himself here as a very paltry ghost indeed?' It may, however, be answered, by some wise one, that every thing should in this world either not exist, or exist as a credit to its Maker. This, indeed, would be very praiseworthy and agreeable; but the courteous reader knows very well that the image of God in this world often reduces himself to a most hideous and foolish caricature of a man; but does any body on that account doubt of the wisdom of the Creator? *Yes: let us look into the mirror, and I am afraid we shall find ourselves very much unlike the original image of God.*"

Kerner then gives a series of well-attested cases of the apparitions of such distorted and degraded spirits, and adds, "It is an incontestable truth which Jacob Böhme ably demonstrates, and which the Seeress of Prevorst confirms; namely, that 'The body being now broken up and dying, the soul retains her likeness as the spirit of her will. Now is it away from the body; for in dying there is a separation. Now the likeness appears in and amid the things which the soul had here imbibed, which she had infected herself with, which she allowed to build themselves up in her, since she has the same well-spring in her. That which she loved here, which was her treasure, and into which the spirit of her will entered, is now expressed in her, and becomes her spiritual image, not as a reminiscence, but as an actual condition.'"

Let us hope that the day is near when a more reverent attention will be lent to facts which are the key to much that confounds our scrutiny in our studies of human nature.

Johann Jung-Stilling, born in Westphalia, in Germany, 1740, was, like Kerner, a devoted Spiritualist. His "Theory of Pneumatology," translated into English by Samuel Jackson, was republished in New York, in 1851, with an introduction from the pen of our revered friend, the late George Bush, whom it was our fortune to introduce to some of the phenomena of somnambulism, which we were investigating at the time. Stilling appears to have been well versed in the facts which the manifestations of 1848 brought so prominently before public attention. The phenomena of rapping and knocking he frequently notices, as modes of spirits announcing themselves. He was convinced of the existence of the spiritual body. "There is a natural body, and there is a spiritual body," says St. Paul; *is* NOW, not *is to* BE.

Stilling was unconsciously a "medium." He announced, more than ten weeks before the occurrence, the tragic fate of Lavater, who was shot by a soldier in Zurich, in 1799. Stilling wrote seasonably to Hess, and begged him to communicate the prediction to Lavater. The warning seems to have been unheeded. Stilling's presentiments of evil were sometimes very strong, and as unerring as they were strong. In his "Pneumatology," he has collected a great number of authentic narratives of apparitions and other phenomena indicative of spiritual powers. The "many-sided" Goethe was Stilling's fellow-student at Strasburg, and became strongly attached to him. "I urged him," says Goethe, "to write his life; and he promised to do so." The promise was fulfilled.

Stilling was well acquainted with the phenomena of animal magnetism. His experiments convinced him, as our own long since convinced us, that the soul does not require the outward organs of sense in order to be able to see, hear, smell, taste, and feel, in a much more perfect state.* "Animal magnetism," he says, "proves that we have an inward man, a soul, which is constituted of the divine spark, the immortal spirit, possessing reason and will, and of a luminous body, which is inseparable

* "The vision that can see through brick walls," says Professor William Denton (1868), "and distinguish objects miles away, does not belong to the body: it *must* belong to the spirit. Hundreds of times have I had the evidence that the spirit can smell, hear, and see, and has powers of locomotion. As the fin in the unhatched fish indicates the water in which he may one day swim, as the wing of the unfledged bird denotes the air in which it may one day fly, so these powers in man indicate that mighty realm which the spirit is fitted eternally to enjoy."

from it. Light, electric, magnetic, galvanic matter, and ether, appear to be all one and the same substance, under different modifications. The light, or ether, is the element which connects soul and body and the spiritual and material world together.

"The ideas we form of the creation, and all the science and knowledge resulting from them, depend entirely upon our organization. God views every thing as it is in itself. For, if he viewed things in space, and as no space can be conceived as really existing unless limited, the views which God takes would therefore also be limited, which is impossible; consequently no space exists out of us in nature, but our ideas of it arise solely from our organization. If God viewed objects in succession and rotation, he would exist in time, and thus again be limited. Now, as this is impossible, time is therefore also a mode of thinking peculiar to finite capacities, and not any thing true or real."

From these principles, Stilling arrives at the opinion that, since time and space are only modes of thinking suited to our present state, it is impossible that rational inferences, though mathematically just, can serve to guide us into the truths of the invisible world, when their premises are founded on modes of thinking adapted to the visible world, but excluding operations from the invisible.

Perhaps this theory may explain why natural science makes such blunders in its attempts to deal with the recent phenomena.

CHAPTER VIII.

SOMNAMBULISM, CLAIRVOYANCE, ANIMAL MAGNETISM.

"Shut your eyes, and you will see." — *Joubert.*

"Whereas the atheists impute the origin of these things to men's mistaking both their dreams and waking fancies for real visions and sensations, they do hereby plainly contradict one main fundamental principle of their own philosophy, that sense is the only ground of certainty and the criterion of all truth." — *Cudworth.*

IN the face of the opposing protestations of a negative materialism, there is one great fact established by the positive testimony of the past and of our own age; this, namely, that there are and have been such individuals as seers, somnambulists, mediums, exhibiting powers which wholly transcend those of our mortal senses, and who must derive such powers either from spiritual faculties of their own, superseding the physical and normal, or from communication with spiritual forces and intelligences external to themselves. The manifestations upon which our convictions of this fact are based are of daily occurrence, and such as may be tested by all who will take a little trouble and exercise a little patience.

More than thirty years ago, by a series of experiments which extended over a period of two years, we satisfied ourselves of the facts of animal magnetism, or mesmerism, including the higher phenomena of lucid somnambulism. Our opportunities of investigation were of daily occurrence, and such as to make imposture impracticable. We made many observations of high psychological significance, as we believe, confirming most of the accounts of similar experiences by Puysegur, De Leuze, Dupotet, Chauncy Hare Townshend, and others.

The interest of these observations has been, to a great extent, merged in the more comprehensive generalizations of modern Spiritualism, including the phenomena of animal magnetism, as well as of witchcraft and sorcery, and thus showing them all to be expressions of one great spiritual or psychical fact.

Moreover, many of the most surprising phenomena of animal magnetism, though ridiculed and denied for a long time by the scientific world, are now admitted by the leading physiologists of the day. Science is just beginning to change its attitude of angry contempt for the less unbecoming position of inquiry and attention. One has only to read the medical and physiological writings of Dr. Carpenter, his admissions on the subject of somnambulism, of brain action without consciousness, and other unexplained mysteries, to be satisfied on this point; for Dr. Carpenter now represents the most advanced school of England in his department of physiology, and few equally high contemporary authorities can be named.

It is true that some of the more surprising facts of clairvoyance are still kept at a distance, on probation, even by Dr. Carpenter; but they are no longer treated with that disdainful vituperation or easy indifference which the magnates of science observed towards them up to the year 1856.

The phenomena of lucid somnambulism are a constant offence and stumbling-block to the modern materialistic school, of which Moleschott, Vogt, Feuerbach, and Büchner are active representatives. With the asperity of partisanship, these able writers deny all evidences of a psychical nature in man, and seem to take it as a personal affront if we credit them with immortal souls.

"It may appear singular," says Dr. Büchner, "that at all times those individuals were the most zealous for a personal continuance after death, whose souls were scarcely worthy of such a careful preservation."

This modest philosopher would seem to look upon the Augustines, Origens, Pascals, Johnsons, and Goethes of the human race as small specimens, compared with Dr. Büchner!

Ludwig Feuerbach (born 1804) has the following remark: "No one who has eyes to see can fail to remark, that the belief in the immortality of the soul has long been effaced from ordinary life, and that it only exists in the subjective imagination of individuals, still very numerous."

That the belief in immortality has been largely effaced from the ordinary life of many educated persons, is, we fear, but too true; but this is owing, in a great degree, to the circumstance, that the class of facts which modern Spiritualism has re-verified has been excluded, by false theories and an imperious ignorance, from scientific consideration. Belief in immortality was more general in ancient times than now, if we except the rapidly increasing body of Spiritualists. Even so good a Catholic as Frederick Schlegel admits this. "Among those nations of primitive antiquity," he says, "the doctrine of the immortality of the soul was not a mere probable hypothesis: it was a lively certainty, like the feeling of one's own being."

One has to go back only to the time of Richard Baxter, to see how largely the convictions of immortality, in his day, were based on a knowledge of spiritual phenomena.

But in what mole's labyrinth can the learned Feuerbach have been burrowing, that he does not know that some of the most eminent anthropologists of the present time — men who build their belief on a patient induction of objective facts — have admitted the phenomena and the hypothesis of Spiritualism?

He has no doubt heard of the Darwinian theory; for it is a favorite one with the materialists, while at the same time it does not in the least disturb the Spiritualists. Among the Spiritualists of England is Mr. Alfred R. Wallace, a distinguished naturalist, who made explorations on the Amazon; and of this gentleman, Dr. Hooker, the president of the British Scientific Association, spoke as follows, in his address at the meeting at Norwich, in August, 1868: —

"Many of the metaphysicians' objections have been controverted by that champion of natural selection, Mr. Darwin's true knight, Alfred R. Wallace, in his papers on 'Protection,' in the

'Westminster Review,' and 'Creation by Law,' in the 'Journal of Science,' October, 1867, &c., in which the doctrines of 'Continual Interference,' the 'Theory of Beauty,' and kindred subjects, are discussed with admirable sagacity, knowledge, and skill; but of Mr. Wallace, and his many contributions to philosophical biology, it is not easy to speak without enthusiasm; for, putting aside their great merits, he, throughout his writings, with a modesty as rare as I believe it to be in him unconscious, forgets his own unquestionable claims to the honor of having originated, independently of Mr. Darwin, the theories which he so ably defends."

Mr. Wallace's testimony to the facts of Spiritualism is therefore that of a competent scientific man of the highest reputation; and for such a man to be complacently set down, by a metaphysician in his closet, as the victim of his "subjective imagination," is a reversal of the order of things.

Mr. Karl Vogt (born 1817) is very intolerant of any facts of a spiritual tendency. He says, "Physiology" (*my* physiology?) "decides *definitely and categorically* against individual immortality, as against any special existence of the soul. The soul* does not enter the fœtus, like the evil spirit into persons possessed, but is a product of the development of the brain; just as muscular activity is a product of muscular development, and secretion a product of glandular development. . . . The fœtus manifests no mental activity: this changes with the periods of life, and ceases altogether at death."

Here is mere dogmatic assertion, without any proof or apology for proof. How does Mr. Vogt *know* that "the fœtus manifests no mental activity"? From the time of that Elizabeth, men-

* In the "Ontology" of Dr. Doherty (Trübner & Co., London), some of the most advanced facts of physiology are harmonized with those which Spiritualism reveals. "The spirit," says this writer, "forms the body *in utero*, by collecting and associating particles of matter from the blood of the mother to form organs; and it sustains the physical organism during life by a constant interchange of atoms with the external world. It is the soul which originates the body, and adapts it to its own special functions."

ioned by St. Luke, down to our own day, there are mothers by he million who will tell Mr. Vogt that his declaration is erroeous.

Mr. Vogt labors through an entertaining volume, in which the language of science is diversified with that of sarcasm, to prove that we need not pass over many links of our genealogy to find apes for our ancestors. We have no special repugnance to the ape-theory. Many Spiritualists are inclined to it. The Darwinian hypothesis might become a certainty to-morrow, and it would not clash with the convictions of a man who knows that the phenomena proclaimed in this volume are substantially true; and Spiritualism, while it encourages us to aspire to the attainments of the loftiest seraph, would, if rightly meditated, teach us a humility that would not shrink from sympathy with the creature that is lowest in the scale of being.

But so far as Mr. Vogt's system rests on his ignorance of spiritual facts, it needs reconstruction, if he would have it conform with the science of the future.

Dr. Moleschott (born 1822), who has acquired high distinction as an anthropologist, imagines, with the sanguine temperament of youth, that he has uttered the last word of science in regard to a certain class of facts, when he says, "Unprejudiced philosophy is compelled to reject the idea of an individual immortality and of a personal continuance after death."

So the philosophy which differs from that of Dr. Moleschott, and which refuses to accept his cheerful doctrine of the soul's annihilation, is a philosophy of "prejudice"! Newton, Leibnitz, and the rest, were men of prejudice!

With equal positiveness, the late Dr. Elliotson (who knew a good deal that Dr. Moleschott has yet to master) taught, for many years, in the "Zoist," a materialism quite as dense and narrow as his. But, after he had lived threescore years and ten, he stumbled on one little fact, demonstrated to him by his senses and his reason, which shivered the "unprejudiced philosophy" of a lifetime as by a lightning-flash, and convinced him that the Spiritualists, with their vulgar intuitions and their stubborn experiences of spiritual intervention, were, after all, in the right and that there *is* "a personal continuance after death."

Dr. Moleschott's philosophy is foredoomed to the same end· for it rests on a repudiation of the positive testimony of a large portion of the human race up to the present time. Unless, like the man who refused to look through a microscope, because it would subvert his pet theory, he stubbornly persists in ignoring the great facts of Spiritualism, he must some day do as Elliotson did, and humbly acknowledge his error. The following lines of Beattie explain the rest: —

"So fares the system-building sage,
Who, plodding on from youth to age,
Has proved all other reasoners fools,
And bound all Nature by his rules;
So fares he in that dreadful hour
When injured Truth exerts her power
Some *new phenomenon to raise*,
Which, bursting on his frightened gaze,
From its proud summit to the ground
Proves the whole edifice unsound."

There are hopeful tokens already in the more recent writings of Dr. Moleschott, that he is reconsidering his barren doctrine of the "all-mightiness of the transmutations of matter," in which he seems merely to have revamped some of the notions of Heraclitus.

The denial of the continuous life of man, after the dissolution of the material body, is a negation that never arises from *knowledge*. It is not the exposition of any positive knowledge, but the mere dogmatic assertion that beyond the line of such knowledge there lies nothing more. This is why we regard as unphilosophical and irrational the position of those *who teach dogmatically that the phenomenon called death is the end of the conscious individualism of man*. All such teaching is as unphilosophical and unscientific as it is arrogant and presumptuous.

The utmost that the materialist can rationally say, is, "I *doubt* the fact of a future life." To say "There *is* no future life," ıe ought to be the spirit whose existence he repudiates. If it requires spirit to reveal the fact of spirit, surely nothing less than spiritual authority is requisite to teach the fact of no-spirit. Thus the dogmatist against a future life is involved in a contradiction. To teach the matter confidently, he ought to have an illumination, the possibility of which his theory utterly denies. No one but a seer has a right to say, "There is no life for man beyond the grave;" and the seer's own seership would give the lie to his assertion. The Pyrrhonist may be a philosopher; but the teacher of annihilation is simply a charlatan.

The Spiritualist, on the contrary, having a knowledge of phenomena, mental and physical, proving to his satisfaction the existence of spiritual powers, would be false to his own legitimate convictions if he did not teach the great fact of immortality *as a certainty*, in view of which our mortal life ought to be shaped, and our thoughts and affections constantly refreshed by the sublime consciousness that death is a mere incident, which leaves the essential part of our being untouched; and that we shall survive to study the infinite works of the Creator in other worlds, and to commune with the loved ones gone before, and the great and good of all ages, in progressive stages of being, with which this rudimental state, and our discipline here, shall be found hereafter to have been in perfect accord.

Dr. Büchner has an easy way of disposing of certain inconvenient facts. He says, "Some of these phenomena, clairvoyance especially, have been laid hold of to prove the existence of the supernatural and supersensual. . . . All these things are now, by science and an interrogation of the facts, considered as idle fancies. . . . What the belief in sorcery, witchcraft, demoniac possession, &c., was in former centuries, re-appears now under the agreeable forms of table-moving, spirit-rapping, psychography, somnambulism, &c. . . . There can be no doubt that all pretended cases of clairvoyance rest upon fraud or illusion. Clairvoyance, that is, perception of external objects without the aid of the senses, is an impossibility. It is a law of nature, which cannot be gainsaid, that we require our eyes to see, our ears to hear, and that the senses are limited in their action by space."

It would thus seem that Dr. Büchner, like his master, Moleschott, bases his whole structure of atheistic materialism upon his ignorance of certain facts, known to be true at this day by several millions of intelligent persons, and publicly proclaimed as true by several thousands. If he will open his eyes, he will find that he is behind the age. Even Dr. Carpenter and the "Edinburgh Review" admit the power of somnambulists to see through opaque substances, and to read without the normal use of their physical organs of sight.

Dr. Maudsley, in his recent work on the "Physiology and Pathology of the Mind," has presented the materialistic view of his subject with exhaustive ability; but in doing this he has to ignore almost entirely the great facts of somnambulism. His reference to the subject is of the most meagre and casual kind. "Perhaps," he says (page 267), "no more fitting opportunity than the present will present itself for referring to the singular state of somnambulism." And then, after attributing the phenomena to the "independent action of the sensorial and corresponding motor centres," he winds up with "a striking instance" that recently came under his observation. It is a story of a young sempstress who got up in the night and finished, in a state of somnambulism, the work on which she was engaged.

And here is the moral he draws from the incident: "Soon the long day's task will be over with her, and she will sleep well where no troubles more can reach her, and no dream of work or sorrow disturb her slumbers." All which is simply a repetition of Chaumette's epigraph in the days of the French revolution: "Death is an eternal sleep." An hypothesis which all the facts of somnambulism confute! And yet to this momentous subject Dr. Maudsley gives less than two pages out of the four hundred and forty-two to which his volume extends.

M. Georget, a much esteemed physiologist of the Paris school, appears to have arrived ultimately at a very different conclusion from that where Dr. Maudsley leaves us. Georget was the author of a much esteemed work on the "Physiology of the Nervous System (1821)." In it he professed opinions, charged with materialism, very similar to those of Dr. Maudsley; but, after numerous experiences in magnetic somnambulism, Georget completely changed his views, and had the courage and good faith to avow it, and to give the avowal an added sanctity by incorporating it in his last will and testament, as follows: —

"I must not conclude without an important declaration. In 1821, in my work on the 'Physiology of the Nervous System,' I boldly professed materialism. . . . This work had scarcely appeared, when renewed meditations on a very extraordinary phenomenon, somnambulism, no longer permitted me to entertain doubts of the existence within us, and external to us, of an intelligent principle, altogether different from material existences; in a word, of the soul and God. With respect to this I have a profound conviction, founded upon facts which I believe to be incontestable. This declaration will not see the light till a period when its sincerity will not be doubted, nor my intentions suspected. As I cannot publish it myself, I request those persons who may read it, on opening this will, that is to say, after my death, to give it all possible publicity."

Anton Mesmer (1734–1815) bears much the same relation to animal magnetism that Miss Kate Fox does to modern Spiritualism. The fact of the influence of one human being by another, under certain conditions, through passes of the hand, or by the simple exercise of the will, was known and practised long before Mesmer introduced the subject anew to public attention. Recent discoveries at Pompeii show that it was a mode of relief known there centuries ago. Plautus, in "Amphitryo," makes one of his characters ask, "How if I stroke him slowly with the hand, so that he sleeps?" These magnetic means of cure were not only practised, but directions for them were inscribed on sacred tables and pillars, and illustrated by pictures on the temple walls, so as to be intelligible to all. Apuleius furnishes similar evidences of the ordinary practice by the Romans of magnetic manipulations, to induce somnambulism and clairvoyance. In Livy alone, there are more than fifty instances in which he refers to the literal fulfilment of dreams, oracles, prognostics by seers, &c.

It was Mesmer's theory, that the universe is submerged in an eminently subtle fluid, which he thought should be named animal-magnetic fluid, because it can be compared to the fluid of the magnet; that this fluid impregnates all bodies, and transmits to them the impression of motion; that it insinuates itself into, and circulates through, all the fibres of the nervous system; and that it may be accumulated, when the magnetizer wills it, in buckets, tubs, &c., and especially in the organs of the magnetizer who transmits it to the magnetized. This hypothetical fluid will remind the classical reader of the "chain uniting all beings" of Hesiod, and the "soul of the world" of Plato.

With Mesmer's operations began the modern interest in animal magnetism, whatever its antiquity may be. In 1778, he arrived in Paris, and for five or six years made a great noise by his experiments. The king appointed a commission, consisting of five members of the Royal Academy, and four members of the Faculty of Medicine, to report upon Mesmer's exhibition. Franklin was a member of the commission; but he was at the time unwell, and unable to attend its sittings.

The commission, in their elaborate Report, allow that in what they witnessed, there was something that seemed the working of a mysterious agent. They reduced Mesmer's exhibitions to four classes: first, those which could be explained on physiological grounds; second, those which were contrary to the laws of magnetism; third, those where the imagination of the mesmerized person was the source of the phenomena; and fourth, facts which led them to admit a special agent. One member of the commission, the eminent Laurent de Jussieu, became a convert to Mesmer's views, and testified to "several well-verified facts, independent of the imagination."

In the year 1826, the French Academy of Medicine appointed a second commission. They labored diligently for five years, and presented a report (June, 1831) through Dr. Husson. It is igned by nine members of the commission, two only, Messrs. Double and Magendie, having declined to assist at the investigations. The commission admit nearly all the important facts of animal magnetism.

"It is demonstrated to us," they declare, "that magnetic sleep has been produced in circumstances where the magnetized persons have not been able to see or gain any knowledge of the means employed to determine it." The magnetizer being in a separate apartment, and the subject wholly unaware of his intention, the sleep was induced through the mere operation of the magnetizer's will. We have ourselves repeatedly tested this phenomenon here admitted by the commission.

The Report speaks of a terrible operation (the removal of the right breast) which was performed by M. Cloquet upon Madame Plantin. During the twelve minutes that the operation lasted, the invalid, previously magnetized, "continued to converse calmly with the operator, giving not the slightest evidence of sensation."

The late Dr. Valentine Mott, of New York, who was present at this operation, added his personal testimony in our presence to the truth of the foregoing statement.

In regard to clairvoyance, the commission report several facts. Among others, they speak of a law student, M. Villagrand, whose eyelids were kept closed by the different members of the commission; but who, nevertheless, recognized cards entirely new, and read from a book open before him. In short, the interior life, the perception of the state of the body, the prevision of crises, the instinctive prescription of remedies, are forcibly attested in the Report.

"The magnetized person," it says, "can not only be acted upon, but he can, without his knowledge, be thrown into and aroused from a complete somnambulic condition, when the operator is out of his sight, at a certain distance from him, and separated by doors. . . . The phenomenon of clairvoyance takes place even with the fingers pressed tightly over the eyelids. The previsions of two somnambulists, relative to their health, were realized with remarkable accuracy."

The Academy was rather astonished at the Report, and for a long time refused to discuss it. But the experiments continued to multiply. Insensibility to pain, during terrible operations, was one of the phenomena that was regarded as most wonderful. Pistols were discharged close to the heads of the somnambulists without making them start; without even interrupting the sentence they had commenced.

Facts like these could not long be ignored, nor could the Report of the eleven commissioners be silently consigned to oblivion. The Academy then decided to discuss it; and the result was, that they refused to print the Report, voting only for the autograph copy, which, as Count Gasparin tells us, remains shut up in the archives of the Academy of Medicine! "To deny these phenomena," he says, "one must also deny natural somnambulism, assuredly not less extraordinary than magnetic somnambulism. Inasmuch as the existence of natural somnambulists cannot be denied (and who will deny it?), little will be gained by contesting mesmerism."

M. Georget, to whom we have already referred, thus expresses himself: "My somnambulists are so insensible to sound, that the very loudest noises, produced unexpectedly to them, do not cause them the slightest emotion. Yet they will always hear the magnetizer." A phenomenon we have ourselves frequently experienced in somnambulists; as we also have the following, described by M. Rostan: "The outward life ceases; the somnambulist lives within himself, completely isolated from the exterior world; this isolation is especially complete for the two senses of sight and hearing. . . . The eyes of the majority of somnambulists are so insensible to light, that the lashes have been burned without their testifying the least impression; if the lids are raised, and the fingers passed rapidly in front of the eye, the immobility remains complete. . . . And yet they are conscious of the objects which surround them; they avoid with the greatest address obstacles in their path."

The French commissioners mention some experiments in which rare powers of detecting disease were manifested by somnambulists. Internal symptoms, inappreciable to the eye, were described by them, and the correctness of the description afterwards verified by a post-mortem examination of the bodies.

M. de Puységur says of a peasant whom he had magnetized, "I have compelled him to move quickly about on his seat, as if dancing to a tune, which, singing mentally myself, I made him repeat aloud. . . . I have no occasion to speak to him; I think in his presence; he understands and answers me."

"Having performed," says Dr. Bertrand, "on my first somnambulist the process by which I usually awakened her, exercising at the same time a firm will to the contrary, she was seized with strong convulsive movements. 'What is the matter with you?' said I. 'Indeed,' she replied, 'you tell me to awake, and yet you do not will that I shall awake.'" Dr. Bertrand says that he has thrown into the somnambulic state a person a hundred leagues from him.

M. Filassier relates that a young somnambulist described at Paris, minute by minute, the various acts, the attitudes, and even the secret thoughts of her mother, who was at Arcis-sur-Aube. "Every possible precaution," he adds, "was taken to ascertain the truth regarding this vision into space. The inquiry was conducted by a family of intelligence and strict integrity, in connection with some conscientious physicians. The lucidness of Mlle. Clarice was in all cases justified by the event."

The celebrated Arago, in an article on Mesmerism, says, "The man who, outside of pure mathematics, pronounces the word *impossible*, is wanting in prudence. . . . Nothing, for example, in all the wonders of somnambulism, is looked upon with more mistrust than an oft-repeated assertion touching the faculty, possessed by certain persons in a crisis state, of deciphering a letter at a distance by means of the foot, the hand, or the stomach. Yet, I do not doubt that the suspicions of even the most rigidly critical minds will be removed, after having reflected on the ingenious experiments in which Moser produced, also at a distance, very distinct images of all sorts of objects on all sorts of bodies, and in the most complete darkness."

"The phenomena we are made to observe in somnambulism," says Deleuze, "demonstrate the distinction of the two substances, the double existence of the interior man and of the exterior man in the same individual: they offer the direct proof of the spirituality of the soul, and the answer to all objections that have been raised against its immortality." "Among the men who have made magnetism their study, there are, unfortunately, some materialists. I cannot conceive how it is possible that many of the phenomena witnessed by them — such as sight at a distance, prevision, the action of the will, the communication of the thoughts without external signs — could have failed to appear in their eyes as sufficient proof of the spirituality of the soul."

"The repose of the outer," says Townshend, "is an absolute condition for the revelation of the inner, sensibility. We all may feel that, in order to call up before our mind's eye the face of a dear friend, or the beauties of a familiar landscape, we must retreat from the obtrusive impulses of the external world. Would we rise to a yet higher discernment of remembered objects, we must yet more calmly check the beating of our pulses, until we pass into that state of mind so beautifully described by Wordsworth, —

'That serene and blessèd mood
In which the affections gently lead us on,
Until the breath of this corporeal frame,
And even the motion of our human blood
Almost suspended, we are laid asleep
In body, and become a living soul:
While, with an eye made quiet by the power
Of harmony, and the deep power of joy,
We see into the life of things.'

"The mesmeric vision, or clairvoyance, has been gravely and grandly pronounced to be 'physically and physiologically impossible.' How can we reply to this? Only, I suppose, as Pascal did to some one who asserted that it was impossible for God, being so great, to busy himself about our little world,—'To decide such a question, one ought to be great indeed.'"

> "Impossible is nowhere to be found,
> Except, perhaps, in the fool's calendar."

Dr. Edwin Lee, in his "Report upon the Phenomena of Clairvoyance" (London, 1843), mentions the case of the prediction of the death of the king of Würtemberg by two different somnambulists: the one having foretold the event four years beforehand; the other, in the spring of the same year, mentioned the exact day, in the month of October, as also the disease (apoplexy).

"The exact coincidence," says Dr. Lee, "of the event with the predictions is not doubted at Stuttgard; and, a fortnight ago, Dr. Klein, who is now in England, accompanying the Crown Prince of Würtemberg, having been introduced to me, I took the opportunity of asking him about the circumstance, which he acknowledged was as has been stated, saying, moreover, that his father was physician to the king, who, on the morning of the day on which the attack occurred, was in very good health and spirits."

Shelley, the poet, appears to have been partially somnambulic on several occasions. He was also sensitive to mesmeric influence. Williams, who was drowned with Shelley, says in a note in his diary shortly before the event, "After tea, walked with Shelley on the terrace. . . . Observing him sensibly affected, I demanded of him if he was in pain; but he only answered, by saying, 'There it is again! there!' He recovered after some time, and declared that he saw, as plainly as he then saw me, a naked child (Byron's Allegra, who had recently died) rise from the sea, and clasp its hands as if in joy, smiling at him. This was a trance that it required some reasoning and philosophy to wake him from entirely, so forcibly had the visions operated on his mind."

Almost every family has its tradition of some event like the following: The Pacific Hotel, in St. Louis, was destroyed by fire in February, 1858; and twenty-one lives were lost on the occasion. On the night of the fire, a little brother of Mr. Henry Rochester, living at home with his parents, near Avon, in the State of New York, awoke some time after midnight with screaming and tears, saying that the hotel in St. Louis was on fire, and that his brother Henry was burning to death. So intense were his alarm and horror, that it was with considerable difficulty he could be quieted. On the following day, at noon, the parents received a telegram from St. Louis, confirming the little boy's dream in every particular.

Well-authenticated instances of spontaneous clairvoyance like this could be collected from the newspapers of the last ten years till the record would fill volumes. Not many years since a New-Orleans merchant, being in Paris, woke up from sleep one night, having heard, as he thought, the voice of his son uttering the words, "Father, I'm dying." So much impressed was he by this, that he got out of bed, lighted a candle, and made a record of the occurrence, stating the exact hour by the clock, in his note-book. When he arrived in New Orleans, a few weeks afterwards, the first friend he met told him of his son's death, and added, "His last words were, 'Father, I'm dying.'" The merchant took out his note-book, pointed to the record, and afterwards learned that his son had died at the precise hour named, after making the proper allowance for difference of longitude between Paris and New Orleans.

Bacon recognizes a natural divination proceeding from the internal power of the soul. "The mind," he tells us, "abstracted or collected in itself, and not diffused in the organs of the body, has, from the natural power of its own essence, some foreknowledge of future things; and this appears chiefly in sleep, ecstasies, and the near approach of death."

"The phenomena of clairvoyance, prevision, and second sight," says De Boismont, "depend on a sudden illumination of the cerebral organ, which calls into activity sensations that have hitherto lain dormant."

Rather do they depend, we should say, on an intromission from latent spiritual forces, called into action by some abnormal conditions affecting the relations of the physical to the spiritual body.

De Boismont, whose work on "Hallucinations" (Paris, 1852) has a high reputation in France, admits that some cases of prevision "appear to spring from an enlarged faculty of perception, a *supernatural intuition*."

To our instances of clairvoyance in dreams, we add the following perfectly well-authenticated case, related (1858) by the Rev. Dr. Horace Bushnell. "As I sat by the fire," he says, "one stormy November night, in a hotel-parlor, in the Napa Valley of California, there came in a most venerable and benignant-looking person, with his wife. The stranger was Captain Yount, a man who came over into California, as a trapper, more than forty years ago. Here he has lived, apart from the great world and its questions, acquiring an immense landed estate, and becoming a kind of acknowledged patriarch in the country. His tall, manly person, and his gracious, paternal look, as totally unsophisticated in the expression as if he had never heard of a philosophic doubt or question in his life, marked him as the true patriarch.

"The conversation turned, I know not how, on spiritism and the modern necromancy; and he discovered a degree of inclination to believe in the reported mysteries. His wife, a much younger person, and apparently a Christian, intimated that probably he was predisposed to this kind of faith by a very peculiar experience of his own, and evidently desired that he might be drawn out by some intelligent discussion of his queries.

"At my request, he gave me his story. About six or seven years previous, in a mid-winter's night, he had a dream, in which he saw what appeared to be a company of emigrants, arrested by the snows of the mountains, and perishing rapidly by cold and hunger. He noted the very cast of the scenery, marked by a huge perpendicular front of white-rock cliff; he saw the men cutting off what appeared to be tree-tops rising out of deep gulfs of snow; he distinguished the very features of the persons, and the look of their particular distress.

"He woke, profoundly impressed with the distinctness and apparent reality of his dream. At length he fell asleep, and dreamed exactly the same dream again. In the morning he could not expel it from his mind. Falling in, shortly, with an old hunter comrade, he told him the story, and was only the more deeply impressed by his recognizing, without hesitation, the scenery of the dream. This comrade had come over the Sierra by the Carson-Valley Pass, and declared that a spot in the pass answered exactly to his description. By this the unsophisticated patriarch was decided. He immediately collected a company of men, with mules and blankets and all necessary provisions. The neighbors were laughing, meantime, at his credulity. 'No matter,' said he: 'I am able to do this, and I will; for I verily believe that the fact is according to my dream.' The men were sent into the mountains, one hundred and fifty miles distant, directly to the Carson-Valley Pass. And there they found the company in exactly the condition of the dream, and brought in the remnant alive."

Dr. Bushnell adds, that a gentleman present said to him, "You need have no doubt of this; for we Californians all know the facts and the names of the families brought in, who now look upon our venerable friend as a sort of savior." These names he gave, together with the residence of each; and Dr. Bushnell avers that he found the Californians everywhere ready to second the old man's testimony. "Nothing could be more natural than for the good-hearted patriarch himself to add that the brightest thing in his life, and that which gave him the greatest joy, was his simple faith in that dream."

Instances similar to the foregoing could be multiplied indefinitely. We have heard of the case of the brother of an ancestor of our own, whose ship was struck by lightning, the consequence of which was that he and his crew were compelled to escape from the wreck in the long-boat, where they were exposed for many days, at an inclement season, in the middle of the Atlantic.

The captain of a vessel sailing from the same port dreamed of seeing them, and was so vividly impressed by the vision, that he determined on altering his course, and going back in search of the boat. This he did, against the expostulations of his mates. On the morning of the third day he fell in with the boat, and rescued the occupants of it.

The phenomena of clairvoyance in the somnambulism induced by mesmerism were first noticed, in modern times, in the year 1784, by the Marquis de Puységur, a disciple of Mesmer. That these phenomena afford conclusive evidence of spiritual faculties latent in man, and developed under certain circumstances even in this life, is a conviction at which most persons, who have given much thought to the subject, have finally arrived. We see no escape from the conviction. The added marvels of Spiritualism are hardly needed to give it force; but let them be none the less welcome on that account.

We need not multiply instances of clairvoyance, clairaudience, &c. The fact is established, if any fact can be by human testimony. It needs but a single experiment with Mr. Charles H. Foster, in pellet-reading, to shatter the most elaborate structure of Sadducean materialism from turret to foundation-stone. If the faculties of sight and hearing, in their highest manifestations, are not dependent on their proper physical organs, who can rationally argue that they are likely to be destroyed by the dissolution of the physical body itself?

Mr. S. B. Brittan, one of the earliest to accept the facts of phenomenal Spiritualism, remarks, "The individuality of man does not belong to his body; but inheres in a supra-mortal and indestructible constitution. . . . Within this corporeal frame there is another body of more ethereal elements. . . . If there were no inward form or spiritual constitution, the molecular eliminations would periodically destroy the identity of man."

"Our soul," says Joubert, "is ever fully alive. It is so in the sick; in those who have fainted; in the dying; it is still more alive after death."

"The soul," says Zschokke, himself a clairvoyant, "has the

faculty directly, and without inference, both of perceiving occurrences at a distance, and of being sensible of future events. The ancients, who knew as much as we do of the properties of the human soul, observed this inexplicable power of perception and foresight, especially in cases of nervous weakness, and in the dying."

That the instances of clairvoyance on the part of the ancient oracles were numerous, no student of history can deny, without rejecting, through simple prejudice, a vast amount of explicit and concurrent human testimony. The genuineness of the oracles was conceded by Justin Martyr, Athenagoras, Theophilus of Alexandria, Tatian, Clemens Alexandrinus, Origen, Eusebius, Athanasius, Chrysostom, Cyril Alexandrinus, and others of the Greek fathers; and by Minucius Felix, Cyprian, Tertullian, Lactantius, Maternus-Firnicius, Jerome, Augustine, and others of the Latin. Thus, Augustine writes, "They [the spirits] for the most part foretell what they are about to perform; for often they receive power to send diseases by vitiating the atmosphere. Sometimes they predict what they foresee by natural signs, which signs transcend human sense; at others they learn, by outward bodily tokens, human plans, even though unspoken, and thus foretell things to the astonishment of those ignorant of the existence of such plans."

The Jews before Christ, and the Fathers after, believed that departed spirits lurked about images, spoke in oracles, controlled omens, and in various ways encouraged men to worship them.

If human testimony is to be taken as of any account, compared with the mere speculations of closet professors, putting forth decisions on matters they refuse to investigate practically, this question of spiritual phenomena is decided. "Why, then," asks Cicero, "doubt the certainty of this argument, if reason consent, if facts, people, nations, Greeks, barbarians, our ancestors, and the universal faith? If chief philosophers, poets, the wisest of men, founders of republics, builders of cities? Or, discarding the united consent of the human kind, shall we wait for brutes to speak?"

"Si divinatio est, dii sunt," *if there is divination, there must be gods* (or spirits), was a common saying of the ancient Romans. One authentic instance of clairvoyance satisfied them of the great fact of spiritual existence.

"That we should rather evolve from our present corporeal elements the body that is to be ours, than begin existence *de novo*," says Townshend; "that, in other words, we should really possess a fundamental life, or body, incapable of passing away with the grosser covering that envelops it; that, at death, we should retain something physically from our actual condition,—seems pointed out to us by all the analogies of nature.

"Everywhere we behold that one state includes the embryo of the next, not metaphysically, but materially; and entering on a new scene of existence is not so much a change as a continuation of what went before. The very rudiments of organs, intended in a higher stage of animal life to be useful, are found, uselessly, as it were, appearing in the lower classes of animated creatures; or, stranger still, lying in embryo in the same creature in one state, only to be developed in another. The wings that form the butterfly lie folded in the worm.

"We should, then, *à priori*, expect to find the principle that individualizes man, and is the true medium of his instruction, attached to him from the beginning, and that the germs of future capacities, physical not less than intellectual, should be discoverable in his constitution.

"The dissolution of this coarser covering is, by us, called death; that is, we seem unto men to die: but with our inner body we never part; and, consequently, by that we still retain our hold upon individual existence. As Leibnitz has remarked, 'There is no such thing as death, if that word be understood with rigorous and metaphysical accuracy. The soul never quits completely the body with which it is united, nor does it pass from one body into another with which it had no connection before: a metamorphosis takes place; but there is no metempsychosis.' *

* *Metamorphosis*, a change of form or shape; transformation. *Metempsychosis*, the passage of the soul from one body into another.

"Man is shown by the facts of mesmerism to be capable of increased sensitive power. To what end, if hereafter this increase of power become not permanent? Would wings be folded in the worm if they were not one day to enable it to fly? We cannot think so poorly of creative power, or of thrifty nature. . . . Wretched, indeed, must be the view of man which confines him to this bank and shoal of time; which does not regard him, and all his glorious endowments, as intended *for a series of existences.*"

"What do we understand by the term *spiritual?*" says the Rev. B. F. Bowles. "May we not all agree upon the common idea that the spiritual is the unseen, and, to our senses, intangible? I think we may. Now, that there is an unseen force within us, constituting our interior personality, and that manifests itself through these outward forms, seems self-evident. It is this that is the source of all outward action, and that receives from without all impressions. It is this that constitutes the *I* or the *me*, and to which we refer when we use these pronouns. We are all conscious of this unseen self. And when we speak of seeing, of hearing, of tasting, of smelling, of feeling, we refer to a being who possesses all these senses, but who exists behind the organs of their outward manifestation. I do not properly say 'my *eye* sees,' or 'my *hand* feels,' but rather '*I* see *through* my eye,' and 'feel *with* or *through* my hand.' Nor do I say, 'my *brain* thinks,' but '*I* think *with* my brain.' And our common consciousness approves.

"And the one who possesses all these senses *is never seen.* I never have seen you, nor you me. We have only seen the manifestations of each other. The individual who dwells in either of the living forms before me, or the one who occupied the form that is dead, has never to material senses been tangible. We have never come *directly* in contact with him or her, but always through the mediation of the outer form. Each of us, then, in our real self, answers to the common idea of spirit: we are intangible.

"And, again, each of us, in his voluntary action, betrays

purposes and desires, intelligence and thought; and surely we cannot attribute these to tangible matter. It would be repugnant to all our sense of fact, to affirm that *flesh* could *think* and purpose. We inevitably refer all such action to the unseen. It is the unseen one that loves, and that we love.

"And now with reference to the spiritual *body*, it seems natural to conclude that these secret powers exist in *combination*, forming an interior *being*. We refer them all to one, and yet each is distinct. The same being sees, thinks, and loves. And yet seeing, thinking, and loving are quite different. There is, then, an *organization* interior to this physical organization, possessing in itself each of the senses, and all of the intellectual and emotional power we see expressed through the exterior form. And, being so, it is in a proper sense a *body*. It is in all things, but its *texture*, like the body we see. Only in this (its texture), can we mention aught that the body possesses that the spirit hath not. Indeed, except this and the *shape* of humanity, the body hath nothing when the spirit hath gone out. It hath no senses, no power. Here, then, we have not only the existence of a spirit, but a spiritual *body*, in the sense of organization.

"But what of its substance? Hath it substance? or, is it without? I have often received the impression from friends, that they supposed a spirit to be without substance. Perhaps they had no clear conception of what a spirit is. Perhaps I was unable to *receive* their conception. But, so far as able, it seemed to be, in the words of another, 'the most definite conception of nothing ever given to mankind.' And yet I think it manifest that spirit hath substance. To see this truth, let us inquire what we mean by *substance*. Do we mean some *particular* thing? No; for every thing is substance. Do we not mean by this term *some*thing, in distinction from *no*thing? Can we mean any thing else? Borrowing an illustration, then, think of the millions of human bodies now being moved about by spirits. They would all stop, were the spirits to go out. Is this immense amount of substance moved without substance, moved by nothing?

"Further, to illustrate, think of the material universe all in motion. Go with the astronomer and count the worlds. Endeavor, then, to conceive of those unseen even by him. Ask yourselves of the immensity of their weight. You cannot answer. Well, they are all upheld; they are all in motion with inconceivable velocity. And by what? By nothing? By no substance, which is nothing? No; but by spirit, which is the greatest of all things. By an immeasurable organization of spirit. By that which constitutes all that is unchangeable in the universe. By God, who is a spirit, 'without variableness or shadow of turning.' And the effort to conceive of God without substance, is perilous to our conviction of his existence. And so of the human spirit. In such an attempt, we grapple with the impossible, and are worsted in the struggle. . . .

"In spirituality, then, I think you must bear me witness, there is nothing to forbid the thought that spirits out of the flesh reach and affect those in the flesh, thus triumphing over the death of the body. It becomes, then, a question of *fact*, to be determined by other data. In the absence of *experience*, this may be *doubted*, but not on this ground *denied*. In the presence of experience, and on the part of such as have the evidence of their own senses to this point, it must be affirmed. By the use of their senses they are to be judged, and must judge.

"Such is some of the evidence I draw from our *common knowledge;* such the inferences from common ground, and which, for this reason, I think should find general acceptance. Evidence that '*there is a spiritual body,*' *indestructible, independent of the physical, and hence immortal.*"

It will be seen, however, as we proceed, that the spiritual hypothesis is not the only one which human ingenuity has invented for the phenomena of clairvoyance and of Spiritualism. Mr. H. G. Atkinson, who was associated with Miss Martineau some years since in the authorship of an atheistic book, in which some of the phenomena of mesmerism were accepted and attributed, as they were by Dr. Elliotson,* to exclusively material causes, professes to be not at all inconvenienced by the added wonders of Spiritualism. He admits them all, but is too uncompromising a Comtean to allow that they point to any thing outside of this barrier of flesh and blood.

Seers and spirits may protest as much as they please; nay, the latter may show themselves in their habits as they lived, — Mr. Atkinson is inexorable.

"I think it can now be shown," he says, referring to the spiritual phenomena, "that there is not *any very essential distinction between these extraordinary facts and the ordinary ones of every-day life!*"

Shut out from the spiritual hypothesis by his whole past philosophy, Mr. Atkinson consoles himself, after the manner of the antediluvian philosopher, who, according to the profane, was shut out from the ark by Noah, and who revenged himself on the patriarch by telling him that "it was no sort of consequence; for he believed it was not going to be much of a shower after all."

A fact of importance, in connection with the history of animal magnetism, has been recently brought to light by the French Spiritualists. This fact is no other than that the magnetists of France anticipated, by at least half a century, the knowledge, since made the world's property by the events

* Dr. Elliotson surpassed even Mr. Atkinson in the enthusiasm with which he sought in a bald materialism for a sufficient explanation of the phenomena of life. But, as we have already seen (page 20), Dr. Elliotson came right at last. The "London Spiritual Magazine" tells us that when modern Spiritualism was introduced, he was one of the most scornful of its opponents. He separated himself on the question from his old friend and colleague in the management of the "Zoist," Dr. Ashburner; to whom it must have been a source of great satisfaction, after years of estrangement, that Dr. Elliotson's conviction of the truth of Spiritualism was the means of re-establishing their former friendship. Spiritualism was not with Dr. Elliotson a conviction barren of results. It revolutionized the philosophy of a lifetime. He bitterly lamented the misdirected efforts he had made, however conscientiously, in the promulgation of materialistic principles. He became a thoroughly changed man, and changed in all respects for the better.

at Hydesville; a fact which is proved by the publication of the correspondence of the two celebrated French magnetic philosophers, Messrs. Billot and Deleuze, in two volumes, in 1836. This correspondence commenced in 1829; and in it we find M. Billot asserting that there are none of these marvellous things that he has not witnessed during the last thirty years.*

This carries his knowledge of spiritual phenomena back as far as 1789, the period of the commencement of the French Revolution; into the period, in fact, of Lavater, Jung-Stilling, Kerner, Goethe, San Martin, &c. These phenomena, not only known to, but avowed by, those distinguished men, were, it now appears, equally well known to MM. Billot and Deleuze, who, as scientific men, had not, however, dared to reveal them. The sects of the Initiated and the Illuminati were well acquainted with these phenomena in the seventeenth and eighteenth centuries; and the only difference to note is, that then they were familiar only to a few who kept the knowledge of them to a certain extent secret, and that now they are familiar to the public at large.

But there is another circumstance especially noteworthy in this discovery of Spiritualism amongst the magnetists, which is, that the class of scientific men among them has been as a body stoutly opposed to the admission of Spiritualism as a fact. In England, we know with what pertinacity Dr. Elliotson and others resisted for many years the conviction that spiritual phenomena underlie those of magnetism; or, in other words, mesmerism. So in France, Dupotet, Morin, and the rest of them fought hard against this conviction; and so much so, that M. Morin, the successor of Baron Dupotet, has constantly resisted the invitations of the Spiritualists to witness spiritual phenomena.

Here, however, we have the curious fact of two of the most

* For this abstract of the correspondence, we are largely indebted to a paper by William Howitt (July, 1868). We can mention no man who has been more earnest, indefatigable, and courageous in his advocacy of the truths of Spiritualism than Mr. Howitt.

celebrated magnetic philosophers of France, avowing after a concealment of the fact through a career of half a century, that they all the time, whilst prosecuting their magnetic inquiries, had become fully aware of other and still more wonderful phenomena supervening and arising out of those inquiries which they prosecuted with no such expectations. These arose like apparitions upon them, startling and astonishing them, like the genius which stood before Aladdin when he rubbed his lamp, meaning only to polish it, and with no idea further from his mind than that his friction was the invocation of a spirit. So MM. Billot and Deleuze, experimenting only in magnetism, and expecting none but strictly natural though abstruse results, found that they were pressing on those secret and mysterious springs and laws of life which awake the attention of the inhabitants of the invisible, and cause them to manifest their presence.

It is still more remarkable that these two great magnetists — who had published, each, work after work, and whose names were famous in that science — did not work in company, or with a knowledge of each other's proceedings. They had each their own avowed theory, differing greatly one from the other; and these they had propounded and defended with zeal and persistency, till they had acquired a certain character of antagonism. All this time, however, their writings bore to the ordinary reader no traces of any thing but the legitimate facts and doctrines of magnetism. But, to these great antagonist magnates of science, there was something in their language which awoke a more than ordinary sensation in each other; and, opening a correspondence, they began to approach each other, putting forth the delicate feelers of an intense curiosity, grounded on a conviction that each possessed secret knowledge that he had not yet laid open to the light, and that this knowledge was, in reality, the property of both. They had each a consciousness that, whilst they had been going along separate and even hostile paths, they had been treading the very same enchanted ground, and were twins in a life which they had hitherto hidden from each other and from mankind.

On the 24th of March, 1829, M. Deleuze wrote to M. Billot, complaining that certain magnetizers made their experiments out of mere curiosity. To this implied censure Billot replied, on the 9th of April, that modern magnetizers had many humiliations to suffer from the jealousies of their *confrères;* but he now abandoned his cause to God, who had done great things for him. "Yes!" said he, advancing more boldly, "I have seen, I have understood all that it is permitted to man to see and know!" Still going further in his enthusiasm, and stimulated by the conviction that Deleuze himself had arrived at discoveries like his own, he says, "Permit me to observe that all that you write seems to me to betray *une arrière pensée* (an after thought). Your theory is only a solemn *ruse* to avoid scandalizing the *esprits forts* who will have nothing of the positive."

The ice was now broken, and the two great magnetists proceed to make a clean breast of it to each other. M. Billot, nevertheless, is by far the more open, and is ready to throw off the cautious disguise that they both had worn for so many years. It turns out, in the end, that they have seen nearly all the phenomena of modern Spiritualism, — apparitions, elevations of the person into the air, the fact of material substances being brought by spirits, obsessions and possessions by spirits, and nearly all the wonders which the ancient philosophers and the priests of different churches have declared as truths; and all this, be it remembered, long before the knockings at Hydesville opened up the great drama of renewed spirit-intercourse in our time. But it will be interesting to trace this remarkable correspondence a little further in its natural course.

On the 27th February, 1830, M. Billot writes to M. Deleuze, assuring him that he stated to him the whole truth regarding the extraordinary phenomena manifested through his clairvoyante, Mademoiselle Mathieu, and that he will never deviate from this in his communication of his experiences; and he proceeds to reveal to him things which, he says, he will probably regard as reveries, and then adds, "You would not have combated the theory of spirits for these forty years, if, like me, you had had under your eyes and your hands the masses of facts which have compelled me to adopt it." He then gives some curious facts concerning a clairvoyante in a state of wakefulness.

Deleuze, on the 15th of May, avows that he has seen lucids in that state. "Dr. Chase," he says, "reports having seen the same;" and then he makes the candid confession, "I have suppressed many things in my works, because it was not yet the time to disclose them." Billot, on the 16th of June, touches on certain particulars of somnambulism, which Deleuze in his writings had affected to treat as inexplicable; but he insinuates that he is quite satisfied that they now understand each other on these points. After referring to various passages in Deleuze's writings, "between us, Monsieur," continues Billot, "what need of so much reserve? In spite of your reticences, I understand you."

In his reply, on the 24th of September, Deleuze treats of matter at great length, and at first professes to think that the only thing which proves the communication of spirits with us, are apparitions; but again, thawing a little more, he says, if his health permit, he will write an article in the "Hermes" on psychical phenomena, in which he will free himself from the reserve which he too, hitherto, imposed on himself, and of which M. Billot has divined the real cause. "These facts," he says, "are now so numerous and so well known, that it is time to speak the truth."

On the 24th of June, 1831, M. Billot wrote to M. Deleuze, that in reading his works, he had seen that certain phenomena had been already familiar to him before he himself had entered on his career, and that there was nothing of the marvellous of which he had not been a witness during the thirty or forty years of his magnetic experience. "If you have not made mention of these things," he added, "you have lost your reason for keeping silence." To this M. Deleuze, on the 9th of July, replied that he had designedly avoided the statement of marvellous facts, considering it not always necessary to show these to the incredulous, as being indeed not the most likely way to convince them.

Billot then went on much further with his cautious correspondent, who, though he did not reveal much, was forced to confess that his friend had penetrated into his secret, and that he knew a great deal. "The time," said M. Billot, "is come when I ought to have no further concealment from you. I repeat that I have seen and known all that it is permitted to man to see and know. I have been witness of an ecstasy, not such as Dr. Bertrand imagines, but I have seen magnetic clairvoyants with stigmata. I have seen obsessions and possessions, which have been dissipated by a single word: I have seen many other things, which others have seen also, but which the spirit of this age has not permitted them to reveal. I am an *esprit fort;* and that which the priests have not been able to do now for many years, magnetism has accomplished. *The truths of religion have been demonstrated by it.*"

He then proceeds to relate some of these revelations, which very much resemble the teachings of the ancient philosophers, mingled with that of Christianity, — doctrines which prepared the way for the inculcations of Spiritualism. Superior intelligences, he says, presented themselves; presided at *séances*, and manifested themselves by the delicious odors which they diffused around them. The ambrosia of the mythologists, the odor of sanctity of the Church were discovered to be realities. Evil and unclean spirits also presented themselves; but the clairvoyants immediately recognized them (July 23, 1831). These and other statements, M. Billot says, which he extracted from the journals of the *séances*, could never have seen the light of day, had he not deemed it for the interest of the great science to confide them to the bosom of prudent and discreet friendship; and, on the 9th of September, he announces that he is about to proceed to more substantial proofs of the apparition of spirits, — such as, he says, it will be impossible to deny or to diminish: for *these spirits were tangible; you both saw and touched them.* Perhaps, he adds, M. Deleuze may think these things a little too marvellous for belief; but his doubt will no longer be pardonable when he may touch them himself, and touch them again. What he says on September 30, must convince the most skeptical: there is neither illusion nor vision. He and his co-secretaries have seen and felt, and he calls God to witness the truth of it.

On the 6th of November, 1831, Deleuze writes, that he is greatly grieved that the state of his health and his great age will not permit him to make a journey to see M. Billot, as he most anxiously desires; that *the immortality of the soul is proved to him, and the possibility of communicating with spirits;* but that, personally, he has not seen facts equal to those cited by Billot. Nevertheless, persons worthy of all confidence have made the like reports to him. "I have this morning," he continues, "seen a very distinguished physician, who has related to me some of your facts, without naming you, and who gave me many others of a like character. Amongst others, his clairvoyants *caused material objects to present themselves.* I know not what to think of all this, though I am as certain of the sincerity of my medical friend, as I am of yours. I cannot conceive how spiritual beings are able to carry material objects."

M. Billot, on the 25th of June, 1832, wrote that in the doctrine of Spiritualism the question is not of *opinions* but of *facts:* these are the things which lead to the truth; but neither the magnetizers nor the magnetized can reproduce these at will.

On another occasion, M. Deleuze remarks that "the clairvoyant seizes *rapports* innumerable. He catches them with an extreme rapidity: he runs, in a minute, through a series of ideas which, under ordinary circumstances, would demand many hours. Time seems to disappear before him. He is himself astonished at the variety and rapidity of these reflections. He is led to attribute them to the inspiration of another intelligence. Anon, he perceives in himself this new being. He considers himself in the clairvoyant sleep a different person from himself awake. He speaks of himself in the third person, as some one whom he has known, on whom he comments, whom he advises, and in whom he takes more or less interest, as if himself in somnambulism and himself awake were two different persons."

M. Deleuze finishes by urging M. Billot to publish his experiences, but with his habitual caution counsels him to suppress the most astounding facts. Billot heroically determines to victimize himself for the truth, to brave the sarcasms of the learned; "For," he observes, "to talk of spirits in France, where the majority of the magnetists hold fast by their accepted theory, of merely material agencies, is to become an object of contemptuous pity."

He was also aware of another difficulty, — the uncertainty of securing successful *séances;* which, whilst the causes affecting them are but partially understood, so often fail in the presence of the determinedly skeptical.

Such was the correspondence of the two celebrated magnetists, at a time when Spiritualism in its present phase was yet unheard of. The great facts of spiritual life thus bursting upon them in pursuance of their scientific experiments in magnetism, and in opposition to all their prejudices, as well as most contrary to their expectations, must be regarded as one of the most curious and most interesting events in the annals of Spiritualism. Besides the transport of material objects by invisible agents, the spirits which appeared to them were solid to the touch, as they have so often made themselves since. Living persons were elevated in the air in their *séances.* Dr. Schmidt, of Vienna, and Dr. Charpignon, of Orleans, also give some striking cases of delicious odors, or cadaverous effluvia issuing from pure or impure spirits which presented themselves: the most startling communications of facts otherwise unknown were made; and they had cases of obsession and possession as well as of successful exorcism.

After all the confessions of M. Deleuze, he afterwards was greatly tempted, like Sir David Brewster, to recover favor with his scientific and incredulous contemporaries. Becoming one of the chiefs of magnetic initiation, he endeavored to weaken or to neutralize the force of his avowals. A gentleman well instructed in these mysteries, wrote to him thus: "You have endeavored to fortify your readers, in your journal, against the system of the magnetists of the North, who admit superhuman powers as intermediates in certain magnetic phenomena. I would take the liberty of observing to you that this is not at all a system with them; but the simple enunciation of a fact, that a great number of their somnambulists, raised to a high degree of lucidity, have asserted that they were illuminated and conducted by a spiritual guide."

The answer of Deleuze is worthy of attention: "The facts which seem to prove the communications of souls separated from matter with those who are still united to it, are innumerable, as I know. These are existent in all religions, are believed by all nations, are recorded in all histories, may be collected in society; and the phenomena of magnetism present a great number of them. Yes: a great number of somnambulists have affirmed that they have conversed with spiritual intelligences; they have been inspired and guided by them: and I will tell you why I have thought it best not to insist on such facts and proofs of spirit communication. It is because I have feared that it might excite the imagination, might trouble human reason, and lead to dangerous consequences."

Deleuze did not, when thus challenged, walk backwards out of his previous avowals, like some on the other side of the water: he was only timid and cautious, not untruthful. The frank bravery of M. Billot, in regard to a truth which he knew would be unpopular, is deserving of the highest praise.

The author of these valuable papers has given a number of other instances amongst the magnetists who have arrived at the same conclusions as MM. Billot and Deleuze, in the same manner. They have found themselves in contact with unmistakable spirits, when they have been expecting merely the operations of magnetic laws. Amongst these were M. Bertrand, physician, and member of the Royal Society of Sciences. Baron Dupotet declared that he had rediscovered in magnetism the spiritology of the ancients, and that he himself believed in the world of spirits.

"Let the *savant*," he says, "reject the doctrine of spiritual appearances as one of the grand errors of the past ages; but the profound inquirer of to-day is compelled to believe this by a serious examination of facts."

Dupotet asserts the truth of all the powers assumed by antiquity and by the church, by all religions, indeed, such as working miracles and healing the sick. "When," he says, "lightning, or other powerful agents of nature, produce formidable effects, nobody is astonished; but let an unknown element startle us, let this element appear to obey thought, then reason rejects it; and, nevertheless, it is a truth; for we have seen and felt the effects of this terrible power." Terrible, however, only when nature is not understood as Spiritualism has revealed it. "If," adds Dupotet, "the knowledge of ancient magic is lost, the facts remain on which to reconstruct it." He exclaims, "No more doubt, no more uncertainty: magic is rediscovered."

He then gives a number of phenomena produced of a most extraordinary kind, and laughs at those brave champions of science who, far from danger, talk with a loud and firm tone, reason on just what they themselves know, and pay no regard to the practical knowledge of others; who, in fact, hug their doubts, as we, with more reason, hug our faith.

These avowals were made in 1840, long before the American phenomena or those of Vienna were heard of. But as Spiritualism began to show itself as a distinct faith, the majority of magnetists took the alarm. Those who, like Messieurs Bertrand, D'Hunin, Puységur, and Seguin, had stood on the very threshold of Spiritualism, began to step back a step or two, and to shroud themselves in mystery, and to shake their heads at the prospect of awful consequences in pushing further on such a path.

"The magnetic forces cannot be explained," said Puységur. "We have no organs," said M. Morin, "for discovering spiritual beings." "The real causes of apparitions, of objects displaced, of suspensions, and of a great portion of the marvellous," said D'Hunin and Bertrand, "are inscrutable."

Seguin, who thought that magnetism would revolutionize the

whole of science, starts, and stands still: he finds himself on the brink of a precipice. Inaccessible to danger, however, M. Seguin would wish to pursue his researches; but wisdom commands him to stop on the edge of an abyss, which no man, he affirms, can ever pass with impunity.

What is the precipice which M. Seguin and his fellow-magnetists see at their feet? Simply, the precipice of Spiritualism. The spiritual world opens before them when they desire only to deal with this. In the words of Baron Dupotet, "There is an agent in space, whence we ourselves, our inspiration, and our intelligence proceed; and that agent is the spiritual world which surrounds us." A step further, and the magnetists were aware that they must cut the cable which held them to the rest of the scientific world, and float away into the ocean of spiritual causation. They must consent to forfeit the name of philosophers, and to suffer that of fanatics in the mouths of the material *savans*.

We find in a late number of the "London Spiritual Magazine," a paper, by Mr. R. H. Brown, on the relations of clairvoyance to the facts of Spiritualism. We do but condense his admirably clear and logical statement in the remarks on the subject, which follow to page 195: —

"It is all clairvoyance!" Such is the objection made by many who have slightly investigated the spiritual phenomena. Thus it is that Spiritualism has come to the aid of clairvoyance. Before the advent of Spiritualism, clairvoyance was denounced as the great "humbug" of the day. Nearly all the scientific men of the land shook their heads, and lamented the credulous, wonder-loving ignorance of poor human nature. Now, as the world moves, and as the phenomena of Spiritualism come up, these same wise gentlemen would use what they denounced as the "humbug" of yesterday as the truth of to-day; that is, to help them to explain these more advanced facts!

"It is all clairvoyance!" But what is clairvoyance? Its phenomena may be briefly described as follows: Persons thrown into the somnambulic trance by animal magnetism, through the agency of an operator, or falling into the same state involuntarily, have been known to see without the aid of the physical or external organs of vision, and *without the assistance of light.* Books are read as well in the darkness of night as in the full glare of noonday. Objects and scenes, at great distances, far beyond the reach of the external organs of vision, are seen and described. The clear sight of the clairvoyant mind not only penetrates through the most opaque and dense substances, but also sees the thoughts that bud and blossom in the inmost recesses of the soul. The past is illuminated, and its most hidden passages revealed; and the future, hidden by an impenetrable veil from the normal eye, prophetically presents its yet unrolled panorama, and stamps upon the clairvoyant mind the impress of its coming form. This is clairvoyance. Now let me ask the candid investigator *what it is that sees without the physical eyes,* and *without the assistance of light?*

It is evident that neither the optic nerves nor the crystalline lens are employed by those who read a book, amid the darkness of midnight, unaided by a single ray of light. The answer to this question is all-important; for therein, hidden, lies the golden key which will unlock all the mysteries of Spiritualism. What is normal sight? What is it that *sees* when the natural or external eye, together with light, are the mediums of perception? It is evident that the mere fluid called light cannot see, neither can the lens or humors of the eye, nor the optic nerve, nor a combination of these; for light and visual organs are only the media by which perception is conveyed to that mysterious something which lies hidden within.

In ordinary or normal sight, three things are employed: the object, the eye, and the light which serves as the connecting link or medium of contact between the eye and the object. The eye, like a beautiful and delicate camera obscura, paints with fidelity the picture of the exterior world upon the retina. *It is the immortal soul which stands behind the curtain, and gazes on the shifting panorama.*

Let the soul be absent, and sight ceases, though the organ be perfect: it becomes but a common camera obscura, — the mere arrangement of parts for the production of a picture. The picture is perfect, but there is no spectator. When a person falls into a state of profound abstraction, the eyes, though open, often cease to convey any idea of sight to the soul. This is because the attention of the spectator behind the curtain is turned in another direction: he does not regard the panorama which moves along the darkened curtains of the eye. The materialists reply to this, that sight is not the result of the attentive perception of the soul to the pictorial sensations of the optic nerve. They tell us that the soul has no separate and distinct existence apart from the body. Light, they claim, is but sensation; and sensation is the result of organization. When the organization ceases, sensation will cease; and when sensation ceases, the whole being ceases to be; for organization and sensation, say they, compose the whole of man: there is no soul.

This method of argument is plausible. But the moment that sight is proved to exist *without the use of either light, sensation, or any of the physical and material organs of vision*, the whole pyramid of their logic falls to the ground.

Thus it is that clairvoyance furnishes the most conclusive answer to the materialists, and presents the most satisfactory proof of the existence of the soul, separate from the body, residing within it, generally employing its organs for the reception of ideas, but at times acting independently of them, and obtaining information without their aid. By clairvoyance, we have thus shown the truth of the first proposition upon which Spiritualism rests, — the existence of a dual nature in man, a soul as well as a body.

The second proposition, which lies at the basis of the new philosophy, is the existence of a spiritual body, interfusing and permeating the physical, material, or natural body.

If, in an obscure field, you should pick up the fragments of the bones of an arm, the inference that there had once been a full and complete organization, of which the fragments before you were a part, would be logical and correct. The naturalist is enabled, from the fragment of the skeleton of an extinct antediluvial animal, to reconstruct the whole, and draw the portrait of a creature which existed before the flood.

Let us apply this method to the subject under consideration.

The clairvoyant mind *sees* without the aid of light, or the assistance of the external or physical eye.

The soul does not leave the body to place itself in direct contact with the object seen; therefore the mind must have some medium of sight. This medium of perception is neither light nor the optic nerve. What, then, is it? It is not the odic* force simply; for there must be some means *whereby the character of the impression conveyed by the odic force is determined and individualized*, — some agency whereby the impression of sight is made distinguishable from that of hearing, or the impression made by an abstract idea. It is the peculiar function of an organ to individualize and characterize the nature of an impression received.

A simple object — for instance, a tree — makes upon the physical body a multitude of impressions; and it is the various organs of the body which individualize these impressions. The impression which the size, form, and color of the tree makes is individualized and characterized by the organs of sight. The impressions which its hardness and impenetrability make are individualized and characterized by the sense of touch. If it were not for this, the mind would receive a mass of confused impressions, without possessing any means to analyze, arrange, or distinguish them. As a prism separates and individualizes the various colors which compose a ray of sunlight, so the senses separate and individualize the combined impressions which an object makes upon the physical organism, and present them in an orderly and defined spectrum to the mind.

* This word *odic* is derived from the Greek ὁδός, a way or passage. Reichenbach gave the name *od* to what he conceived to be the force producing the phenomena of mesmerism, and developed by various agencies, as by magnets, heat, light, chemical or vital action. The terms *odyle*, or the *odyllic* or *odic* force, were thought preferable by his English disciples.

If the reader has followed with close attention our train of reflection, he will be prepared for the conclusion at which we have arrived, to wit: If the mind *sees* without the aid of light or the assistance of the optic nerve, it must have *some other medium* by which the simple impression of sight can be individualized, and presented separate and distinct from all other impressions; or, in other words, there must be a *spiritual* organ of sight, distinct and separate from the *physical* organ of sight.

The remainder of our task is now simple and easy; for if there is a spiritual organ of sight, there must also be a spiritual organ for the individualization of all the other impressions. In nature, each part is adapted to all the other parts, and the existence of one part presupposes the existence of all the other parts. If there is a spiritual organ of sight, there must also be a complete spiritual organization or body, interfused with and permeating the physical body.

Nature, our wise and powerful mother, fore-adapts every thing for the conditions amid which she intends it shall live. How shall we escape the conclusion, that by adapting the soul to another state of being, and endowing it for that purpose with the power to exist, act, think, see, and hear, without the aid of the body, and separated from it, *Nature has given us her solemn and sacred guarantee that we shall live hereafter?* To arrive at any other conclusion, is to charge Nature with the weakness of creating that which is useless, and God of the folly of adapting man to a sphere of existence which he does not intend him to enjoy.

All the arguments which have ever been made against the immortality of the soul are based upon the idea, that the soul has no identity of being separate from the body. From which premise the conclusion is correctly drawn, that the soul and body, being one in substance, must perish together. But clairvoyance demonstrates to us that *this premise is false*, and teaches us that the soul and the body are not one in substance; but, on the contrary, that the former can think, act, see, and hear without the aid of the latter, and independent of all its organs. It is thus that clairvoyance, with a mighty hand, crushes to powder the labored logic of the materialists, and places the belief in our immortal nature upon a firm and scientific basis.

But again, clairvoyance, by demonstrating the truthful character of the teachings of *intuition*, has afforded conclusive proof of a higher sphere of existence. God has given man two methods of attaining a knowledge of truth, — intuition and reason. The one is intended to prove the correctness of the other, thus affording man the highest evidence of truth, by giving him the power to arrive at the same results by two distinct and totally diverse mental operations. What intuition and reason both affirm to be true, no man need doubt.

It is true that neither is infallible; and he who expects to find any *human* faculty infallible in its nature, only betrays his own ignorance of the laws of mind and matter. Nevertheless, intuition is a faculty of the soul, just as reliable as that of reason, and the teachings of the one may be reposed upon with as much confidence as those of the other. Clairvoyance has demonstrated beyond all cavil the truthful character of intuition.

What does intuition say in regard to the immortal nature of the soul?

There is not a clairvoyant in the world, no matter what may be his *normal* belief, who does not affirm the existence of the soul after death has destroyed the clay-built palace wherein it dwells during its brief residence upon earth.

Many philosophers have puzzled themselves about the theory of "*innate ideas*." And the belief in our immortality has been classed as an "innate idea." But the philosophers may learn a lesson from clairvoyance. It is no "innate idea," but only the divine voice of intuition, which, deep within each man's soul, proclaims a life to come.

We must look to intuition for the true cause of that faith in a future beyond the grave, which has prevailed in all nations and all ages.

Clairvoyance, then, in demonstrating the truthfulness of intuition, has also demonstrated the immortality of the soul.

We have now arrived at the last of the propositions which is to be considered: the proof which clairvoyance affords of the *power* of spirits who have left the earth-form to communicate with those who remain behind.

As a matter of course, this portion of the argument, as well as the former, is addressed only to such as believe in the phenomena of clairvoyance. To those who are yet so far behind the great age in which they live as to doubt or sneer at magnetism and psychological science, all that has been said or will be said by the writer can be of no use. Such persons have yet to learn the *a b c* of that great science which lies at the basis of all others, and is the most important of them all.

In order to make it plain that clairvoyance does afford scientific and conclusive proof of the power of spirits to communicate with us, it will be necessary to refer to some of the familiar and ordinary phenomena of animal magnetism. Those phenomena may be divided into three classes: —

1. Profound abstraction, magnetic sleep, and insensibility to all external influences. 2. Sympathetic clairvoyance. 3. Independent clairvoyance.

Attention is more particularly requested to the second class; namely, sympathetic clairvoyance. The *subject*, while in this state, is almost entirely under the control of the *operator*. No vocalization of the will of the positive *operator* is required to induce obedience in the negative *subject*. The simple concentration of the unspoken will is all that is required to direct and control the subject. So great is the sympathy induced between the two, that the will of the one acts freely upon the muscular system of the other, and compels him to rise up, sit down, walk, stand, or talk, according to the volition of the operator. The nervous systems of the two are united by a constant interchange of the odic fluids. The result of this intimate union and sympathy between the operator and the subject is, that the thoughts of the one are known to the other. An idea evolved in the mind of the operator, *though unspoken*, immediately becomes present in the mind of the subject. But you will remember that the will of the operator also has control of the muscular system of the subject. Hence, no sooner is the idea of the operator present in the mind of the subject, should the operator will that idea to be spoken by the subject, than the subject is compelled to speak it. In other words, the operator, for the expression of his own silent thoughts, can use the vocal organs of the subject.

EXAMPLE. — A, in the presence of C, magnetizes B, and throws him into the *sympathetic* clairvoyant state. This being done, A silently thinks in his own mind these words: "Good-evening, friend C." Now, by virtue of the sympathy established between the *operator* A and the *subject* B, those words are immediately impressed upon the mind of B, and become present there. A now silently wills B to speak those words, which B is compelled to do; and so he turns to C, and says, "Good-evening, friend C." Thus you perceive A, instead of using his own organs of speech, has employed those of B. In other words, A has been speaking to C *through a medium*. This is an experiment which we have repeatedly performed with success.

It will be observed that the *body or physical organism of the operator was not employed in the above experiment*. The operator used two things only: first, his will; second, an odic force, which was controlled and directed by his will, and made the agent for the transmission of his thoughts and commands to the subject.

It is evident, therefore, that though the operator *be deprived of his body*, he will not lose the power to control and speak through B, provided he yet retain the power of volition and the command of the odic force.

It needs no argument to show that the escape of the soul from the body will not deprive the soul of the power of volition. The will is an essential attribute of the soul. Without volition, a soul would not be a soul; and nothing short of a total annihilation of the soul can destroy its volition. The whole is equal to the sum of its parts. If the whole is immortal, all the parts must be immortal. Hence, we see that the immortality of the will is just as certain as the immortality of the soul.

But will the disembodied volition still retain command of the odic force? There is a natural body, and there is a spiritual body. This spiritual body is very rare and refined in its nature, but is yet less refined than the soul enshrined within it. The soul, therefore, needs some agent by which it can put itself in connection with that spiritual body. The soul cannot come in direct contact with that body: it requires an agent which may transmit its commands to the various parts and members of the same.

What Nature requires, Nature supplies; and such an agent exists. The agent which serves to put the soul in connection with its new spiritual organization is an etherealization of what we term the odic force or vital fluid. It has been termed spiritual magnetism, in contradistinction to animal magnetism. Hence, we have surviving the destruction of the human form the only two conditions needed to enable A to control and speak through B.

This, then, is the true philosophy of the method by which spirits speak through media. It is sympathetic clairvoyance in both cases. In the one case, the *operator* is a spirit *in* the form; in the other case, the operator is a spirit *out* of the form. In both cases the subject is the same. In the former case, the spirit *in the form* uses his will, and the odic force evolved from his physical organism. In the latter case, the spirit *out of the form* uses his will, and the odic force flowing from his spiritual organism. The analogy between the two is perfect, and the means used are the same; and this view being true, it anticipates and answers all the objections to the spiritual hypothesis which we present in the chapter which follows.

We have offered some positive physical reasons for our rejection of the theories of the physiological materialists of our day. Let us give a little space to a metaphysical analysis of their arguments. "Moleschott, Vogt, Büchner, Wiener, and others," says Professor Reubelt, "maintain the following propositions: It is not the mind that thinks: thoughts are the secretions of the phosphoric brain. There is no liberty of the will, but a man *is* what he *eats;* there is no immortality, but a resurrection, of the body when it is used for the manuring of the fields; there is no personal God, who would be as much as a gaseous, vertebrated animal; but the universal law of causation is God. There is no *à priori* knowledge. There is no knowledge without sensual impressions, and no such impressions without a material object. The human mind is no spontaneous and productive, but only a receptive and digestive, organ."

Of this coarse materialism, Professor Gustave Franck, of Vienna, lately said, in his inaugural address, "Scientific criticism has first to take in hand the principle of materialism; and that is, *all is matter, or there is nothing but matter.* But this leading idea is not met with in matter. Materialism is thus based on an immaterial principle, which cannot be proved from matter, and which thus contradicts itself. If materialism could account for every thing else in the universe, it could not account for its own first principle, and thus rests on the belief of a dogma or a prejudice.

"If all is matter, thought is likewise a product of matter, an accidental conglomeration, as Vogt says, of atoms in the brain. Each sphere of thought is, therefore, an accidental phenomenon; each lacks the character of logical necessity. If two men think the same thoughts, it must be owing to the accidental sameness of the substance of their brains. Universal and necessary truths, that is, truths which each and every one has, by necessity, to recognize, there cannot be.

"But if this is so, what right has the materialist to proclaim his idea of the world as the only true one, and what interest prompts him to attack opposite views? If he is consistent, he cannot do any thing else than complain bitterly of fate or accident, by which, in the brains of others, atoms conglomerate in a manner so vastly different from that in his own.

"Now what is the position of materialism when tested by mathematics? Are its propositions and axioms universal and necessarily true, or are they accidental? To admit the first part of this question involves a denial of the very first principle of materialism; and to assume the second is absurd.

"Philosophically neither proved, nor capable of being proved, and perfectly unable to account for the most common phenomena, modern materialism has sought its main support in natural science. The materialist reasons thus: —

"'The most minute and thorough examination and observation of nature has not yet been able to discover a spirit,* and there is consequently no spirit.'

"But with the same right a man may say, I have never seen music with my eyes; and there is, therefore, no music. All that natural science can do is to confine itself to a relative negation, and to say, '*With the means at my command, I cannot discover a spirit.*' As soon as it oversteps this limit, and makes its negation absolute, it is pretentious: it has left its own legitimate sphere, and enters another of which it knows nothing, and of which it has, therefore, nothing to say.

"Materialism is atomistic: it accounts for the universe and all the phenomena taking place therein, by assuming the existence of eternal infinitesimal bodies that are endued with force. But as these atoms cannot be perceived by the senses, materialism, to be consistent, has nothing, can have nothing, to do with them. Again the forces of cohesion and expansion, supposed to inhere in the atoms, cannot possibly produce any connection conformable to design; and the materialistic philosopher must, therefore, deny the existence of any thing of the kind in organisms.

"As these atoms are entirely destitute of intelligence, the origin of a self-conscious intelligence and the identity of this self-consciousness during the whole life, amid the constant changes of matter, cannot be accounted for on materialistic principles; and the materialist has to doubt his own self-consciousness in order to be consistent.

"But as thinking, so is volition, a purely physical mechanism; and there is, therefore, no freedom of will, but every apparently free act is the necessary result of a chain of mechanically act-

* As far as the faculties of man are capable of discovering a spirit, by sight, touch, speech, and the joint efforts of the reason and the senses, modern Spiritualism claims that this has been done, although no chemical test may yet have been found.

ing causes: there is no moral self-determination. Materialism has, consequently, no morality; but leads consistently to the doing away with all moral and human order.

"Where there is no room for morality, there is of course none for religion. Thus materialism is everywhere a sad negation of every thing ideal; yea, a mere negation itself, a heap of ruins."

In dismissing the materialism of Moleschott, Büchner, Maudsley, and the rest, we are happy to quote Professor Tyndall, who, though he has shown, like some other men of high scientific culture, a lack of courtesy, if not of courage, in dealing with the spiritual phenomena, discourses well in regard to the unphilosophical attitude of the German Sadducees. Speaking of the connection between physical and mental processes, he says, "Were our minds and senses so expanded, strengthened, and illuminated as to enable us to see and feel the very molecules of the brain; were we capable of following all their motions, all their groupings, all their electric discharges, if such there be; and were we intimately acquainted with the corresponding states of thought and feeling, — we should be as far as ever from the solution of the problem, 'How are these physical processes connected with the facts of consciousness?' The chasm between the two classes of phenomena would still remain intellectually impassable. Let the consciousness of *love*, for example, be associated with a right-handed spiral motion of the molecules of the brain; and the consciousness of *hate* with a left-hand spiral motion. We should then know when we love that the motion is in one direction, and when we hate that the motion is in the other; but the 'why?' would still remain unanswered. In affirming that the growth of the body is mechanical, and that thought, as exercised by us, has its correlative in the physics of the brain, I think the position of the materialist is stated as far as that position is a tenable one. I think the materialist will be able finally to maintain this position against all attacks; but I do not think, as the human mind is at present constituted, that he can pass beyond it. . . .

"The problem of the connection of body and soul is as insoluble in its modern form as it was in the pre-scientific ages. Phosphorus is known to enter into the composition of the human brain; and a courageous writer has exclaimed, in his trenchant German, '*Ohne Phosphor keine Gedanke.*'* That may or may not be the case; but even if we knew it to be the case, the knowledge would not lighten our darkness. On both sides of the zone, here assigned to the materialist, he is equally helpless. If you ask him whence is this 'matter' of which we have been discoursing, who or what divided it into molecules, who or what impressed upon them this necessity of running into organic forms, he has no answer. . . . Science, also, is mute in reply to these questions."

Long before these remarks were made, in some comments upon the physiological writings of Mr. Bain, Mr. James Martineau expressed similar ideas, as follows: "If modern cerebral researches were ever so successful; if we could turn the exterior of a man's body into a transparent case, and compel powerful magnifiers to lay bare to us all that happens in his nerves and brain, — what we should see would not be sensation, thought, affection, but some form of movement or other visible change, which would equally show itself to any being with observing eyesight, however incapable of the corresponding inner emotion. Facts thus legible from a position foreign to the human consciousness are not mental facts, are not moral facts, and have no place in the interior of a science which professes to treat of these, and reduce them to their laws."

"Whatever force," says Mr. Shorter, "there may be in the argument from metaphysics for the soul's immortality is unaffected by Spiritualism, save in the way of confirmation to its conclusion. Spiritualism converts what before was but probability into certitude; it supplies the missing link; it makes good that embarrassing defect in the evidence which has perplexed so many, leading them to question or reject the belief in immortality

* No thought without phosphorus; an assertion which the facts of somnambulism and Spiritualism would seem to render rather questionable, to say the least.

as not adequately sustained. Let, then, the metaphysician marshal all his forces, and do what service he may in the cause of this great truth: I would only say, in the language of an elder Spiritualist, 'Yet show I you a more excellent way.'"

To many minds, familiar with the facts of Spiritualism, all arguments in proof of the soul's immortality will seem as superfluous as it would be to argue to a photographer that pictures can be made by the aid of light. To them the question is no longer an open one; for to them the fact of spiritual existence has been proved, as far as it can be to our limited human faculties. Enough has been given to satisfy them that to give more might be to cross some of the purposes of this disciplinary mundane existence. And so they wait serenely for the dawn of the great morning.

"Soon the whole,
Like a parchèd scroll,
Shall before our amazèd sight unroll;
And, without a screen,
At one burst be seen
The presence wherein we have ever been."

CHAPTER IX.

MISCELLANEOUS PHENOMENA.

"Oh, hearts that never cease to yearn!
Oh, brimming tears that ne'er are dried!
The dead, though they depart, return,
As if they had not died!

The living are the only dead;
The dead live—never more to die;
And often when we mourn them fled,
They never were so nigh!"

WELL authenticated accounts of apparitions of the departed may be found in Mr. Owen's "Footfalls on the Boundary of Another World," and in Mr. Howitt's comprehensive "History of the Supernatural."

"The department of apparitions alone," says Mr. Howitt, "is a most voluminous one, and that on evidence that has resisted all efforts to dislodge it. Amongst those of recent times is that which warned Lord Lyttleton, in a dream, of the day and hour of his death; the truth of which has been assailed in vain. Equally well attested is that which appeared to Dr. Scott in Broad Street, London, and sent him to discover the title-deeds of a gentleman in Somersetshire, who would otherwise have lost his estate in a lawsuit with two cousins. That which drove Lady Penniman and her family out of a house in Lisle at the commencement of the French Revolution, is well known and authenticated. That which announced to Sir Charles Lee's daughter at Waltham in Essex, three miles from Chelmsford, her death that day at twelve o'clock, and which took place then, is related by a bishop of Gloucester. That of Dorothy Dingle, related by the Rev. Mr. Ruddle, a clergyman of Launceston in Cornwall, occurring in 1665, is well known. Still more celebrated is that of Lord Tyrone to Lady Beresford, to warn her against a most miserable marriage, and to predict the marriage of his (Lord Tyrone's) daughter with Lady Beresford's son, and her own death at the age of forty-seven. In proof of the reality of this ghostly visit, the spirit took hold of her ladyship's wrist, which became marked indelibly, so that she always wore a black ribbon over it. The apparition to Dr. Donne of his living wife, when he was in Paris, representing the death of his child, is related by Dr. Donne himself; that of the father of the Duke of Buckingham, warning his son of his approaching fate, is well attested. Baxter relates several cases as communicated to him at first hand. But of all cases, ancient and modern, none are better authenticated than that of Captain Wheatcroft, who fell at the storming of Lucknow in 1857."

In this last case, the apparition presented itself to two different ladies, one of them the wife of Captain Wheatcroft. Nor could it be said that the recital of one lady caused the apparition of the same figure to the other. Mrs. Wheatcroft was at the time at Cambridge, and Mrs. N—— in London; and it was not till weeks after the occurrence that either knew what the other had seen. Those who would explain the whole on the principle of chance coincidence, have a treble event to take into account: the apparition to Mrs. N——, that to Mrs. Wheatcroft, and the actual time of Captain Wheatcroft's death, each tallying exactly with the other.

Examples of apparitions at the moment of death might be multiplied without number. In the case of the Wynyard apparition, which took place Oct. 15, 1785, at Sydney, in the island of Cape Breton, off Nova Scotia, Sir John Sherbrooke and General George Wynyard, then young men, both witnessed it at the same moment. "I have heard," said Sherbrooke, "of a man being pale as death; but I never saw a living face assume the appearance of a corpse, except Wynyard's at that moment." Both remained silently gazing on the figure as it passed slowly through the room, and entered the bed-chamber, casting on young Wynyard a look of melancholy affection. The oppression of its presence was no sooner removed, than Wynyard, grasping his friend's arm, exclaimed, "Great God! my brother!"

They instantly proceeded to the bedroom, searched, but found it untenanted. The case was made known to their brother-officers. With the utmost anxiety they waited for letters from England. At length came a letter to Sherbrooke, begging him to break to Wynyard the news of the death of his favorite brother, who had expired on the 15th of October, and at the same hour at which the friends saw the apparition.

Recently, while in England, Mr. Owen took pains to authenticate this narrative. "It will not, I think, be questioned," he writes, "that this evidence is as direct and satisfactory as can well be, short of a record left in writing by one or other of the seers, — which it does not appear is to be found. A brother-officer, the first who entered the room after the apparition had been seen, testifies in writing to the main facts. Sir John Sherbrooke himself, when forty years had passed by, repeats to a brother-officer his unaltered conviction that it *was* the spirit of his friend's brother that appeared to them in the barracks at Sydney, and that that friend was as fully convinced of the fact as himself."

Colonel Swift, late keeper of the crown jewels in the Tower, London, communicates to "Notes and Queries" of Sept. 8, 1860, an account of a singular apparition witnessed by himself and family in October, 1817, in his room in that ancient fortress, famous for so many royal murders and executions; and adds, that, soon afterwards, a sentinel on duty before the door of the jewel-office was so frightened by an apparition, that he died.

The Cambridge Association for Spiritual Inquiry, familiarly called the Ghost Club, have stated that their carefully conducted researches on the subject of apparitions have led them to regard such appearances as a settled fact. A member of this association informed Mr. Owen that he had collected two thousand cases of apparitions.

Dr. Garth Wilkinson, in his "Life of Swedenborg," says truly, "The lowest experience of all time is rife in spiritual intercourse already; man believes it in his fears and hopes, even when his education is against it; almost every family has its legends; and nothing but the wanting courage to divulge them keeps back this supernaturalism from forming a library of itself." This was also the candid confession of Kant.

In "Recollections, Political, &c., of the Last Half Century," by the Rev. J. Richardson (London, 1856), there is a circumstantial account of the appearance of Mr. John Palmer (an actor, who died suddenly on the stage at Liverpool on the 2d of August, 1798), on the night of his death, to a person in London, named Tucker. "The fact of his absence from London was known to Tucker, but he was not aware about his arrangement for his return. On the night just mentioned, Tucker had retired at an earlier hour than usual; but the company in the drawing-room was numerous, and the sound of their merriment prevented him from falling asleep. He was in a state of morbid drowsiness produced by weariness, but continually interrupted by noise. As he described the scene, he was sitting half upright in his bed, when he saw the figure of a man coming from the passage which led from the door of the house to the hall. The figure paused in its transit for a moment at the foot of the couch, and looked him full in the face. There was nothing spectral, or like the inhabitant of the world of spirits, in the countenance or outline of the figure, which passed on, and apparently went up the staircase. Tucker felt no alarm, whatever: he recognized in the figure the features, gait, dress, and general appearance of John Palmer, who, he supposed, had returned from Liverpool, and, having the *entrée* of the house, had, as usual, availed himself of his latch-key. . . . Next morning, in the course of some casual conversation, he informed Mrs. Vernon that he had seen Mr. Palmer pass through the hall, and expressed a hope that his trip to Liverpool had agreed with his health. The lady stared at him incredulously; said he must have been dreaming or drinking or out of his senses, as no Mr.

Palmer had joined the festivities in the drawing-room. His delusion, if delusion it were, was made a source of mirth to the people who called in the course of the day. He, however, persisted in his assertion of having seen Mr. Palmer; and on the arrival of the post from Liverpool on the day after he had first made it, laughter was turned into mourning, and most of the guests were inclined to think there was more in it than they were willing to confess.

"It should be added, that this 'Tucker' was a sort of hall-porter in Mrs. Vernon's house, and slept on a couch in the hall; and 'those who entered the house, and were about to go up stairs, had to pass by the aforesaid couch.'

"It is very curious, also, that Palmer dropped down dead on the stage, while performing the part of the 'Stranger' in Kotzebue's well-known play of that name, and immediately after uttering these memorable words, '*There is another and a better world!*' A benefit was got up in Liverpool for his children, which produced £400."

The positive statements of hauntings are so numerous, that, to deny them, or set them down as delusion, requires a skepticism akin to credulity. It turns out, on a thorough re-examination by Mr. Shorter of the celebrated "Cock-Lane ghost-story," for his belief in which Dr. Johnson has been so repeatedly ridiculed, that the phenomena of that case were in accordance with laws now familiar. The girl, a child of thirteen, was simply a medium. To learn how the raps were made, she was tried in all sorts of ways, and with tied-up hands and feet, from the supposition that she made the noises herself; but in vain. The noises went on, and that in different rooms, and even different houses. Floors and wainscots were pulled up; but no trick was discovered, though the search was made under the supervision of Dr. Johnson, Bishop Douglas, James Penn, and Stephen Aldrich. "That such a deception," says Howitt, "should be carried on by a family on which it only brought persecution, the pillory, and ruin, was too absurd for the belief of any except the so-called incredulous."

Beaumont, in his "Gleanings of Antiquities," published in 1724, mentions the *rapping* phenomena, and says, "There is a house in London, in which, for three years last past, have been heard, and still are heard, almost continual knockings against the wainscot overhead, and sometimes a noise like telling money, and of men sawing, to the great disturbance of the inhabitants; and often lights have been seen, like flashes of lightning; and the person who rents this house has told me that when she has removed eighteen miles from London, the knockings have followed her."

Glanville says that there were knockings, and that a hand was seen at old Gast's House in Little Burton in 1677. The knockings were on a bed's head, and the hand was seen holding a hammer, which made the strokes. Our times do not have the exclusive experience even of knockings. Bishop Heber says that the evidences of such things, which Glanville gives, are more easily ridiculed than disproved.

The cases on record of *direct spirit-writing*, when no medium was near enough to co-operate in any known way, are very numerous.* A work by Baron L. De Guldenstubbé, a Swedish nobleman, resident in Paris, entitled "La Réalité des Esprits," and published a few years since, contains numerous *fac-similes* of writings made on paper by some invisible and intelligent force. The names of ten distinguished persons who witnessed the phenomenon are given. The Baron is a gentleman well known to personal friends of our own; and his character gives all possible weight to his testimony.

"The absurd fear of demons," he says, "has incapacitated our orthodox priests and theologians from combating the materialists and the Sadducees with effectual experimental weapons. This *demonophobia* has unfortunately grown to be a veritable *demonolatry*. The priests having fear of demons, and, consequently, not wishing to occupy themselves with these spiritual

* At the house of Mr. Daniel Farrar in Boston, some years since, we were present at some very curious experiments of this sort. The late Charles Colchester was the medium.

phenomena, have unwittingly formed a pact with the devil, by virtue of which the reign of incredulity and materialism, that reign of the demon *par excellence*, continues to subsist in all its *éclat*. . . .

"The two fundamental ideas of Spiritualism — namely, that of the immortality of the soul, and that of the reality of the invisible world which reveals and manifests itself in different ways in our terrestrial world — are but the necessary corollary of the idea of God or the Absolute, and *vice-versâ*. We may even assume that the idea of the immortality of the soul, and of its relations to the supernatural world, is more intimate and primitive than that of God, Creator and Supreme Author of the universe. . . .

"The Bible does not formally teach the idea of the immortality of the soul, graven by the Eternal himself on the heart of man, but it supposes it everywhere. (Job xix. 26, 27; Num. xxiii. 10; Isa. xxvi. 19.) . . . The practice of necromancy, according to Samuel (1 Sam. xxviii. 3-25), and according to Deuteronomy (xiii. and xviii.), necessarily presupposes the doctrine of the immortality of the soul; and so with the visions and apparitions, of which the Bible is full."

Dr. Henry More gives a remarkable story touching the stirs made by a demon in the family of one Gilbert Campbel, by profession a weaver, in the old parish of Glenluce in Galloway, Scotland, in November, 1654. Among other phenomena in this case, we read that "presently there appeared a naked hand and arm from the elbow down, beating upon the floor till the house did shake again."

Certain surprising occurrences, which took place in 1806 at Slawensick Castle, Silesia, are thoroughly well authenticated. Councillor Hahn, in the service of Prince Hohenlohe, had gone to Slawensick, and with an old friend, a military officer named Kern, had taken up his abode in the castle. "Hahn, during his collegiate life, had been much given to philosophy; had listened to Fichte, and earnestly studied the writings of Kant. The result of his reflections, at this time, was a pure materialism."

He had been reading aloud to his friends the works of Schiller. when the reading was interrupted by a small shower of lim which fell around them: this was followed by larger pieces; bu they searched in vain to discover any part of the walls or ceiling from which it could have fallen. The next evening, instead o the lime falling, as before, it was thrown, and several pieces struck Hahn; at the same time they heard many blows, sometimes below and sometimes over their heads, like the sound of distant guns. On the following evening, a noise was added, which resembled the faint and distant beating of a drum. On going to bed, with a light burning, they heard what seemed like a person walking about the room with slippers on, and a stick, with which he struck the floor as he moved step by step. The friends continued to laugh and jest at the oddness of these circumstances, till they fell asleep; neither being in the least inclined to attribute them to any supernatural cause. "But, on the following evening, the affair became more inexplicable: various articles in the room were thrown about, — knives, forks, brushes, caps, slippers, padlocks, funnel, snuffers, soap, — every thing, in short, that was movable; whilst lights darted from corner to corner, and every thing was in confusion. At the same time the lime fell, and the blows continued. Upon this, the two friends called up the servant, Knittel, the castle-watch, and whoever else was at hand, to be witnesses of these mysterious operations. Frequently, before their eyes, the knives and snuffers rose from the table and fell, after some minutes, to the ground." So constant and varied were the annoyances, that they resolved on removing to the rooms above. But this did not mend the matter: "the thumping continued as before; and not only so, but articles flew about the room which they were quite sure they had left below." Kern saw a figure in the mirror interposing, apparently, between the glass and himself; the eyes of the figure moving, and looking into his.

It is unnecessary to recount the means employed to trace out these mysteries. Hahn and Kern, assisted by two Bavarian officers, — Captain Cornet and Lieutenant Magerle, and all the aid they could assemble, — were wholly unsuccessful in obtaining the slightest clew. And Hahn, from whose narrative this account is taken, declares, "I have described these events exactly as I saw them; from beginning to end, I observed them with the most entire self-possession. I had no fear, nor the slightest tendency to it; yet the whole thing remains to me perfectly inexplicable."

Those who have read Mrs. Poole's "Englishwoman in Egypt," will recollect her curious account of the hauntings and apparitions in the house of her brother, Mr. Lane, at Cairo. The account is fully confirmed by Mr. Bayle St. John. He relates having seen a ghostly Sheik enter the house at noon, where he himself lived; having had the doors immediately closed, and the visitor actively hunted up, but to no purpose. He relates also, that, in Alexandria, cases of throwing of stones from the roofs are of no unfrequent occurrence, where no one can discover the perpetrators.

M. Joseph Bizouard, in a work published in Paris, under the title of "Des Rapports de l' Homme avec le Démon," relates some details, given by Görres, of strange events at Münchshofe, situated a league from Voitsberg, and three leagues from Gratz. They occurred in the house of a Herr Obergemeiner, and were observed and recorded by Dr. J. H. Aschauer, his father-in-law, a very learned physician and professor of mathematics at Gratz. They commenced in October, 1818, by the flinging of stones against the windows on the ground-floor, in the afternoon and evening. The noise generally ceased when they went to bed. As nobody could discover the cause, towards the end of the month, Obergemeiner, without saying any thing to his family, engaged about thirty-six of the peasants of the environs, and placed them in cordon all round the house well armed, and with orders to allow no one to go in or out of the house. He then took into the house with him Koppbauer and some others, assembled all his people to see that none were missing, and thoroughly examined every apartment, from the attics to the cellar. It was about half-past four o'clock in the afternoon.

The peasants formed their circle, and saw that no one was concealed within it, nor was able to pop in or out; notwithstanding, the throwing of stones commenced against the windows of the kitchen. Koppbauer, placed at one of them, endeavored to ascertain their direction. Whilst Obergemeiner was in the kitchen with the others, a great stone was launched against the window where he stood, and broke many of the panes. It was previously thought that the stones were thrown from the interior; and it was in effect from that direction that they now continued to come till half-past six in the evening, when the whole ceased. Every place in the house where a man could possibly conceal himself was visited; and the guard without continued its position.

At eight o'clock in the morning, the stone-throwing re-commenced before more than sixty persons; and they were convinced that, issuing from beneath the benches of the kitchen, they struck the windows in a manner inexplicable. Pieces of limestone, weighing from a quarter of a pound to five pounds, were seen flying in all directions against the windows; and immediately afterwards all the utensils, spoons, pots, plates, full and empty, were launched from the midst of the spectators against the windows and the doors with a velocity inconceivable. Some broke the glass, some remained sticking in the broken panes; and others, only appearing to touch the glass, fell into the interior. The spectators, when struck by the stones, felt only a slight blow. Whilst utensils were being carried from the kitchen, they were forced from the hands of those who bore them, or they were knocked over on the table on which they were placed. The crucifix alone was respected: the lights burning before it were forcibly flung down. At the end of two hours, all the glass in the kitchen and all the fragile objects were broken, even those which they had carried away. A plate full of salad carried up to the first floor, in the act of being carried down again, by a servant, was snatched from her hands and flung into the vestibule. The disorder ceased at eleven o'clock. We omit many particulars which took place at this time.

M. Aschauer, having heard this strange news from his son-in-.aw, desired to know when any thing further took place; and, being sent for, as he entered he saw his daughter, with the man named Koppbauer, picking up the fragments of a pot, which had been thrown on the floor just as he entered. Then, all at once, a great ladle was launched from the shelf on which it lay, and, with incredible velocity, against the head of Koppbauer, who, instead of a severe contusion, only perceived a very light touch. M. Aschauer saw nothing further till the next day; when, issuing from the kitchen on account of the smoke, some stones were thrown against the windows. This physician examined the lightning-conductor, and every thing else, with an electrometer; but neither he nor Obergemeiner, who had offered a reward of a thousand francs to any one who could discover the cause, could detect any thing. On the second day, about four o'clock in the afternoon, Aschauer, troubled at these strange occurrences, was standing at the end of the kitchen, having opposite to him a shelf on which stood a large metal soup-tureen, when he saw the tureen suddenly dart towards him in a nearly horizontal position, and with surprising velocity, and pass so near his head that the wind of it raised his hair; and the tureen then fell to the earth with a great noise.

Curiosity caused people to hasten from all parts, who were struck dumb with astonishment at these phenomena, and others of a similar nature. Towards five o'clock came a stranger, who pretended that a man must be concealed in the chimney. This ridiculous explanation excited the anger of M. Aschauer; and he led him towards the door, whence nothing could be seen from the chimney, and, pointing to a copper dish upon a shelf, he said, "What would you say, monsieur, if that dish should, without any one touching it, be thrown to the other side of the kitchen?" Scarcely were the words uttered, when the dish, as if it had heard them, flew across. The stranger stood confounded.

We omit many particulars, because they are of the same kind. A pail of water, weighing fifteen pounds, which had been set on the floor, fell from the ceiling without any one being able to conceive how it got there; for there was nothing to hang it upon. As they were seated round the fire, a pot, which none of them could touch, was suddenly turned over, and emptied itself little by little, contrary to the law of such a fall. Then came egg-shells flying from every corner, nobody being there to throw them, and no one being able to imagine whence they came. After the departure of M. Aschauer, the wheels of a mill, about six minutes' walk from the house, stood still from time to time; the miller was thrown out of his bed, the bed turned over, the lights were extinguished, and various objects were thrown to the ground.

After this, nothing more is said to have happened; at all events, M. Obergemeiner, who did not love to speak of these things, made no report of any. They made a great sensation, however, amongst the government officials; and the district of Ober-Greiffenneck sent its report to the circle of Gratz. "Although it is said that we exist no longer in the times of ignorance, when phenomena which could not be comprehended were attributed to demons, &c., it is remarkable that, at an epoch in which civilization and the progress of the natural sciences have put them to flight, we yet see extraordinary things which the *savans* cannot explain." The report accords with the recital of M. Aschauer, and a mention is made in it of an inquiry by order of the magistrates, conducted by M. Gayer, with his electric apparatus; and the report concludes by recommending a further inquiry, "as a natural solution can alone combat the hypocrisy of some and the superstition of others."

We do not ask the reader to imagine the conclusion to which the government came on this matter, for he never could divine it. It was "that a man concealed in the tunnel of the chimney was probably the cause"! These professors of natural science were, however, charged to proceed to a further inquiry; but they considered it beneath their dignity, and refused. Afterwards an agent of the police visited the house; and Görres says, that, amongst the various causes that he imagined, the most amusing was that M. Aschauer had only astonished the people by a series of scientific tricks. Görres, however, stating that his account is literally found in a letter of M. Aschauer to a friend, dated Jan. 21, 1821, and in details communicated to himself, at a later period, assures us that M. Aschauer was not only a man of the profoundest science, but of the profoundest regard to truth, and one who feared no ridicule in stating it, however strange it might be. On this occasion, he asserted that no master of legerdemain was capable of producing the things which he saw. Neither was the force employed a mere scientific or physical force: it was a force free and reasoning; and these effects were the sport of a spirit or spirits, immaterial or invisible.

M. Babinet, in an essay in the "Revue des Deux Mondes," reasons, like Faraday, that these, and similar phenomena attested by Spiritualists, are impossible, because they contradict the law of gravitation. Dr. Brownson urges in reply, that when he sees a fact of this kind, he does not pretend that it is in accordance with the law of gravitation, but the essence of the fact — that which constitutes its marvellousness — is precisely that it is not. "Now, to deny the fact for that reason," he says, "is to say that the law of gravitation cannot be overcome or suspended, and precisely to beg the question. How," he asks, "does M. Babinet know that there are not invisible powers who can overcome this force as easily as we ourselves can do? The fact of the rising of a table, or a man to the ceiling, is one that is easily verified by the senses; and, if attested by witnesses of ordinary capacity and credibility, must be admitted. That it is contrary to the law of gravitation, proves not that it is impossible, but that it is possible only preternaturally." In the words of Mr. Mill, there must be an "adequate counteracting cause."

"Scientific men," says Mr. Shorter, "should learn from experience to be cautious in affirming the limit of the possible. The more completely they prove that the phenomena in question are not due to, and are impossible by, any physical agency, the more completely do they establish the necessary spiritual causation of such phenomena. Those men of science who have erected theories about the impossible, have not unfrequently built a monument to their own folly and shame. The circulation of the blood, the prevention of small-pox by vaccination, the fall of meteorolites, the lighting of towns by gas, conveyance by steam, painless surgery, clairvoyance, — these, and many other things now familiar to us, have, each in its turn, been pronounced impossible by high authorities. One age laughs at an idea; the next adopts it. The impossible of yesterday is the familiar fact of to-day. In an age when steam is our conductor, and electricity our messenger, and the sun our portrait-painter; when the every-day facts of life would have been a fairy tale a hundred years ago; who, especially with the knowledge that spiritual forces are working around and within us, will have the presumption to affirm that it is impossible for spiritual beings so to operate upon ourselves and surrounding objects as to make their presence evident even to our senses? Lord Bacon says, 'We have set it down as a law to ourselves to examine things to the bottom, and not to receive upon credit, or reject upon improbabilities, until there hath passed a due examination.'"

The late Thomas Starr King was intuitively a Spiritualist. "What more arrogant and presumptuous folly can there be," he says, "than that which a person exhibits, who makes *his experience* of nature the measure of the *possibilities* of nature? Yet this is what all of us do who object to the doctrine of the soul's immortality, that we cannot conceive *how* it is released from its fleshy bondage, nor what are the methods of its disembodied life. If we should hear any man soberly affirm that he did not believe that any process could go on in this universe, or any thing be true, which baffled his powers of comprehension, we should probably think that the application to him of Paul's apostrophe to the Corinthian doubter involved no dangerous lack of charity. It has pleased God to endow us with five senses, through which we hold conversation with the created realm. We do not know that five other media of communication might not be opened that would make the physical universe seem as different and as much higher than it now does, as if we were transported into another sphere. Who has told us that there cannot be any other avenues between the soul and matter than the touch, the taste, the ear, and the eye? Who has told us that all which *exists right about us* is reported by the limited apparatus furnished to our nerves? . . .

"It has been truly said by another, that we should 'easily believe in a life to come, if *this present life* were the wonderful thing to us which it ought to be.' Here is the point. Not that there are startling difficulties in the way of conceiving a future existence, but that we lose the fine sense and the nice relish of the mystery and miracle that invest us here. There are a thousand scientific facts that would seem as marvellous to a cultivated mind, if they had not been demonstrated and published in veracious treatises, as the continued existence of the body. What would Plato have said, could he have seen a man, without using any flame in the experiment, cause fire to burst out of a lump of ice? Suppose that Newton had never heard of a loadstone, what would he have thought, could he have seen an iron weight, in defiance of the law of gravitation which he had just demonstrated, spring from the floor to the wall? Before seeing the fact for the first time, would not the proposition have seemed as surprising to him, and as difficult to be believed, as the return of a dead man to life before his eyes, or the appearance of a spirit? And after he had seen it, how could he explain it? How can any man explain the phenomenon now?

"Is the statement that there is an enduring spirit within us, entirely distinct from the corporeal organization, and which the cessation of the heart liberates to a higher mode of existence, any more startling than the statement that a drop of water, which may tremble and glisten on the tip of the finger, seemingly the most feeble thing in nature, from which the tiniest flower gently nurses its strength while it hangs upon its leaf, which a sunbeam may dissipate, contains within its tiny globe electric energy enough to charge eight hundred thousand Leyden jars, energy enough to split a cathedral as though it were a toy? And so that, of every cup of water we drink, each atom is a thunder storm?

"Is the idea of spiritual communication and intercourse, by methods far transcending our present powers of sight, speech, and hearing, beset with more intrinsic difficulties than the idea of conversing by a wire with a man in St. Louis, as quickly as with a man by your side, or of making a thought girdle the globe in a twinkling? And when we say that the spiritual world may be all around us, though our senses take no impression of it, what is there to embarrass the intellect in accepting it, when we know that, within the vesture of the air which we cannot grasp, there is the realm of light, the immense ocean of electricity, and the constant currents of magnetism, all of them playing the most wonderful parts in the economy of the world, each of them far more powerful than the ocean, the earth, and the rocks, — neither of them at all comprehensible by our minds, while the existence of two of them is not apprehensible by any sense."

"Sweep away the illusion of Time," says Carlyle, "compress the threescore years into three minutes," and what are we ourselves but ghosts? "Are we not *spirits*, that are shaped into a body, into an appearance? This is no metaphor: it is a simple scientific fact. We start out of Nothingness, take figure, and are Apparitions: round us, as round the veriest spectre, is Eternity; and to Eternity minutes are as years and æons. . . .

"O Heaven! it is mysterious, it is awful to consider that we not only carry, each a future Ghost within him, but are, in very deed, Ghosts! These Limbs, whence had we them; this stormy Force; this life-blood with its burning Passion? They are dust and shadow; a Shadow-system gathered round our ME; wherein, through some moments or years, the Divine Essence is to be revealed in the Flesh. That warrior on his strong war-horse, fire flashes through his eyes; force dwells in his arm and heart: but warrior and war-horse are a vision; a revealed Force, nothing more. Stately they tread the earth, as if it were a firm substance: fool! the Earth is but a film; it cracks in twain, and warrior and war-horse sink beyond plummet's sounding. Plummet's? Fantasy herself will not follow them. A little while ago they were not; a little while and they are not, their very ashes are not.

"So it has been from the beginning; so will it be to the end. Generation after generation takes to itself the Form of a Body; and forth issuing from Cimmerian Night, on Heaven's mission APPEARS. What Force and Fire is in each he expends: one grinding in the mill of Industry; one, hunter-like, climbing the giddy Alpine heights of Science; one madly dashed in pieces on the rocks of Strife, in war with his fellow: and then the Heaven-sent is recalled; his earthly Vesture falls away, and soon, even to Sense, becomes a vanished Shadow. Thus, like wild-flaming, wild-thundering train of Heaven's Artillery, does this mysterious MANKIND thunder and flame, in long-drawn, quick-succeeding grandeur, through the unknown Deep. . . . Earth's mountains are levelled, and her seas filled up, in our passage: can the Earth, which is but dead and a vision, resist Spirits which have reality and are alive? On the hardest adamant some foot-print of us is stamped in; the last Rear of the host will read traces of the earliest Van. But whence?—O Heaven, whither? Sense knows not; Faith knows not; only that it is through Mystery to Mystery, from God and to God."

Carlyle reveals to us the spiritual side of man whilst in this world and fettered to his clog of flesh. The great facts of Spiritualism reveal man to us as he is when he emerges into "a purer ether, a diviner air," with his individualism unimpaired, and all that he has gained of good, through his affections and his understanding in this life, left whole as the vantage-ground of future progress.

CHAPTER X.

THEORIES IN REGARD TO THE PHENOMENA.

"It is only since the middle of the eighteenth century that Spiritualism began to cease to be the prevalent faith of Christendom; and parallel with this decline has been the denial of all revelation and the spread of atheistical philosophy. God, however, has not left himself without a witness; and in our day, when Sadduceeism most abounds, evidences of a spiritual world have been multitudinous." — *Thomas Shorter.*

WE have seen what the first theories were in explanation of the phenomena of 1848. It was soon found that these theories were insufficient. Like Faraday's notion of an unconscious exercise of muscular force, they did not cover the new facts as they came up and multiplied.

So long as the manifestations were confined to raps and table-tippings, it was surmised that they might proceed in some mysterious way from animal electricity, put in operation by the unconscious will of the medium or of other persons present.

The late Dr. E. C. Rogers, a gentleman personally well known to us at the time the Rochester phenomena began to excite public attention, was the author of a work bearing the following title: "Philosophy of Mysterious Agents, Human and Mundane, or the Dynamic Laws and Relations of Man." His theory is that the whole body of phenomena, physical and mental, are referable to cerebral or mental action, through the medium of "a physical force associated with the human organism; and, under peculiar conditions, this physical force is made to emanate from that organism with a most terrible energy, and without any necessary conjunction with either spiritual or psychological agency." This agent may be the *od*, or *odic force*, of Reichenbach. It is not under the general control of the will, but is the mere agent of the unconscious organs, playing its part automatically, as the brain is affected.

The material agent is thus put in operation by the peculiar changes that take place in the cerebral organs. That every thought, emotion, or passion is accompanied with a change of the motion of the brain, is assumed as one of the undisputed facts in physiology. It is the prerogative of every man's mind or spirit to control the motions, and, consequently, the changes of his brain, within prescribed limits. But, in certain conditions of the latter, such as mesmeric trance, catalepsy, sleep, cerebral inflammation, passiveness of mind and will, and many others, the man's own personality is suspended in its prerogative action. The predominant influence upon it, then, becomes material or sensuous; and here, according to Dr. Rogers, the reflex action of another's brain will readily take effect. Another's wish or request will act like a law; and a fictitious personality may be induced in the brain, and represented independently of the conscious personality, reason, and will of the individual. It therefore follows that the specific action of one person's brain may be unconsciously propagated to another's brain, and there be exactly represented in a second cerebral action. This may propagate itself to the automatic centres in the spinal axis, and thus the involuntary play of the muscles may produce the rappings, movements of furniture, and the other phenomena.

In view of the many evidences of unconscious cerebral action, Dr. Rogers regards it as precipitate to attribute to the influence of disembodied spirits that which may lie within the sphere of the human organization and of mundane agencies. He then proceeds to show how the human organism may be influenced by drugs, so as to alter its conditions; and argues that, inasmuch as the agent, the substance on which it acts, and the new condition, are purely physical, the results must be physical also. It follows, therefore, that visions, somnambulism, ecstasy, which are pathematically produced, and also produced by the influence of drugs upon the organism, are the results of the material conditions of that organism, and do not require the spiritual hypothesis for their explanation.

Dr. Rogers's conclusion is, that the whole body of phenomena of Spiritualism, including the past and the present, "offer to the philosopher a new view of man and his relations to the sphere in which he lives, by neglecting which the deepest mysteries of the human being are left unsolved."

This ingenious writer died before the more advanced phenomena recorded in this volume were made known to the world. Had he lived to become acquainted with them, he might have found that, whatever there may be of truth in his theory, is not inconsistent with the fact of the agency and appearance of disembodied spirits.

Professor A. Mahan, Mr. Charles Bray, Dr. Samson, of Columbia College, and others who have adopted the apneumatic or *no-spirit* view in regard to the phenomena, have done little more than either to put in new and expanded form the arguments of Dr. Rogers, or to substitute for his notion of an odic force the simple hypothesis of nervous action. None of these opponents of the spiritual theory deny the facts. Professor Mahan says, "We shall admit the facts claimed by Spiritualists. We admit the facts for the all-adequate reason that, after careful inquiry, we have been led to conclude that they are real. *We think that no candid inquirer who carefully investigates can come to any other conclusion.*"

The facts being admitted, Professor Mahan finds in Reichenbach's odic force the mysterious agent by which they are manifested. But it is somewhat remarkable that Reichenbach himself, the original hypothetist of this odic force, modestly disclaims for it all such power as these writers attribute to it. He avowedly regards it merely as the means by which spiritual intelligence manifests itself; as the channel through which it sends its forces. That it is in itself an intelligent, personal principle, able to take the shape of the human body, and to conduct itself like an individual in the flesh, makes no part of his hypothesis; and this notion certainly demands as great an effort of credulity as any theory of direct spiritual action.

President Samson is of opinion that all the manifestations, supposed to be spiritual, are really natural, the working of an agent intermediate between mind and matter, for which agent he can give no better name than the nervous fluid.

He tells us that, when, in 1848, Arago witnessed the attraction and repulsion of heavy bodies at the presence of Angelique Cottin, a nervous factory-girl, who, having begun suddenly to exhibit this wonderful derangement, was carried up to Paris, to appear before the Academy, that great philosopher, being asked his opinion about it, remarked, "That is yet to be settled. It seems to have no identity with electricity; and yet, when one touches her in the paroxysms, there is a shock, like that given by the discharge of the Leyden jar. It seems to have no identity with magnetism proper, for it has no re-action on the needle; and yet the north pole of a magnet has the most powerful re-action on her, producing shocks and trembling. This is not effected through the influence of her imagination, as the magnet has the same influence, whether brought secretly near her, or otherwise.* *It seems a new force.* At all events, whatever it be, time and research will determine, with a sufficient number of cases. One thing, however, seems to be certain: the phenomena of this case show, very plainly, that whatever the force is which acts so powerfully from the organism of this young girl, it does not act alone. It stands in mysterious relation to some mundane force, which acts and re-acts with it. This is witnessed in the re-action which external things have upon her person, often attracting her with great power. It is a curious inquiry, and may open to us new resources in the nature of man and of the world, of which we have little dreamed."

In two bulky volumes, published in 1855, Count Agenor de Gasparin, takes a view of the question not dissimilar to that of President Samson, whom he quotes and commends. The Count

* This is no proof, however, that her imagination may not have operated in the case; for her clairvoyance may have enabled her to detect the instances in which the magnet was *secretly* brought near her.

is a leading Protestant writer of the evangelical school, and is well known to Americans. He avows his belief in the reality of the early phenomena, gives an extended narrative of facts elicited by himself at a series of sittings, in 1853, and shows the fallacy of Faraday's attempted explanation. He replies, at length, to the suggested fear that to admit the facts will give ground for superstition and credence in false miracles. He shows the marked line between just confidence in undeniable facts and the perversions of imagination, by reference to Ammianus Marcellinus, the old Roman historian, who refers to *table*-revelations the perfect counterpart of those of 1848. The people of Rome were expecting that *Theodorus* would become the emperor; and, of course, when the *tables* were consulted, they gave the letters of that name; whereas it proved that *Theodosius* became the emperor. He quotes, also, Tertullian's mention, in these words: "Mensæ divinare consueverunt:" *Tables are accustomed to divine.*

He quotes a case examined by Chamillard, doctor of the Sorbonne, in the seventeenth century, in which the same result was reached as that reported by the French Academy's commission to report on Mesmer's experiments, which prior result was thus sententiously recorded: "Multa ficta, pauca vera, à dæmone nulla:" *Many things fictitious, a few true, from a demon none.* Coming to the consideration of the natural cause of the phenomena, Gasparin ascribes them to the excess of nervous susceptibility. All that is real in such as are regarded as supernatural is to be found, he thinks, in an undue and diseased action of the nervous organism. He quotes from Arago what that philosopher says on the subject of Mesmer's experiments: "Effects, analogous or inverse, might evidently be occasioned by a fluid subtle, invisible, imponderable; by a sort of nervous fluid, or of magnetic fluid, if this be preferred, which may circulate in our organs." He also quotes from Cuvier, who was of opinion that the effects of mesmerism are clearly due "to some sort of communication established between the nervous systems" of the subject and the operator.

His conclusion is substantially like that of the Report of the French Commission on Mesmerism; namely, that the reported phenomena of the so-called spiritual manifestations are to be referred partly to errors of testimony, arising from the natural spirit of man to exaggerate the character and number of the facts; partly to the hallucination of an excited imagination, which suggests an exaggerated idea of the cause as supernatural; and chiefly to the *real action of the nervous fluid*, by which phenomena analogous to those in electricity and magnetism are wrought.

The new and irreconcilable facts that have come up since Gasparin arrived at these conclusions, make his theory wholly unsatisfactory at this time. It will not do to attribute to hallucination the results of the calm scrutiny of hundreds, nay, thousands, of competent observers, free from all undue excitement or bias, investigating the phenomena with the perfect composure which continued familiarity must always give, and actuated by no sectarian or anti-sectarian preconceptions. There are a multitude of witnesses now to the extraordinary, as well as to the ordinary, facts of Spiritualism; and some other hypothesis must be resorted to than that of "errors in testimony." When such men as De Morgan, Wallace, Varley, Denton, Owen, Wilkinson, Shorter, Howitt, Leighton, Coleman, Gunning, Gray, Mountford, Ashburner, Bell, Farrar, Livermore, Brittan, and hundreds of others in all the various professions, testify to a certain class of phenomena, the pooh-pooh argument, in reply, has lost its power, and falls flat, except on the ears of the uninformed.

To admit all the marvellous facts of Spiritualism, and still to reject the spiritual hypothesis in accounting for them, seems to require, at the first thought, a greater stretch of credulity than the wildest spiritual belief. But Mr. J. W. Jackson, of England, an experienced mesmerist, a man of science, and a full believer in spiritual realities, admits the most startling of the recent phenomena; but, like Sir David Brewster, will not "give in" to the theory of spiritual agency in their production. He assumes that mesmerism, *minus* spirits, explains all. He treats modern Spiritualism as Comte treats all religious creeds, simply as a new illustration of the same tendency of mind which induced the human race, in earlier ages, to attribute great natural phenomena, such as thunder, eclipses, volcanoes, &c., to the intervention of spiritual beings, angry deities.

"The spiritual hypothesis," he says, "is the product of a law of the human mind, in virtue of which it is impelled to supplement knowledge by superstition; and so, when there is no assignable cause for a phenomenon, it is at once relegated to the realm of miracle." — "Originating in a mental necessity for assigning some cause, real or imaginary, for every clearly recognized effect, the spiritual hypothesis is an inevitability with minds at the theologic stage, whenever a phenomenon transcends the range of recognized scientific knowledge." — "In earlier ages, the spiritual hypothesis, or, in other words, a theory of the miraculous, amply sufficed as an explanation of all otherwise inexplicable phenomena."

In reply to these views, Mr. Andrew Leighton remarks, "I doubt not every competent and patient investigator will find that, after the most careful discrimination of facts, after discounting all that is clearly mundane, and all that is *not* clearly, but only possibly, mundane, there will remain a residuum, which, if we are to attempt the resolution of the facts at all, will *necessitate* the supramundane hypothesis, and thus render it, so far from being 'inadmissible,' really the only rationally admissible one, since it will be found to be the only hypothesis adequate to cover all the facts."

The rival hypothesis he sets down as this: "That the brain has in it active potentialities unknown to consciousness, — not only unknown, but opposed to consciousness; to which potentialities, as a last resource, must be referred the otherwise inexplicable and indomitable facts."

"Notwithstanding," adds Mr. Leighton, "what has been said as to the rationality, and indeed necessity, of the spiritual hypothesis, it is not meant that this is to be held, *except* as an hypothesis, ready to be yielded up immediately that another capable of more perfectly explaining the facts, in accordance with all other truths of science, can be produced. Until the scientific mind, *par excellence*, shall produce that, it had better suppress its scorn and its supercilious condescensions."

With respect to the *facts* of Spiritualism, Mr. Jackson makes as large admissions as any Spiritualist could desire; yet his explanation is substantially the same as that of other upholders of the anti-spiritual hypothesis.

"Spiritual manifestations," he says, "are divided into mental and physical; and the spiritual hypothesis presupposes that, under each, there are phenomena to whose production Nature is inadequate. Let us now test this in reference to the first class, where it may be freely admitted that you not only have intelligence, but supersensuous intelligence; that is, you obtain information beyond the ordinary cognition of the medium, and sometimes beyond the knowledge or experience of any one present at the circle, and this, too, in reference to things past, distant, or future. It is in this way, indeed, that you have obtained a very large moiety of your converts, and those too often of a rather superior order of intellect; and yet there is nothing here but a manifestation of that clairvoyant power with which the mesmerist has been long familiar.

"After more than twenty years' experience, in which I have employed *lucides* of various ages and of both sexes, I could not fix the limits of this extraordinary faculty, and say, Here the natural power of the medium terminates, and there spiritual aid must have supervened. This probably reveals to you the key by which I propose to unlock the mysteries of the circle. The latter, when rightly constituted, is a most powerful mesmeric battery, of whose nervo-vital current the medium is the duly susceptible recipient. Now, in the present very imperfect state of our knowledge, it is quite impossible to predicate the maximum of result obtainable under such conditions, and unless we can do so, the assumption of spiritual aid, in any particular case, is perfectly gratuitous; quite permissible as a soothing succeedaneum to undisciplined minds, but altogether inadmissible as a scientific hypothesis. The same remark applies to spontaneous exaltation, whether of a literary, artistic, or even prophetic character, on the part of a medium. Such unusual displays of mental power are simply manifestations of ecstatic lucidity, taking that particular form; and, in the present state of our knowledge, it is quite impossible to say what are the unaided limits of a gifted human mind in this direction."

On the subject of levitation, elongation of body, and other phenomena, Mr. Jackson says, "But when we find lightness of body frequently recorded as an accompaniment of ecstatic illumination, not only in Christian, but also Brahminical and Buddhistic legends, the idea is at once suggested that it may be the result, in certain temperaments, of unusually exalted nervous function. Such facts suggest the institution of further experiments, rather than the hasty formation of a spiritual hypothesis; for they seem to indicate that nervo-vital power has in it an element antagonistic to the action of gravitation; and lightness of body may be only an extreme manifestation of this force, the accompaniment of a crisis, or the effect of consentaneous action in a well-constituted and harmonious circle of human organisms."

As to the movements of ponderable articles, these are referred, by Mr. Jackson, to "the intervention of life-power under conditions not yet known to science."

Upon this Mr. Leighton remarks, "We hold that the intelligence and will implied in the physical manifestations are not those of the passive media in whose presence they occur, but are demonstrably those of beings distinct from the members of the mundane company. Sometimes, as Mr. Jackson knows, they are said to be actually visible to one or more of the company, though invisible to the rest. The moral argument of the integrity of the seers — not to be got over by mere psychological imputations — has therefore to be met, besides the evidence of seers and non-seers alike, when the physical manifestations alone are considered. That 'there is really nothing more miraculous in the apparently spontaneous ascent of a table to the ceiling than in the corresponding ascent of a needle under the influence of a magnet,' is quite as firmly asserted by the Spiritualist as by the Non-spiritualist. Why should Mr. Jackson imply, and so constantly iterate, the implication to the contrary? His notion of the intervention of a vague 'life-power'—an unconscious efflux of the company, accomplishing all the intelligent voluntary motions imposed upon the table or other passive piece of furniture, sometimes according to the desire of those present, sometimes *against* their wishes, and in defiance of their every effort to prevent them—approaches far more nearly the 'miraculous' than the hypothesis he so persistently attempts to identify therewith.

"Repeating the sophism already exposed in other relations, Mr. Jackson says, 'As we are ignorant of the power of a life-circle, it is impossible to assign limits to its effects; and until these are reached, spiritual intervention is a needless accessory.' Was it the *life-power of the circle*, which on one occasion concentrated itself in my presence, seized a slate-pencil, and wrote out a sentence which was certainly not in the mind of any who were visibly present? Was it the same power which manipulated the keys of an accordion, and played, with artistic ability and feeling never surpassed, the tune of 'Home, Sweet Home,' in opposition to the expressed wishes of several present, who asked for other tunes? Talk of the miraculous in Spiritualism! Can any thing be more miraculous or gratuitous than the conceptions of this votary of science in his endeavors to escape the only hypothesis which, without straining, naturally and completely covers all the facts? To assume that the mesmeric power of the circle, in any form or degree, is capable of accounting for such facts, appears to us as gratuitous, not to say ridiculous, as to apply Faraday's unconscious muscular hypothesis in explanation of the movement of physical objects upon which there was no muscular impact, or upon which the muscular impact was strenuously exerted the opposite way."

The remarks of the "London Spiritual Magazine" (May, 1868), in relation to Mr. Jackson's theory, deserve to be quoted in this connection. We here subjoin them:—

"Mr. Jackson, the author of 'Ecstatics of Genius,' and of various lectures on mesmerism, has long, like other magnetists, found a great difficulty in accepting the phenomena called spiritual as actually proceeding from spirits. Some years ago, a friend of ours, on reading Mr. Jackson's mesmeric publications, told him that he saw exactly where he was, — that he was on the staircase leading to the chambers of Spiritualism, but had not reached the rooms for which the staircase was built. Mr. Jackson is on the staircase still, and, to all appearance, likely to remain there. In an address delivered some time ago to the Glasgow Spiritualists, he assured them that he fully admitted the reality of the phenomena which they attributed to spiritual influence, but that he was quite satisfied himself that spirits had nothing whatever to do with them. In this assurance we are persuaded that Mr. Jackson is perfectly sincere; and, still more, that he cannot possibly come to any other conclusion. It is the result of the pre-occupation of his brain with lucid magnetic theories, from which he can no more escape than the bird that is once enclosed in the net of the fowler. That he will ever persuade a single Spiritualist, however, to adopt his convictions, we cannot encourage him to hope. Louis Büchree, in his 'Natur und Geist' and 'Kraft und Stoff,' and Carus Sterne, in his 'Naturgeschichte,' have gone over the whole of his ground most elaborately and ably, but with the discouraging result of convincing nobody who had come to the examination of these phenomena with a mind free from professional theories.

"Many men, eminent for their habits of metaphysical research; many men of profound science, — have tested the character of these phenomena, and have been compelled to adopt the spiritual theory as the only one capable of explaining them. Professor Hare, of America, entered on this inquiry with as strong a persuasion as any man has ever entertained, that he should rout the spiritual theory altogether. As a man of practical science, a profound electrician, and an avowed disbeliever in revelation, he entered on the inquiry with the utmost care, and pursued it with the utmost pertinacity for two years; but he came out of it a firm believer in the spiritual agency, concurred in the manifestations, and proclaimed himself a thorough Christian. Judge Edmonds, as a lawyer, went through the same laborious inquiry with the same result. Professor Mapes and Dr. Gray, of America, are also examples of philosophers as accomplished and as practical as those who are likely to follow in the same track. If philosophers, as Mr. Jackson affirms, be the only men capable of unravelling the mystery of these phenomena, here we have a number of them; and their decision is adverse to his position.

"Mr. Jackson in a stately and *ex cathedra* style assures us that, in his opinion, physical laws will explain the whole of the phenomena. That such laws, and others yet little known, are at work in these matters, every one knows; but it seems to us to require very little acquaintance with these things, to perceive that the laws which operate in them are conjointly resident in spirits incarnate and spirits de-carnated. Mr. Jackson refers to the great fact, that the intelligences involved in these phenomena have uniformly asserted that they are individual and actual spirits, and not mere laws and forces; have asserted this in every country and to every class of people; and he thinks he has an answer to this rather strong fact. In all ages and countries, he says, communications, professing to proceed from spirits, have reflected the creeds and opinions of those to whom they came. Pagans, Greek, and Roman philosophers, Buddhists, Brahmists, Chinese followers of Fohi and Lootse, Christian, Catholic, and Protestant, all have received communications in accordance with their own beliefs. Nay: mythologic gods have appeared to mythologists; the Virgin Mary and Catholic saints, to Catholics. Mr. Jackson's conclusion, therefore, is, that all these communications and apparitions are the objective results of the subjective powers and spirits of those who indulge in these occult practices and speculations.

"The fact is correct and historical; but the explanation, in our opinion, comes from a very different quarter. It is the result of a fixed law, — 'like draws to like.' Beyond this, we know enough now to understand that spirits carry with them into the other world the views, opinions, habits, creeds, prejudices, and self-wills which had taken possession of them here. The immense hosts of spirits, 'gone before,' are always anxious to perpetuate their peculiar faiths and opinions amongst their successors on earth, and spare no pains or disguises to effect this. To the old Greeks and Romans they came in the shape of their gods; they delivered oracles to them as their gods; to the Roman Catholics they came as the Holy Mother, and as saints and saintesses. To those who think themselves philosophical, they still come as Socrates, Bacon, Shakespeare, Franklin, and the like, though with very little evidence of the intellect or genius of those great souls. As the Romans believed that, at the battle of Cannæ, their soldiers and those of the Carthaginians still continued the conflict in the air after they were slain; and as the hosts of Attila, in the battle of the Huns, were said to do the same, — we believe and have no doubt, that every species of departed spirit, and that in hosts and countless battalions, are still zealously infusing their own views, and the views of their partisanships, into the minds of their successors on earth, and endeavoring to rule here still, and thus stir up the worst passions and practices of this afflicted world.

"Now, though the forces operating in these phenomena, profess themselves to belong to different churches and religions, different creeds and philosophies, they all agree in one point; namely, that they are individual spirits, and not mere forces, or laws physical or spiritual. Their evidence regarding this fact is clear, uniform, and persistent; and for this universal and unvarying expression there must be a cause, and that cause cannot be a lie. Why should mere laws, physical or spiritual, be lies? How can they be lies, if they are laws and forces impressed upon the living cosmos by its Creator? Mr. Jackson, on reflection, must perceive the dilemma into which his theory has led him. And let him for a moment suppose that these powers, whatever they be, had as uniformly, as clearly and persistently declared themselves to be merely laws and forces;

suppose, in fact, that they had declared themselves on the side of the philosopher, — does he not see with what an Io Pæan of triumph they would have been received? with what a clamor the philosopher would have denounced all attempts to declare them not laws and forces, but spirits?

"Mr. Jackson is of opinion that scientific men are the only ones qualified to judge of these phenomena, and to bring to light what they really are. No idea can be more delusive. That scientific men are the best judges of their own natural laws and processes, we readily admit; but in these phenomena there are laws in operation which they are totally ignorant of, and which they cannot possibly test by any apparatus or materials in their laboratories. Beyond and besides this, they are, from their prejudices and adopted theories, totally disqualified for a clear and effective examination of this question. Their minds have become stereotyped in particular theories, to which the phenomena of Spiritualism run counter. Mr. Jackson himself is a living proof of such men being totally disqualified for the free and penetrating examination of such a subject. He believes in all the phenomena, but denies the conclusions drawn by the common sense of many millions of men, and can bring himself to believe that intelligences which can come, and reason acutely, and make themselves seen, heard, and felt avowedly as individual spirits, are *mere laws and forces* emanating from, or existing in, the persons who perceive them.

"And what is really astounding is, that Mr. Jackson, whilst uttering so decided an opinion, shows that he has totally misunderstood the nature of the phenomena on which he discourses. He puts into the same category the 'flowers, fruit, birds,' &c., 'which form the stock wonders of the circle.' He imagines them to issue from the vital forces of the circle itself, and to disappear and dissolve again rapidly. This may apply to the hand which appears at the Davenport *séances*, and to the flowers which were brought by the apparition wife of Mr. Livermore, of New York; but the flowers, fruits, &c., which are produced at the *séances* of Mr. Guppy, and the birds which have appeared at these *séances*, are real earthly flowers and birds, which are brought through walls and doors of closed rooms, and remain. One of the birds remains in a cage to this day. Some of the fruits are kept by those who received them. They were not produced by any physical power of the circle. They came whence no one knew; and they could not, therefore, come in consequence of any internal power exercised by the party assembled. They must be brought by beings, reasoning beings out of the flesh; and no philosopher can possibly propound a more simple or palpable theory than the universal one, that they are brought by spirits who affirm themselves to be spirits.

"Again, the iron collar, which we now hear is made to pass over the head of a youth in America, though seven inches less in interior circumference than the head, is not a collar evolved magically from the minds or the latent forces of the persons of the circle, but is an actual collar, made without any hinge or opening by the blacksmith. The philosopher, who shall explain this phenomenon, must know a great deal more about matter than the most profound physiologist who ever lived; and, in our single opinion, it can never be explained, except on the hypothesis that matter, under the influence of spirit, is in a condition totally different from its condition when operated upon solely by natural laws, however subtle and potent.

"We are so far from entertaining Mr. Jackson's idea that scientific men are the best qualified to examine these singular phenomena, that we feel sure that so soon as they are compelled, like himself, to admit the reality of the facts, their scientific prejudices will lead them vehemently to endeavor to treat them as the results of material laws, as he himself does. This will assuredly become the philosophical phase of the question, whenever the denial of the *fact* is at an end. We cannot hope, that, on having made this step of advance, the philosophers will have got much nearer the truth, because they will, from habit, persist in seeking for the solution of the mystery in a direction in which it is not to be found. The plain sense of mankind will still march on far ahead of them."

Another critic asks, "Has not Mr. Jackson resuscitated the theories of Democritus and Epicurus, peopling the universe with *Εἴδωλα*, or imagery the objective world has mirrored forth into space? Epicurus tell us that our brain imagery is constantly flitting about, distinguishable from the reflected forms of an objective reality, by its greater subtileness and evanescent character. He says, 'The imagery of the senses, and of our phantasy, are realities (Ἐναργῆς ἄλογος), and cannot be denied.'"

We do not see a difficulty in admitting both the pneumatic and apneumatic solution for these manifestations. It is not unlikely that many of the minor phenomena, attributed with sincerity by many partially developed mediums to spirits, may be produced by the unconscious exercise of spiritual powers latent in the individual; while other phenomena are of so extraordinary a character that the more rational explanation may be found in the theory of the application of an *external* spiritual intelligence or force.

The narratives of apparitions of living persons are very numerous, and the facts collected in this volume are not inconsistent with the possibility of such phenomena. The Germans have a familiar word to designate persons of whom they are related; calling them *doppelgangers* or *double-goers.* Jung Stilling says, "Examples have come to my knowledge in which sick persons, overcome with an unspeakable longing to see some absent friend, have fallen into a swoon, and during that swoon have appeared to the distant object of their affection."

In his "Footfalls on the Boundary of another World," Robert Dale Owen gives a number of narratives which he personally took pains to authenticate in relation to this subject. We select the following: —

"In May, 1840, Dr. D——, a noted physician of Washington, was residing with his wife and his daughter, Sarah, near Piney Point in Virginia. One afternoon the two ladies were walking out in a copse-wood not far from their residence, when, at a distance on the road, coming towards them, they saw a gentleman. 'Sally,' said Mrs. D——, 'there comes your father to meet us.'

'I think not,' the daughter replied: 'that cannot be papa; it is not so tall as he.'

"As he neared them, the daughter's opinion was confirmed. They perceived that it was not Dr. D——, but a Mr. Thompson, a gentleman with whom they were well acquainted, and who was at that time, though they then knew it not, a patient of Dr. D——'s. They observed also, as he came nearer, that he was dressed in a blue frock-coat, black satin waistcoat, and black pantaloons and hat. Also, on comparing notes afterwards, both ladies, it appeared, had noticed that his linen was particularly fine, and that his whole apparel seemed to have been very carefully adjusted.

"He came up so close that they were on the very point of addressing him, but at that moment he stepped aside, as if to let them pass; and then, *even while the eyes of both the ladies were upon him*, he suddenly and entirely disappeared.

"The astonishment of Mrs. D—— and her daughter may be imagined. They could scarcely believe the evidence of their own eyes. They lingered, for a time, on the spot, as if expecting to see him re-appear; then, with that strange feeling which comes over us when we have just witnessed something unexampled and incredible, they hastened home.

"They afterwards ascertained through Dr. D——, that his patient Mr. Thompson, being seriously indisposed, was confined to his bed; and *that he had not quitted his room, nor indeed his bed, throughout the entire day.*

"It may properly be added, that, though Mr. Thompson was familiarly known to the ladies, and much respected by them as an estimable man, there were no reasons existing why they should take any more interest in him, or he in them, than in the case of any other friend or acquaintance. He died just six weeks from the day of the appearance.

"The above narrative is of unquestionable authenticity. It was communicated in Washington in June, 1859, by Mrs. D—— herself, and the manuscript being submitted to her for revision, was assented to as accurate."

Our friend, Mr. Benjamin Coleman, supplies the following emarks on this subject: "Among the most intelligent inquirers with whom I conversed at Brighton, was a lady of title. She told me that she was one of those present at the Davenport *séance*, held at the residence of Sir Hesketh Fleetwood. She was seated in the dark *séance* by the side of a gentleman, whose previous skepticism, he confessed to her, was fast disappearing in the face of the facts they were witnessing, when a light was suddenly struck, and both of them distinctly saw the form of Ira Davenport glide close past them. This incident very much disturbed the confidence of Lady L——, and entirely satisfied the skeptic that imposition was practised; and he left the room a confirmed unbeliever. I told Lady L——, that, on his return to London, Mr. Ferguson* spoke to me of this very fact, as one of the most curious that had yet occurred at any of the *séances*. He was holding, he said, the box of matches, as he usually does, when the box was snatched from his hand, and a light was struck by the invisible operator; and, during the momentary ignition of the match, he plainly saw a form, apparently of a human figure. He said nothing at the moment, but whispering the fact to Mr. Fay, *he* confirmed it; and afterwards several of those present admitted that they, too, had seen it. Mr. Ferguson, however, was not aware that any one present supposed it to be the actual person of Ira Davenport, as no observation to that effect was made; and, as Ira Davenport was seen instantly afterwards, when the light was restored, fast bound to his chair, it was simply impossible that the suspicions of Lady L—— and her friend could have been well founded. But admitting that two competent witnesses did actually see the form of Ira Davenport on that occasion, it is corroborative of a very important and interesting fact, and distinct phase of these puzzling mysteries of spiritual appearances; namely, *the duplication of individual form*.

* The Rev. J. B. Ferguson, of Tennessee, a gentleman who has given a good deal of attention to the spiritual phenomena, and whose testimony is believed to be above suspicion. He was with the Davenports for a time in England.

"Mr. Ferguson, who did not on that occasion recognize the resemblance to Ira Davenport, nevertheless has, as he solemnly asserts, seen at other times, when alone with them, the entire duplicated form of Ira Davenport, and a part of Mr. Fay; and, in my first conversation with the Davenport Brothers, they told me, among other curious facts of their extraordinary history, that persons had said they had met one or the other of them in places where they had not been. On one occasion their father went to a neighboring shop to order some fruit, when he was told by the shopkeeper that his son Ira had just been there, and had already ordered the fruit. It was, however, satisfactorily proved that Ira had not left the house, and that the man must have seen his 'wraith' or 'double.'

"I may as well anticipate the question that will no doubt arise in the minds of many: 'That supposing the spirit of a living person can assume a natural form and become an active intelligent agent, producing mechanical effects, may not that account for much of what we are accustomed to attribute to the presence of the spirits of departed persons?'

"I answer, '*Yes!*' but not all. We have too much evidence of spiritual individual identity, and too many instances of direct intelligence, perfectly independent of surrounding witnesses, to admit the possibility of our own spirits acting on all occasions the double, and deceiving our senses.

"Again it may be asked, 'Do you think that *any* of the phenomena which we are accustomed to attribute to spirits of the dead may be produced by the spirits of the living?' and again, I answer, 'Yes!' After close observation and calm reflection upon the whole range of these Davenport manifestations, I am inclined to believe that the rope-tying and untying, the handling and carrying about of musical instruments, &c., are partly effected by their 'doubles,' and it may be that these are in part assisted by other spirits. The unerring certainty with which the same phenomena are produced in the presence of the Davenports day after day tends to confirm the opinion that their own 'spirits,' or 'doubles,' produce many of the mechanical effects which we witness. On one occasion when they were bound in the usual manner within the cabinet, and the test of filling their hands with flour was applied, a group of four hands was seen; *and one of them I plainly saw was covered with flour.*

"And another idea occurs to me: as it is certain that four instruments are played upon at one time, requiring the agency of six or eight hands, it may be that the medium's hands are not only duplicated, but that they are triplicated and multiplied according to the necessities of the case, and the existing conditions and strength of the medium-power. We know that there is upon record ample evidence of apparitional appearances of persons still living, sometimes seen at the point of death, sometimes days before, and held to be death warnings; and at other times of persons in health, and remaining so for an indefinite period, and again there are instances of persons seeing themselves.*

"From these, and many other sources, much corroborative evidence may be obtained to establish the fact that the spirit-forms of living persons have been seen at various times and places, and the theory, which I now venture to suggest, is, *that many manifestations which Spiritualists are accustomed to attribute to the spirits of the departed are, in truth, effected by their own doubles.*

"This idea can in no degree destroy our cherished belief in the power of departed spirits to communicate with us. On the contrary, it tends to confirm it; for if spirits in the flesh can assume a tangible form and actually produce certain mechanical effects, why may not spirits out of the flesh be able to do all this and much more? Let it be once recognized that spirit is a living entity when separated from the fleshly body, having a dynamic power over matter, and the great difficulty which enshrouds the materialistic mind vanishes. I am not wedded to a dogma on this or any other subject. I am only concerned to

* Kerner relates a case in which Mrs. Hauffé, who was ill in bed at the time, suddenly perceived the appearance of herself seated in a chair. As Kerner himself saw nothing, the vision will of course be set down by the incredulous as purely subjective.

uphold, in opposition to the arrogant assumptions of ignorant skeptics, that the phenomena of which we speak are not to be attributed to delusion, to legerdemain, or to any recognized natural cause."

If in the human organism there are powers which enable a man to see without eyes, and to do the work of the corporeal senses without the aid of those senses, then we may infer that it is through the exercise of a faculty, independent not only of the particular organ of sense, but of the whole physical body. Mr. Jackson admits that, "in virtue of our being spirits, we possess the powers manifested by spirits," and that "there is not the least necessity for going outside of ourselves for these things." Why, then, should the mere dropping of our material husk at death disable us from producing, as disembodied spirits, the same effects we could produce, through our purely spiritual faculties, while we were in the flesh?

Undoubtedly, many phenomena referred by inexperienced observers to the agency of spirits do not require a supramundane solution. Whether in or out of the corporeal form, the human spirit may have certain powers; and its phenomenal manifestations, whether it be in its embodied or disembodied state (and when we speak of *body* we mean only *the visible earthly body*), may have many points of similarity. It may sometimes be difficult to trace the origin of facts occurring along that mysterious border-land, where the visible and invisible seem to blend.

The advocates of the no-spirit theory have much to say of "unconscious cerebration" and the controlling agency of the will; but may not this be only another name for that spiritual contact of our souls with the spiritual world, from which, according to Swedenborg, we get so many of our impressions?

The puerile character of many of the communications for which a spiritual origin is claimed; the reckless assumption of the names of great men and women by pretended spirits; the author of some imbecile doggerel, claiming to be Shakespeare; the designer of some atrocious picture, signing himself Michael Angelo; and the utterer of some stupid commonplace asking us to believe he is Lord Bacon, — of course make the spiritual pretensions of the communicants ridiculous in the estimation of most persons of taste. But when it is realized that spirits are not a kind of minor gods; that they carry with them the characters they formed in this, or, it may be, in anterior lives; that there are among them the frivolous, the vain, the mendacious, and the malignant, with all their imperfections on their heads, just as they left this world, — the fact that a worthless communication may yet be spiritual in its origin does not seem so difficult of belief.

These indications that the next life is a state similar in kind to this present life, and only a step higher in an ascending series of existences; one into which we carry our human nature, and in which progress * is but gradual, — are contrary to the general theological conceptions of the next stage of being, and are distasteful to the feelings of many, whose notions of the hereafter, of the "saved" and the "elect," are of a state of passive beatitude. But perhaps the views of modern Spiritualism on this subject derive some support from analogy, harmonizing as they do with those facts of physical progress taught by geology and by the study of organic forms from primeval times.

Since we have an eternity before us, in which to grow in knowledge and in virtue, why should we expect to mount at once, without any merit or effort of our own, to the summit of all possible bliss and wisdom? Spiritualism, rightly understood, might teach us that the true kingdom of heaven is not *without* man, either in this present or in any other home, where his spirit may successively dwell in those "many mansions," the scenes of the divine bounty and power; but, as Christ tells us, *within*, in the will, the affections, and the mind.

Our sketch of the noteworthy theories that have been put

* "Mortal progress," says H. J. Slack, "and, for aught we know, part of immortal progress also, is accompanied by occasional retrogression." Or, perhaps, our course of ascension being, as Goethe tells us, *spiral*, what may seem retrogression may be one of the conditions of progress.

forth on the subject of these phenomena would be incomplete without a mention of that of Professor Daumer, whose work "Das Geisterreich," appeared in Dresden in 1867. According to his pneumatology, "Ghosts are neither bodies nor souls, but a third entity which he calls *eidolon*, by which he understands the direct self-manifestation and representation of the *psyche* (soul). The soul is restricted to the corporeal exhibition only so long as it animates the body. Once released, by the death of the latter, it can manifest its immanent reality in any way it pleases. It can even reproduce whole episodes from its former life, including any number of figures of itself or of other persons. It can also produce sounds, and perform other material acts."

We have already seen that Baron Reichenbach, a distinguished German chemist, and the discoverer of creosote, finds in what he calls *od* or the *odic* force the medium of many phenomena. He reports that his sensitive subjects saw, at the poles of the magnet, odic light, and felt, from the near contact of large free crystals, odic sensations, which by Reichenbach himself, and others as insensible as he to odic impressions, were wholly unperceived.

At first distrustful of the spiritual significance of certain phenomena, Reichenbach, if we may believe Mr. D. Hornung, of Berlin, now entertains views not opposed to Spiritualism. While in London in 1861, at the residence of Mr. Cowper, son-in-law of Lord Palmerston, he attended a spiritual circle.

"On that occasion," says Mr. Hornung, "two media, Mrs. Marshall and her niece were present, who did not understand a word of German. Reichenbach therefore, after the rapping had commenced, put his questions intentionally in German; and they were answered correctly by raps on the table, and he had the names of several members of his family correctly given. In regard to one name, however, he began to doubt the capacity of the table to give it; the name to be spelled being 'Friedericke,' while it spelled the letters 'R. I.' But when the name 'R I C K E' was completed, the baron was much surprised, as his sister had been wont to be called 'Ricke.'

"Now comes the most remarkable part of the performance,

and I give it in the baron's own words. He says, 'The answers were rapped by the foot of the table in a brightly lighted room. I wished to ascertain whether the rapping could not be prevented, and for this purpose I leaned with my breast against one of the feet of the table, taking hold of two others with both hands, and pressing them down. The rapping of the feet ceased; but the rapping continued above me, on the top of the table. All at once, by a sudden jerk, the table dragged me forward, with the carpet on which it stood; and I lay prostrate in the middle of the room.'

"This experiment convinced the baron that, besides the emanation of the odic element, higher spiritual powers can manifest themselves; and these he now no longer ignores, but recognizes them as facts of experience, for which, however, he as yet knows no explanation." He regards "the great influences of *od* upon the human spirit" as the mere "physical side of the matter," — "the roots by which it adheres firmly to the ground;" and he is thankful to see the day when all his former discoveries show themselves as the portal through which it is possible for him "to go forward into the spiritual department."

A writer in "Human Nature," under the signature of "Honestas," is of opinion that the transition brought about by death, though carrying with it a vast change, does not so completely alter our nature as to render mundane intercommunication impossible. The laws governing the physical conditions of the next sphere must be in harmony with those that rule this, to us, natural world; these laws being only an outgrowth from those of our present condition, and correlatives of them.

Why then is the intercommunication restricted to the limited bounds of a medium's presence? The writer aphoristically replies, Within our coarser earth-body dwells an ether-body, which derives its elementary sustenance from the ether or odic element, from out which this visible, ponderable world has grown forth, with its plastic, centralizing tendency. Our ether-body manifests its presence in the nerve aura, or odic element (first noticed by Reichenbach), in the streaming forth of a mediated, organically centralized ether element, which element sustains this ether-body, — in the same manner as the food and earth elements, which the organism assimilates, support our bodily condition. A double action is thus carried on in the animal organism; namely, a drawing of supply from the centralized earth elements, simultaneously with that from the primary ether or odic element. In the mesmeric fluid which passes from the mesmerizer to his subject, the odic force is transmitted; and a connection is established between the two, sufficiently primary to mediate a physical correspondence between them. Here is the key to the solution of the problem of spiritual manifestations.

These are divisible into psychical and physical. The psychical effects are produced by an action akin to the mesmeric action; that is, the mind of the operating agent, by an action of the will, throws a current of the odic power of its nerve aura on to the nerve aura of the terrestrial being, and an effect similar to that of the mesmerizer upon his patient results; a phenomenon too well known to need explanation.

The second, or physical effects, arise from an action upon the organically mediated free nerve aura of the body of the medium, which aura enables the spirit to create an organism or mechanism, rendering action upon our ponderable matter possible, and allowing of the production of the physical phenomena of sound, movement of bodies, &c.; appearances familiar to the observer of spiritual manifestations. This centralization can only, however, take place by means of the mediating presence of the nerve aura, enabling a condensation into ponderable matter to be effected. The visible, ponderable world is but a phase in the great chain of ever-continuing progress and development. The imponderable, and, to us, invisible world, is, in reality, the permanent and lasting state, from out which the soul brings with it its principle of life, that which is continuous and imperishable, the power of mediating for its own use the supplying element. It has, too, the power, by right of its earth-born state and bodily organism, of mediating the coarser, ponderable elements of our present condition. But the terrestrial mediation can only be effected by the aid of an organism fitted for that special object and use. This mechanism our earth-body furnishes. The spirit-soul does not, however, possess this: its organism is different, finer, undoubtedly more complex than ours.

By the transition called death, the soul has parted with this, for mundane purposes, adapted organism. But to enable a spirit to operate upon material things, an organism has to be formed adapted for that function: this embodying cannot, however, take place unless aided by the mediating presence of the organic nerve aura of a living being. In the embryonic evolution, the mediating element is the maternal one; and here, too, in obedience to laws of development, the embryo being, once having attained its growth, takes its place on earth with an independent central self-existence. The spirit-soul, when incarnating itself in a material envelope, can only do so by the aid of the nerve aura of a living being, upon which it only momentarily acts, which action is rendered possible by the accident of an affinity, enabling a temporary use to be effected, — this use being restricted, however, within the narrow limits prescribed by the supply which the organism of the medium furnishes; and, further, subject to endless interruptions from external causes; as, for instance, over-excitement, or alarm, or atmospheric changes.

The extreme uncertainty of spiritual phenomena; the difficulty, even when produced, of prolonging their duration beyond a few minutes; and more especially the difficulty of giving a continuity to the more developed forms of spirit appearances, — confirms this view of the dependence of visible, tangible, spiritual manifestations upon our organism, and the necessity of an agreement of our natures with the spirit operating upon the nerve aura of the medium.

This writer gives the name of *pre-development* to that change of organic form of our ether-body taking place during life, and by which the transition to the next state is mediated, prepared.

This change is always in accord with the sphere we have to join after death. And the centre — second centre — is the organism thus changed to adapt itself to the onward and next sphere. Decay and death follow this change as a necessary sequel; that is, as pre-development proceeds, we cast off the organism adapted for this life; it becomes old, not nourished by the supplying elements that hitherto sustained it.

According to Leibnitz, every germ has its pre-existence. Every grade or plane of development of phenomenal life is the outgrowth of a pre-existing state of things, which has prepared the elements from which it has been evolved. This is a fundamental law of nature. The grade beneath and the grade above are intimately connected with the gradation in which we exist. In every grade, the next and superior grade exercises its influence, creates, or rather renders the growth possible, of an organism adapted for existence in the next sphere, plane, or grade.

We owe to the law of *pre-development* continuance of our individuality. Were it not for a growth preparatory to the entering into a next sphere or state, such grade not being mediated, rendered by prior growth fit for our organism, continuous life would be impossible. Step by step, mediated by prior growth, the soul progresses onward and onward in never-ending ascent to the highest conceivable unfoldment of our natures. The past is everlasting: the phenomenal life of the present is but an unfolding of the past; and the future, of which this state will be the past, will be again only an unfoldment of the present.

Accepting the theory of progressive growth, as proved, the writer maintains that the forms of the world beyond this existence, must have developed from the forms of the antecedent grades out of which they have been evolved, and that preservation of the type of the human form follows the soul in its onward step into the next world.

Our next organism is mediated, prepared, by our mundane organism; and, this being so, it must depend in its development upon two conditions, physical and psychical; must carry with it, as it passes into the next sphere, the impress of the character of its progress on earth. Thus our sins and shortcomings impress themselves on our very organism; and the life that now is shapes the life that is to be.

Such is an imperfect sketch of the theory of this ingenious writer, who brings to the discussion a full, practical acquaintance with the most remarkable of the phenomena obtained through Mr. Home and other mediums.

Dr. John Ashburner, the translator of Reichenbach's "Dynamics of Magnetism," and who was one of the first men in England to investigate and accept the phenomena of 1848, in his latest work, entitled "Notes and Studies in the Philosophy of Animal Magnetism and Spiritualism," argues that every law in the natural or physical world depends on the "grand trunk force of universal gravitation," which being divisible into centripetal and centrifugal, in other words, attractive and repulsive forces, is, as the active principle, traceable through all the changes which take place throughout the realm of nature. In the author's words, "All change is necessarily dependent on these forces; no chemical compositions or decompositions can take place without them; they regulate the great orbs in space, as well as the form of the minutest of the primitive crystalline globules, of which every *crystal* in existence is built up." "In vegetable existence, it determines a law of evolution when it decrees the folding up of embryonic forces in those minute spherules or germ-cells which develop *vegetable* crystals;" and, "proceeding with these laws, we observe the law of evolution regulating more complicated germ-cells in *animal* existence, but still obedient to magnetic laws of polarity;" for "human beings, as well as all other animals, vegetables, and minerals, within the magnetic sphere of this magnetic earth, must nécessarily partake of the magnetic influences emanating from the grand trunk force of universal gravitation." The author shows that all the phenomena of the so-called forces of heat, light, and electricity, are dependent on attraction and repulsion; and that these simple antagonistic forces are the sole principles by which every change, atomic or otherwise, is effected, under Almighty guidance throughout the universe.

The author is a stanch opponent of the materialistic notion that *brain thinks*, and consequently an assertor of the absolute inertia of matter, which the Creator has made subject to the attractive and repulsive principles involved in that which is called gravitation or magnetism. Force, therefore, is the life and soul of matter, which, controlled and regulated by it, manifests the phenomena which are continuously taking place in the form, size, weight, and color of objects, from the least unto the greatest.

The condition of sleep and the cause of pain are attributed to the state of the magnetic currents in the animal economy: "Sleep is the result of an attractive force, analogous to the attraction of gravitation; and wakefulness results from a repulsion, analogous to the centrifugal agency constituting a part of the phenomena attendant on the great trunk force." The facts adduced in evidence of the truth of this position are highly illustrative; and the author contends that cases recorded by many surgeons justify the conclusion that the molecules of the brain being subjected to a central attractive force, is the cause of sleep; as the brain, when exposed, is seen to become smaller in that state; and that a repellant action among its particles precede the wakeful condition. The cause of pain is summed up in the following: "The whole body, being a congeries of magnetic molecules, must necessarily be subject to the laws regulating polarities. Any change in the relations of the poles of living animal molecules must be productive of a change in the sensibilities of the part. Whether the change be the cause of pleasure or of pain, must depend upon the faculties of the individual. Endowed with a nervous system, the animal is susceptible of sensations, without which, the idea of pleasure or pain becomes absurd. The inference then remains, that pain is the result of an extreme disturbance of the polarities of a part."

Dr. Ashburner accepts the spiritual hypothesis to the fullest extent, and thinks that any other is wholly unsatisfactory in view of all the facts and phenomena which he has tested.

Another theory, not undeserving of mention, is that put forth in a work published in London, in 1863, and bearing the following extraordinary title: "Mary Jane; or, Spiritualism Chemically Explained." The author's hypothesis, audacious as it may appear, is urged with a certain show of scientific learning. He gives us the following summary of his conclusions: —

1. Man is a condensation of gases and elementary vapors.

2. These vapors are constantly exuding from the skin.

3. They charge (to use an electrical term) certain things; viz., The sensitive plant, — and it droops. The human body (as in mesmerism), — and it becomes insensible to pain. A table, — and ——

4. When these vapors (which Reichenbach calls odic) emanate from certain persons, who appear to have phosphorus in excess in the system, *they form a positively living, thinking, acting body of material vapor, able to move a heavy table, and to carry on a conversation*, &c.

5. That the other persons sitting at the table affect the quality of the manifestations, although the odic vapors from them are not sufficiently strong to move the table, or act intelligently alone.

6. That we do not see the odic emanations from their fingers, has nothing to do with the question; for we can neither see heat nor electricity, — and yet we admit the existence of both from their effects.

7. Thus, if the medium knows nothing of music, and holds a guitar, the sounds given out will be discordant, or such as might be expected of a person knowing nothing of music; but, if a good performer sits at the table at the same time as the medium, the sounds will be harmonious. So, if a medium understands nothing of drawing, and paper and pencil be put under the table, scribbles will be produced; but if an artist sits at the table, flowers or other artistic drawings will be produced; although, in neither case, could the artist produce the slightest movement of the table, or manifestation whatever, without the medium.

8. That this odic being thinks and feels exactly as the persons from whose body it emanates; that it possesses all the senses, — seeing, hearing, smelling, tasting, feeling, and thinking; that it makes up for the want of the muscular organs of speech, by either an electrical power of rapping, or by guiding the medium's hand, or by direct writing with pen or pencil.

9. That its power of sight is electrical; for it can see under a domino, or what is in the adjoining room, — in short, where the human eye cannot.

10. That its power of hearing is also electrical or superhuman.

11. That it is highly sensitive to odors, delighting in those of flowers, and expressing repugnance to some.

12. That it can rap in two, and probably more, places simultaneously.

13. That it can carry on different conversations with different individuals at the same time.

14. That its conversations with different persons will be responsive to the affections, the sentiments, and the religious belief of each person it is talking with, although they are drawn from one common source, — the odic vapor concentrated at, or with which the table is charged, — and although those religious creeds are entirely at variance. And if asked for the name of the (presupposed) spirit, it will give the name either of the desired relative, or of some high authority (on religious matters) in the specific creed of the person making the inquiry.

15. That, from various concurrent testimony, it appears fully proved that this odic vapor possesses the power of taking the shape of hands, arms, dress, &c., and even of an entire person, dressed; and, such fact being certain, the statement that in America photographs of both dead and living persons have been obtained, ceases to be preposterous; but that the souls of those persons produced, or had any thing to do with those shapes, does not appear to be any more proved, than that if a good Turk received a message signed, "Mahomet," it would be accepted as proof, either of the truth of the message, or that the deceased Mahomet had any thing to do with it.

16. That, nevertheless, the high thought, philosophy, independence, conciseness, and deep reflection evinced by many of the answers and sentiments expressed by the odic fluid, point to its connection with a general ***thought-atmosphere***, as all-pervading as electricity, and which possibly is in itself, or is in intimate connection with, the principles of causation of the whole universe.

Such is the bold theory of this chemical investigator. That the emanations of the human body "may form themselves, without our knowing any thing about it, into a distinct personality, with the faculties of perception, memory, reason, and conscience, — a personality that may rap, write, draw, carry on general conversation, make witty and moral observations, and not only think, but 'think deeply and profoundly,' and take to itself a name (as, in the author's fanciful experience, it took the name of 'Mary Jane'), and, in short, in every way conduct itself like an educated and well-behaved member of society, — is certainly an astounding instance of the prodigious capabilities of 'odic vapor.' It is an hypothesis which, if it does not merely amuse, is likely to startle men of science even more than the spiritual theory itself; and their surprise is not likely to be diminished on learning that the odic vapor is convertible into intellect; that the odic emanations actually create life and intelligence; and that there is a universal thought-atmosphere, resulting, we presume, from the phosphorescent and other chemical emanations from the collective brain of humanity, from which these vaporous personages get the information and ideas which at the time they may not in themselves possess.

"Admitting the extravagant assumption of a being evolved from the chemical emanations of our physical substances; nay, more, admitting even that these emanations are imbued with our special idiosyncrasies, — with our mental and moral qualities, — still, as a derivative being, it could have only the knowledge, ideas, and qualities of those from whom it proceeded. That cannot come out of a man which is not in him. Hence, as our author very consistently says in the words we have quoted:

'This odic being thinks and feels *exactly* as the persons from whose bodies it emanates.' Of course, if the hypothesis were true, it *must* do so. But then, unfortunately for the hypothesis, this 'odic being' will not do as he ought to do. He will sometimes think and feel *differently* from the persons from whose bodies he is an out-birth. No fact in this inquiry is better known or more firmly established than that spirits exhibit powers, and maintain opinions surpassing, different from, and sometimes even antagonistic to those of both medium and circle.

"In some instances mediums will give information altogether outside the knowledge of themselves, or of any person present,* and exhibit a mental force transcending their own natural powers, as in others it will be equally below their natural capacity.

"We might pursue our argument from every phase of the manifestations: from vision and prevision; from dreams and apparitions; from impressions, presentiments, and warnings; from clairvoyance and trance; from prediction, possession, and personation; these all demonstrate the same conclusion, — that the acting power is no way a part of ourselves, but is wholly discreted from us, with independent thought, affection and volition. The fact is, that our author confounds conditions and causes. Certain conditions are found necessary to certain effects; *therefore*, he reasons, they are the efficient cause of them. This is just such a mistake as it would be to attribute a telegram to the wires, instead of to the operator at the end of them."

William Howitt, of whom we may say, as Coleridge said of Baxter (another Spiritualist), "I could almost as soon doubt the gospel verity as his veracity," in a letter published in 1862, and

* Professor Hare testified to a message having been sent by a supposed spirit, from a circle at Cape May, to one in Philadelphia, and an answer, giving assurance of actual communication, having been returned in half an hour. The Rev. J. B. Ferguson, of Tennessee, testifies to having heard native Americans, who never knew a word of German, discourse for hours in that tongue in the presence of native Germans, who pronounced their addresses pure specimens of the power of their language. Facts of the same sort without number could be given.

commenting on the odic theory of the Rev. Mr. Mahan and others, writes as follows: —

"They who ascribe the powers exercised by spiritual agency to odic force, betray an equal ignorance of the real properties of that force, and of the present *status* and facts of Spiritualism. Search through Reichenbach's essay on this force, and you will find no trace of a reasoning power in it. He ascribes no such properties to it. He says it throws a flame in the dark, visible to sensitive persons, such as the Spiritualists call mediums; that this flame is thrown from magnets of great power, from crystals, from the light of the sun, &c. That by passes made with magnets, or crystals, or by water impregnated with the sun's rays, certain sensations, agreeable or disagreeable, as the power is applied, are induced, but not a trace of any reasoning in this power, of any revelation of facts, of any pictorial vision, of any faculty of prognostication. It cannot tell you what will take place to-morrow, much less at the Antipodes, or in the spiritual world. But spirits do all this, and more. It does not attract iron, or other physical substances, which, as far as iron goes, its cognate, magnetism, does. But spirits lift iron, or any other body of very great weight, and not in one direction only, but carry them about from place to place. Spirits lift heavy tables: I have seen dining-tables, capable of accommodating more than a dozen people, lifted quite from the ground. Spirits play on all musical instruments: they can carry about hand-bells, and ring them in the air, *as I have seen them*. The music which they produce is often exquisite. Spirits will draw or write directly upon paper laid for them in the middle of the floor, or, indirectly, through the hands of people who never took a lesson, and never could draw. *I am one of them.*

"These are things which are not only going on in England, and amongst my own friends every day, but have been going on for these forty years; ten years in America, and thirty before that in Germany. But, in America, the wide diffusion and constant repetition of these phenomena have convinced some millions of people, and some of them the first men of scientific and legal ability in the country. Those persons have not believed on mere hearsay, or mere hocus-pocus and delusion, but upon the familiar evidence of facts; and, as I have observed, for thirty years before that, in Germany there existed a considerable body of the most eminent philosophers, poets, and scientific men, familiar with most of these things. Amongst these no less a man than Emanuel Kant; and also Görres, Ennemoser, Eschenmayer, Werner, Schubert, Jung Stilling, Kerner; and, pre-eminent amongst women, Mrs. Hauffé, the Seeress of Prevorst, who professed, not merely to have spiritual communications, but to see and converse daily with spirits; and she gave continual proofs of it, as any one may see who reads her story.

"Now it is useless to tell us that the odic force, acting somehow mysteriously on the brain, can produce these results. It cannot enable people to draw, and write, and play exquisite music, who have no such power or knowledge in their brains; for on the old principle *ex nihilo nihil fit*, no such things being in, no such things can come out. It cannot come from other brains, for there are often no other brains present. If it could do such things, it would be *spirit*, endowed with volition, skill, and knowledge; and there would be an end of the dispute. The condition, therefore, of those who ascribe these powers to odic force, is that of one ascribing the telegraphic message to the wire, and not to the man at the end of it. Odic force may be the wire; for spiritual communications are, and ever have been, made through and under certain laws, as all God's works always are: but it certainly is not the intelligence at the end of it. . . .

"Whilst the *odists* and *automatists* speculate about an action on the brain, we cut the matter short, and say, There stand the spirits themselves, seen, heard, felt, and conversed with.

"More than six years ago I began to examine the phenomena of Spiritualism. I did not go to paid nor even to public mediums. I sat down at my own table with members of my own family, or with friends, persons of high character, and serious as myself in the inquiry. I saw tables moved, rocked to and fro, and raised repeatedly into the air. . . . I heard the raps; sometimes a hundred at once, in every imaginable part of the table, in all keys, and of various degrees of loudness. I examined the phenomena thoroughly. . . . Silly, but playful, spirits, came frequently. . . . I heard accordions play wonderful music as they were held in one hand, often by a person who could not play at all. I heard and saw hand-bells carried about the room in the air; put first into one person's hand and then into another's; taken away again by a strong pull, though you could not see the hand touching them. I saw dining and drawing room tables of great weight, not only raised into the air, but when placed in a particular direction, perseveringly remove themselves, and place themselves quite differently. I saw other tables answer questions as they stood in the air, by moving up and down with a marvellous softness. I heard sometimes blows, apparently enough to split the table, when no one could have struck them without observation; and I breathed perfumes the most delicate. I saw light stream from the fingers of persons on the table, or while mesmerizing some one. As for communications professedly from spirits, they were of daily occurrence, and often wonderful. Our previous theological opinions were resisted and condemned, when I and my wife were alone. This, therefore, could be no automatic action of our own brains, far less of the brains of others, for they were not there. We held philosophical Unitarian opinions; but, when thus alone, the communications condemned them, and asserted the Divinity and Godhead of our Saviour. When we put questions of a religious nature to the spirits, they directed us to put all such questions to the Divine Spirit alone. . . .

"Many persons that we know, draw, paint, or write under spiritual agency, and without any effort or action of their own minds whatever, some of them having never learned to draw. Several of my family drew and wrote. I wrote a whole volume without any action of my own mind, the process being purely mechanical on my part. A series of drawings in circles, filled up with patterns, every one different from the other, were given through my hand, one each evening: the circles were struck oft as correctly as Giotto or a pair of compasses could have done them; yet they were made simply with a pencil. Artists who saw them were astonished, and, as is generally the case in such matters, suggested that some new faculty was developed in me; when, lo! the power was entirely taken away, as if to show that it did not belong to *me*. The drawings, however, remain; but I could not copy one of them in the same way if my life depended on it. A member of my family drew very extraordinary and beautiful things, often with written explanations, but exactly in the same mechanical, involuntary manner. In fact, most of these drawings are accompanied by explanations spiritually given, showing that every line is full of meaning.

"I may add that I have never visited paid mediums; but I have seen most of the phenomena exhibited through Mr. Home, Mr. Squire, and others. *I have seen spirit-hands moving about; I have felt them again and again. I have seen writing done by spirits, by laying a pencil and paper in the middle of the floor*, and very good sense written too. I have heard things announced as about to come to pass; and they have come to pass, though appearing very improbable at the moment. I have seen persons very often, in clairvoyant trances, entering into communication with the dead, of whom they have known nothing, and giving those who had known them the most living description of them, as well as messages from them. . . .

"Now it is idle talking of odic force in the face of facts like these, which are occurring all over America, and in various parts of Europe, and which accord with the attestations of men of the highest character in all ages and nations. In Greece, Plato, Socrates, Pythagoras, and numbers of others asserted this spirit-action; in Rome, India, Egypt, Scandinavia, and aboriginal America, as well as in Judea and amongst the most eminent Fathers of the Church. The leading minds of every age but this have but one voice on the subject.

"It is the last, vain clutching at shadows to avoid coming to the substance, which makes those educated in the anti-spiritual theories of the past century, seize so eagerly on the odic force as their forlorn hope. It will be torn by advancing truth from their grasp. The cry that all is imagination is gone already: odic force is the present stage, and it must go too.

"And here I could give you a whole volume of the remarkable and even startling revelations made by our own departed friends at our own evening table; those friends coming at wholly unexpected times, and bringing messages of the most vital importance, — carrying them on from period to period, sometimes at intervals of years, into a perfect history. But these things are too sacred for the public eye. All Spiritualists have them; and they are hoarded amongst the treasures which are the wealth of the affections, and the links of assurance with the world of the hereafter.

"Now, I ask, what right have we, or has any one, to reject the perpetual, uniform, and voluntary assertions of the spirits; to tell them that they lie, and are not spirits, but merely *od*, or some such blind and incompetent force? Nothing but the hardness and deadness of that anti-spiritual education, which has been growing harder and more unspiritual ever since the Reformation, could lead men to such absurdity. Protestantism, to destroy faith in Popish miracles, went, as is always the case, too far in its re-action, and, not content with levelling the abuses, proceeded to annihilate faith in the supernatural altogether."

The Rev. Charles Beecher, in his able review of the apneumatic theories, says, "That mind, separating itself partially from the body, even during this life, should be able to energize at a distance, though mysterious, is not incredible. Cicero recognizes it. Jamblichus builds on it. It is easy to conceive a law by which it should be. But to say that *brain* can push a door open at a distance, project odic spectra, visible and audible to distant observers, perform on distant musical instruments; and, in short, do whatever the person would do, if physically present; or that every particle of the body is a miniature of the whole; and that these, constantly exhaling, remain for years, and coming in contact with sensitive brains, produce visions of the person, and his precise sensuous and mental state at the time the particle was elaborated,—these, though stated as *facts* in a scientific treatise, are not only unsustained by evidence, but shocking to the common mind."

The theory of ***automatic mental action***, Mr. Beecher regards as equally objectionable. It is an attempt to prove that intelligent manifestations can be produced unintelligently; thus overthrowing the foundation of all argument from design to a designer.

Admit that the phenomena are the work of spirits at all, and the conclusion cannot be resisted that they are disembodied spirits.

For what do the facts conceded imply that the embodied spirit can do? It can, by some means, appear at a distance from its own body, speak audibly, hear answers, move bodies, perform on instruments, and do whatever it would do through the body if that were present. It can obtain access to the contents of other minds, reveal distant events, past, present, and future. But if so, the further concession of a temporary going forth of soul from body cannot long be withheld. Mrs. Hauffé firmly declared that her soul left the body and returned. Gilbert Tennent, to the day of his death, believed that during that long and death-like trance his soul left the body. All clairvoyants testify to the same. In this way Cicero accounts for prophetic dreams: "In dreams, the soul hath a vigor free from sense, and disinthralled of every care, the body lying death-like. And since she hath existed from all eternity, and been acquainted with innumerable minds, she beholdeth all things that are in ***rerum natura***." *

All the writings of antiquity are eloquent with this grand idea.

But once admit this of the soul before death, and how can it be denied after?

* Cicero, it will be seen, was inclined to the doctrine of metempsychosis, or successive re-incarnations.

Take, for example, the instance given by Cicero, as a favorite with the Stoics: Two Arcadians stopped at Megara, one at an inn, the other at a friend's. At midnight, the former appeared to the latter, asking help; for the innkeeper was about to murder him. Roused in affright, the latter thought it a dream, and again slept. His friend again appeared, asking him, as he had not come to him alive, to avenge him dead, as the innkeeper had now slain him, and concealed his body in a cart under dirt. In the morning he met the cart as directed, found the corpse, and the innkeeper was executed.

"Here, if it be admitted," says Mr. Beecher, "that the soul appeared at a distance from the body before death, how can it be denied that it did the same after?

"Furthermore, if the soul do, after death, come in contact with the spirit throngs that environ us, how deny that it does the same when severed from the body before death?

"How resist the firm persuasion of Gilbert Tennent, and others, that he did actually converse with spirits? Why should not a sleep, so deep as to be like death, produce in part death's results, in introducing the spirit to scenes behind the veil?

"Is there no weight in the impressive declaration of the almost dying Mrs. Hauffé, that while all sorts of ocular illusions passed before her eyes, yet '*it was impossible to express how entirely different these ocular illusions were to the real discerning of spirits; and she only wished other people were in a condition to compare these two kinds of perception with one another, both of which were equally distinct from our ordinary perception, and also from that of the second sight.*'

"Yet if such converse with the dead be admitted, in even one well-authenticated instance, the whole apneumatic argument falls. With all the gross consequences, then, of the cerebral hypothesis, it is the only alternative.

"If, then, such difficulties embarrass the apneumatic hypothesis, why not adopt the pneumatic? It is an admitted principle of science, that that theory is preferable which accounts most naturally for all the facts known. The pneumatic theory accounts for all facts alleged by the other theories as well as either of them; for some, better; and for many, which they cannot account for at all without absurdity.

"One of the facts most relied on by the apneumatic argument is the misspelling, which, it is asserted, *always* follows the habit of the medium. Such, however, is not the fact. Cases are on record of misspelled communications coming through mediums who could spell correctly, much to their chagrin. But even if the fact were, as claimed, it might be accounted for, either by supposing that illiterate mediums attracted illiterate spirits, or by supposing that spirits, in order to communicate, are obliged partially to incarnate themselves in the body of the medium, and to take on, in part, its organic and mental habits.

"So, also, of the influence of drugs, manipulations, diseases. The pneumatic theory is, that as the soul may by these means be assisted, or disabled, in the use of its own brain, so disembodied spirits may, in the use of an invaded brain. When the odyllic conditions are by these means prepared, the spirit can insinuate itself; when they are by these means destroyed, it is compelled to forego its hold; so in regard to nervous epidemics. The theory is, that these *may* exist without the agency of disembodied spirits; but that *when* they exist, developing proper odyllic conditions, spirits may be expected to take advantage of them. Hence, to find cases of nervous epidemics, where no indications of spiritual agency are apparent, proves nothing, except that the odyllic conditions were not favorable.

"While, then, the pneumatic hypothesis accounts for all the facts adduced by the other theories, as well as they, it also accounts naturally for other facts by which they are embarrassed. It is, therefore, probably the true hypothesis. And before rejecting it, let that saying of Isaac Taylor's be well pondered, that we ought not to reject the almost universal belief of occasional supernatural interference, till we can prove an *impossibility*. 'An absolute skepticism on this subject can be maintained only by the aid of Hume's oft-repeated sophism, that no testimony can establish an alleged fact which is at variance with common experience; for it must not be denied that some few instances of the sort alluded to rest upon testimony, in itself, thoroughly unimpeachable; nor is the import of the evidence in these cases at all touched by the now well-understood doctrine concerning spectral illusions.'

"Now the apneumatic argument virtually implies an *impossibility* of establishing the reality of spiritual communication by any amount of evidence. Suppose a departed spirit, the wife of Oberlin,* for example, were permitted to attempt to converse with her husband, — not to establish a new revelation, not to display divine power, but merely to exercise such potentiality as might pertain to a disembodied spirit, for her own and her husband's edification and satisfaction. How could she do it, in the face of the apneumatic theories under consideration? She speaks to him, moves his furniture, touches his dress, his person, — all automatic action of some brain *en rapport* with that locality! She sings, plays the guitar or piano, takes a pencil and writes, and he sees the pencil in free space tracing his wife's autograph, — automatic still! She shows him a cloudy hand; nay, a luminous form, and smiles and speaks as when in life; that is, an optical illusion, or hallucination, or a particle exhaled from her body has impinged on his sensitive brain, and created a subjective vision. She communicates facts, past, present, and future, beyond the scope of his knowledge; that might be clairvoyance or cerebral *sensing*. Alas! then, what could she do more? She must retire baffled, and complaining that he had become so scientific that all communication with him was *impossible*.

"But if the denial of the pneumatic hypothesis be unphilosophical, it is no less unscriptural."

* The philanthropic Oberlin (1735–1806) was a Spiritualist, and claimed to have frequent interviews with the spirit of his departed wife. When asked how he could distinguish his wife's appearance from dreams, he said to his inquirers, "How can you distinguish one color from another?" He told them that they might as well try to persuade him it was not a table at which they sat, as that he did not receive these visits from his wife. At the same time, Oberlin was remarkably free from any trace of mysticism or fanaticism. He was, in the best sense of the word, a practical man.

In a review of Faraday's exploded argument against the spiritual phenomena, Mr. Isaac Rehn remarks as follows: "The doctrine of the Correlation and Conservation of Forces is based on the indestructibility of matter and force, or, as by some stated, on the indestructibility of matter and the *persistence* of force. From this it is argued that all forms, however diversified, are but the re-appearance of the primitive atoms of elementary matter in new shapes; and, analogous to this, the *powers* of matter are but the re-appearance of the stored forces of the universe, as they are translated into heat, electricity, chemical affinity, gravity, light, vitality, mechanical force, &c. According to this theory, wherever mechanical force is expended, the given amount of this force must *quantitatively* appear as some other form of force; it may be heat or light, or both, or in some other form of force than either; but yet, in whatever form or forms it may appear, it must be quantitatively the total of the initial force, however much it may differ qualitatively from that, and can be no more and no less.

"It is still further urged that the varied forms of matter and force, as they affect the transformations in the world, are also the efficient and only powers through and by which all *vital phenomena* are produced, these vital phenomena being interpreted in that large sense which includes all intellectual or other power, by whatever names called. Now, it is another postulate of the doctrine of the correlation of the forces that every form of force made to appear, may also be made to appear in any other given form of force. Thus, if heat is made to appear as electricity, electricity may, in turn, be made to appear again as heat; and so on through the chapter.

"The point sought to be made against the spiritual theory is, that, under the doctrine of the correlation of the forces, vitality, or vital force, is the re-appearance of some other form of force. According to the law, it may also be made to appear as the initial force or forces engaged in its production, and so can have no continuity of existence beyond the physical duration of the present life; and we are referred to the fact, as a confirmation of this, that, in the retrograde decompositions of the organic compounds of high chemical formulæ back to the binary states of matter, all the forces appear in the putrefactive chemical changes of decomposition. And if *spirit*, therefore, exists in man, it, too, must be *but a form of force;* a translation of some other force which, in its turn, shall *also* be translated, and, therefore, cease to be *as spirit*.

"It presumes all forces *physical*, and in no state can they ever appear in which they may not re-assume the initial form; that is to say, that if all the world, its furniture and people, were, and are, the evolutions of transformed nebulæ, and the forces thereof, then they may, by the law, be nebulæ again.

"But to the point: If it be maintained, as it has been by some, that 'the forces are *indestructible*, *convertible*, *imponderable objects*,' it is not yet settled *how many* such forces there are. Or, if it be assumed that all forms of force are but the translation of *one primal force*, it is no better settled whether there are not *permanent residuary forms*, not convertible by any knowledge we possess, or that all force is, *per se*, *physical*, and that there can be no force but such as appears in transformations of matter, or in the phenomena of heat, electricity, gravity, &c. These points, I say, are not by any means settled; and, until they are, it is but begging the whole argument to declare all spiritual phenomena impossible in view of them.

"The whole argument might, therefore, be rested here, since it is the business of those who urge the argument, founded on the forces, against us, to show in what way they can demonstrate by the 'rigid test of fact and experiment,' that all phenomena are resultant experimentally and logically from the physical forces.

"We simply deny that such demonstration has ever been made, or that even the *vital force* has by any such means been made to appear as a translation of the other forces. The most that can be said upon this point is, that where vital force exists, there the other forces are brought into play, and this nobody pretends to deny. We may also admit that vital force nowhere appears in the absence of the others; and Mr. Faraday, or anybody else, is welcome to all the use that can be made of this admission.

"But who ever heard of *consciousness* being translated into heat, gravity, mechanical force, &c.? Where are the demonstrations that the treasury of the memory, with the thousand incidents which make up the record of our *experience*, and give us the incontestable proof of *personal, individual existence* is convertible into electricity or chemical affinity? For, if the doctrine of the correlation of forces is to be brought against us, we have a right to insist upon the terms upon which its demonstrations are had, which are, in brief, that *any form of force correlated to another form, is susceptible of translation forward and backward, at the will of the demonstrator.* With heat, electricity, chemical affinity, mechanical power, and magnetism, this may be done. With the affections, memory, consciousness, intelligence, and vitality, it has *not* been done, and, in all probability, never will be done. Until this latter has been accomplished demonstratively, our Spiritualism is in no danger of annihilation from arguments founded on the correlation of forces, any more than from damage by the other futile arguments of the learned Professor Faraday."

A psychological theory, for which the writer does not claim entire originality, but which he states with unexampled clearness, is that contained in a little volume published by Trübner & Co., London (1868), and entitled "Chapters on Man; with the Outlines of a Science of Comparative Psychology. By C. Staniland Wake, Fellow of the Anthropological Society of London." Though the theory is not based to any extent on the recent surprising phenomena of Spiritualism, the writer, by a course of scientific reasoning, arrives at results not inconsistent with the great fact of spirit existence, and which accord with the teachings of St. Paul, who, it is contended, distinguishes between the soul, or *psyche*, and the spirit, or *pneuma*, of man.

According to Mr. Wake, the principle of being on which man's superior mental development depends, is the spirit of reflection, or simply, as distinguished from the soul essence, or *psyche*, the spirit, or *pneuma*. "It is by the addition of such a spiritual agent we can alone account for the superior phenomena of the human mental life. Founded, as those phenomena are, in the simple sensational perceptions which the lower animals also possess, we see in them the gradual development of a perception so different in its objects as to be necessarily due to the activity of a superior principle of being. The final result of this perception is the knowledge of the intuitions of truth, which are the very life of the soul essence; a knowledge which requires the operation of a spiritual principle existing beyond the soul, although intimately connected with it. Having no such external principle of spiritual activity, the lower animals can never obtain any knowledge of the soul's intuitions, or of those general truths which are the expression of them in relation to external nature.

"It is thus that the brute creatures are the mere instruments of the soul's activity, operating through the bodily organism; whilst man, having discovered the intuitions which are thus active, realizes them, and makes them instruments for his advancement in knowledge, and for the subjection of the forces of nature to his own purposes.

"The relation between the soul and spiritual essences, or between the *psyche* and *pneuma*, is clearly seen from the nature of the spiritual activity, which leads, not to any change of mental operations, but merely to the improvement of thought *objectivity*. The soul can of itself perceive only the individual objects presented to the eye; but when joined to the spirit, it takes cognizance, not only of the ever-varying phenomena of nature, but also of the qualities of objects on which the changes in such phenomena depend, and even creates those symbols which, as objects of thought, give it so increased a range and activity. The spirit, having to do only with the object, and not with the thought itself, may be classed with the bodily eye, as an instrument of soul vision, — the one giving perception of the material forms of nature, the other of its spiritual forces; and in this relation, although having a much enlarged objectivity, it may be identified with that faculty of reflection which, according to Locke, is a chief source of our ideas.

"As, however, the soul essence, or *psyche*, is indebted to its union with the spirit, or *pneuma*, for all its actual *knowledge*, both of external nature and of its own being, the spirit is entitled to claim a higher nature than that of the soul essence to which it is joined; and it must be recognized as the true principle of *spiritual life*, although not the actual source of *being*.

"That the spiritual life, like the soul activity, has its several phases or stages of development, is evident from the phenomena observable in the mental life of the child, of the woman, and of the man.

"The *child*, in its ceaseless inquiries, shows the first unfolding of the spiritual perception; but that perception being as yet imperfect in its operation, the child is limited in its activity to the imitation which is the result of simple thought.

"In the *woman*, we see the activity of the spiritual principle, in combination with that of the soul essence, in an intuitive recognition of modes of action, without the actual perception of the qualities on which their value depends, which is necessary to the generalizations of reason. We see here the activity of the instinctive soul, vivified by contact with the spiritual principle, resulting in that almost intuitive perception of simple relation, the possession of which by woman is her peculiar distinction.

"In *man*, on the other hand, instinct giving place to reason as the stimulating principle of action, the spiritual perception is employed in supplying objects of thought for the activity of the mind; the final result being the pure reasoning, which is the peculiar attribute of man. In *genius*, we have the crowning glory of man's mental development; the intuitive operation of the emotional soul essence being so perfectly combined with the keen perception of the reflective spirit, that reason itself becomes intuitive, and the mind operates by a process of spiritual instinct." . . .

As to the questions of moral responsibility and immortality,

Mr. Wake thinks it cannot be denied that the soul is the responsible, immortal portion of man's being. As the emotional, thinking, and willing essence, it is the real principle of being, and that which performs, through the physical organism, those actions to which moral responsibility has relation. But the soul is responsible for these actions only because it has a knowledge of their nature as being good or evil. This knowledge depends, however *on the activity of the spiritual perception*, on which the whole special intellectual development of man is founded, and of which conscience itself, the test of responsibility, is one of the fruits.

As the lower animals have not the spirit, or *pneuma*, they can have no knowledge of the nature of actions as being in themselves good or evil; and, therefore, they are not responsible creatures. The question of brute immortality can receive a similar solution. As the soul, or *psyche*, is the principle of being, *it must be the soul which is immortal.* The lower animals, therefore, have within themselves the principle of eternal existence. We cannot believe that any substance, either material or spiritual, can be annihilated; and, therefore, the brute soul, after death, must continue to exist.

By immortality, however, is usually understood *eternal existence in a state of separate identity.* This state does not depend on the possession of the soul essence, or *psyche*, but on that of the higher spirit, or *pneuma*, the activity of which can alone give the self-consciousness on which, apart from the bodily organism, *separate identity* is itself dependent. The brute soul, therefore, according to Mr. Wake, must exist eternally, but not in a separate state.

When, however, it is asked, "In what state, then, does the animal soul exist after death?" the only answer which can be given is, *that it must return to the great source of being from which the soul first had its origin.* As matter is one and eternal, although its grosser forms are ever changing, so it is with the soul essence, whose phenomenal *forms*, numberless as those of matter, are equally changeful, but which in its substance ever continues one and unchangeable. The noble privilege of man, however, is to be individualized as a distinct and immortal spiritual existence.

The tendency of modern scientific thought is to correlate all the phenomena of nature as the manifestations of one simple energy, of which the inorganic and the organic are but more or less complex phases. The professed advocates of the doctrine of material development ultimately reduce all things to an eternally existing and infinitely extended matter, of which force is the phenomenal activity.

"Such would appear to be the conclusion to which the hypothesis of Mr. Darwin tends. Stated in the words of Professor Huxley, it is, 'Given the existence of organic matter, its tendency to transmit properties, and its tendency accordingly to vary; and, lastly, given the conditions of existence by which organic matter is surrounded, — these put together are the causes of the present and the past conditions of organic nature.'

"The existence of matter in an organized form is here assumed; but from Professor Huxley's supposition, that in fifty years' time, science will be able 'to produce the conditions requisite to the origination of life,' we are justified in considering that 'organization' is the *accident*, while the existence of matter in its simple, inorganic form, is the only fundamental requirement. This is, moreover, confirmed by the assertion of a late writer, Mr. David Page, the most recent advocate of the development hypothesis, that man, like the animal, springs from inorganic elements.

"If we turn to the positive philosophy, we see that it has the same material basis. Mr. Lewes, while affirming that there is no real distinction between vital and psychical phenomena, the latter being themselves vital, defines vitality as 'the abstract designation of certain special properties manifested by matter under certain special conditions.' We have here the same fundamental idea as that on which the hypothesis of Mr. Darwin reposes. Mr. Lewes adds, 'Life is known only in dependence on substance: its activity is accelerated or retarded according to the conditions in which the elemental changes of the substance are facilitated or impeded; and it vanishes with the disintegration of the substance.'"

This is the necessary conclusion of materialism.

It is apparent that if this conclusion were established, it would furnish an insuperable objection to the spiritual theory as to man's nature, enforced by Mr. Wake. He, therefore, proceeds to examine the grounds on which the materialistic argument is based. No objection, he contends, can be made to the existence of *spirit* on the ground that it is not capable of *direct* proof. "Positive science allows the existence of matter in so attenuated a condition, that it can be known only by the effects of its motion, and on the 'disintegration of the substance' which attends the destruction of life: the substance itself still remains, although it may take a form which cannot be recognized. *The mere 'non-perceptibility' of spirit is, therefore, no proof of its non-existence.* But, further, supposing the animal organism possesses such a principle of being as this, its real life may continue, notwithstanding the disintegration of the bodily substance, without its existence being perceived. It is extremely probable that the ether can be rendered knowable to us, under the conditions of the present life, only by virtue of its action on the matter of the earth's atmosphere; and if, therefore, this medium were removed, there would be no possibility of our guessing its existence. In like manner, the disintegration of the bodily organism may destroy the only means by which the principle of animal life can reveal itself to us in our present state, *except, it may be, under certain special conditions.*

"Notwithstanding the fact that there is no *primâ facie* objection to the spiritual view of life, the advocates of the material hypothesis may still assert that materialism is quite sufficient to account for all the phenomena of organic matter, without calling in the agency of any special principle of being.

"When, however, we ask what beyond the mere fact of complexity, which itself requires explanation, determines the passage of matter from the inorganic to the vegetable, and from thence to the animal form of organization, the positive philosophy is silent. It does, indeed, declare that there is no 'essential distinction between organic and inorganic matter,' nor yet 'any essential (noumenal) separation' between life and mind; but, at the same time, it admits that it has no other object of inquiry than that of laws. Treating solely of the *laws* of phenomena, it does not concern itself with their cause; and, so far, therefore, as positivism is concerned, any of those phenomena may be due to the activity of an immaterial principle, the presence of which may be the *cause* of the complexity of structure that furnishes the special conditions necessary for such phenomena, and which can perhaps reveal itself only through matter."

The Darwinian hypothesis requires consideration, according to Mr. Wake, only so far as it affects to derive man, equally with both the animal and vegetable kingdoms, from a common and single progenitor. As to the former, Professor Huxley says, "There cannot be the slightest doubt in the world that the argument which applies to the improvement of the horse from an earlier stock, or of ape from ape, applies to the improvement of man from some simpler and lower stock than man."

The same argument may be used to explain the origin of the animal from the vegetable organism. On examination, however, we find that the conclusion cannot be sustained. When it is said that "the structural differences which separate man from the apes are not greater than those which separate some apes from others," we have, independently of the fact that there is no evidence of the past or present existence of any such links between man and the ape, as there are between ape and ape, a statement which is not correct. This may, indeed, be proved by Professor Huxley's own admission. He is constrained to admit "the width of the gulf in intellectual and moral matters which lies between man and the whole of the lower creation," although he explains it as the result of "variation in function" rather than of variation in structure.

According to Professor Huxley, it is *language* which "constitutes and makes man what he is;" and this language depends on "the equality of action" of the two nerves which supply the muscles of the glottis; a change in the structure of which, although imperceptible, might have a result which would be "practically infinite."

"But how can a change of structure, which has so marvellous a consequence be a slight one? The fact is, that its insignificance is merely apparent; for it is associated with a *general* superiority and refinement of nervous structure and sensibility, which give a higher form and tone to the human organization, being the conditions on which the special action of the nerves connected with the muscles of the glottis altogether depends.

"It is, however, a fundamental error to ascribe man's superiority over the animal world to 'language.' The faculty of speech is a most important instrument for the education of man's mental faculties; but it is *merely* an instrument, and one without which man would still be vastly superior to the creatures below him. How strange that man's civilization — and may we not add, his responsibility and immortality? — depend wholly on 'the equality of action of the two nerves which supply the muscles of the glottis'! Surely, the talking parrot must also have a capacity for civilization!

"The Darwinian hypothesis, which Mr. Herbert Spencer accepts as reducible to the 'general doctrine of evolution,' gives no satisfactory explanation of the *origin* of the primitive cell; and thus leaves unsolved the chief problem presented by organic nature in its several phases.

"No ground is assignable, consistent with the hypothesis of evolution, why the only wide gap in the series *should be between the highest ape and man*. The only explanation which can be given by those of its advocates who admit the possession by man of 'special endowments' — 'that nature can produce a new type without our being able to see the marks of transition' — is in reality fatal to the hypothesis itself, seeing that the exercise of such a power bespeaks the operation in nature of some fresh principle of vitality.

"But, secondly, it is evident that the minute modifications of function and structure, supposed, cannot result in the formation of something *fundamentally different* from that which has been thus modified. It has been shown, that it is not the possession of speech which constitutes man's superiority over the animal world, but *the faculty of spiritual perception;* the exercise of which underlies both human language and every other phase of culture by which man is distinguished. This is a power wholly dissimilar from any the animal world possesses; and no modification, therefore, of the animal organization could evolve it.

"Reference to 'a plan of ascensive development' will not meet the difficulty when 'new and special endowments' are admitted; for, according to the principle laid down by Herbert Spencer, that 'function is antecedent to structure,' those endowments can exist only in response to a preceding functional tendency. This principle, moreover, directly contradicts the reasoning of Professor Huxley, that a functional difference which is 'vastly unfathomable, and truly infinite in its consequences,' has arisen from a small structural change. The modification of the organism must have been preceded by that of the function; and as the latter is itself dependent on something which the lower animals do not possess, it is absolutely impossible that either the function or the structural differences which it precedes can have been evolved simply out of an animal organization. . . .

"There must be an antecedent functional tendency, or there can be no formation of organic material, much less of a specialized organism. The very fact of the existence of organisms, so different in their vital phenomena, as the animal and the plant, both of which are made up of the same chemical elements, proves the existence of *two different fundamental tendencies,* which cannot be explained by any peculiarity of combination of those elements, since the function is antecedent to all such combination, and directive of the form it shall take. Supposing, then, specific organized forms are accompanied by peculiar arrangement of their chemical elements, which take the form of 'physiological units,' the tendency of the primitive organic matter, to take this arrangement, has to be accounted for; and it can be only by its dependence on some still more ultimate fact.

This ultimate fact Mr. Wake finds in spirit, deity. The phenomena of life in man are quite distinct from those of either organic or mere animal vitality; and, although intimately related to, and, it may be, necessarily connected with them, the union is one of actual *addition*, as by superposition of a perfectly fresh and independent faculty.

"The universe may be described as an infinitely extended and eternally existing organism. The possession, however, by man of the principles of animal and spiritual life requires the prior existence of *something analogous in nature to them from which these principles can have been derived.* There must, in fact, according to the reasoning of the *materialistic* argument, be an eternally existing principle of being, from which the soul of the animal organism can have had its origin; and thus must it be to enable us to account for the existence of the higher *spiritual* principle which we see in man.

"But, as in phenomenal nature, we see the three discrete degrees of life co-existing in a certain relation, — the lower being essential to the existence of the higher, and the higher again giving a new direction to the activity of the lower, — we are justified in affirming that a similar relation exists between the several co-existing, eternal principles of being which thus reveal themselves. These three degrees of Absolute Life cannot be independent of each other; and, therefore, that Eternal and Infinite Existence from which all phenomenal nature has been evolved, must, although manifesting his activity through a material organism, yet be essentially a spiritual being, as possessing, not only the principle of animal vitality, but also that of the spiritual life.

"As, however, nature is an evolution from the Divine Organism, — man being the final result of such evolution, — we must see in man and nature a representation of God; who, therefore, is not the Unknowable Existence which the hypothesis of evolution, as stated by Mr. Herbert Spencer, requires. God cannot be unlike that which has sprung from himself, — except only so far as he is infinite and perfect, while *it* is finite, and, as such, imperfect.

"Moreover, knowing man and nature, we have a conception — incomplete, because limited — of God himself; and this conception must widen, and therefore become more nearly perfect with every increase of our knowledge. Hand-in-hand, therefore, with the development of science, there should be an ever-increasing veneration for that Being, the laws of whose relative existence science expresses."

And here, according to the system of Mr. Wake, we have the only ground for reconciliation between science and religion.

The argument which we have thus presented, in an abridged form, is worthy the reader's study; and it will, we hope, call attention to the book itself, where some omitted links will be found supplied.

As a fitting termination to our review of the principal theories which the phenomena have called forth, we quote from the London "Morning-Star and Dial" the following remarks: —

"The egotism which sets up its own finite comprehension as the test of possibility, rejects with scorn every thing alien to its experience, or antagonistic to its preconceived ideas. It can scarcely be necessary to urge, that such a mode of dealing with alleged facts is not only grossly unphilosophical, but would, if generally adopted, prove a positive barrier to the elucidation of important truths.

"When a very large number of independent and respectable witnesses testify that they have repeatedly seen phenomena wonderful in their character, identical in their nature, and occurring always under certain fixed conditions, it is obviously our duty to sift their evidence, in order that we may either crush an imposture, dispel a delusion, or establish a new and, possibly, most important truth.

"This is the position which the controversy with regard to

Spiritualism has unquestionably assumed. In England and in America thousands of men and women esteemed for their piety, their intellectual ability, and their social worth, aver that they have been eye-witnesses, not once, but repeatedly, of very strange manifestations, which can scarcely be accounted for by the operation of any known natural agency.

"They tell us that they have seen heavy tables lifted up a foot or more from the ground, and held for some moments suspended in the air; men raised from their chairs and floated across the ceiling of the apartment; accordions and guitars, held in the hand, played upon by unseen fingers; bells carried about a room and rung at intervals by an invisible power, and passed from hand to hand of the quiescent circle; intelligible sentences written upon slates and slips of paper placed beyond the reach of any present; luminous hands appearing in the air, lifting articles from the floor and placing them upon the table; and a host of other marvels, to all appearances equally beyond the grasp of ordinary credibility.

"Mr. Coleman, a gentleman whose word would be unhesitatingly taken on any ordinary matter, tells us of a drawing medium, who has the power of sketching perfect portraits of deceased persons whom he never saw, and with regard to whose personal appearances he had no means of forming any idea. He relates his visit to another medium, to whom he was personally unknown, who, in answer to his mental question, wrote a communication to him from his step-son, sometime deceased, signing it with the young man's full name, and adding his own residence in London; and he states that he listened to some speaking mediums, persons in their ordinary state wholly illiterate, who, under what was asserted to be spiritual influence, spoke in public for more than an hour at a time, with very remarkable eloquence and intellectual power. He recounts an instance, which he declares was certified to him on excellent authority, in which a communication was received through a medium, leading to the discovery of a lost document essential to the success of an important lawsuit; and he recites an example of an opinion obtained by the same means, which brought to light a new point, and put a stop to a harassing litigation.

"But, putting aside all that he gives on the authority of others, his narrative of his own personal experience is strange enough to satiate the most ravenous appetite for the marvellous. At one *séance*, for example, at Boston, he states that a guitar was carried rapidly about the room above the heads of those present, a melody being accurately played upon it as it moved through the air; that bells were similarly floated about, ringing all the while; that the medium, in her arm-chair, was lifted on to the centre of the table, from which position he himself removed her; that his own name was pronounced in a loud voice through a horn; and that, when he complained of the heat of the room, a fan was taken from a drawer and waved before him, and a tumbler of water was raised and placed to his lips.*

"All this, no doubt, is passing strange; and those who have never with their own eyes seen any thing of the sort, may be well excused for shaking their heads in doubt. It is true that the striking singularity of some of the phenomena reported induces us sometimes to forget that, if we concede the possibility of one of them, we may without much difficulty admit that of all. Grant that a power exists which can raise a heavy table from the ground and hold it suspended in the air, it is clear that the same agency may just as easily lift a man from his chair, carry a bell, wave a fan, or play upon a guitar. The simple rapping upon the table, if not fraudulently produced, is intrinsically, though not apparently, quite as marvellous as any of the most elaborate manifestations.

"But these physical effects are by far the least interesting of those which the Spiritualists allege to be of every-day occurrence in their circles. They complain, indeed, that the use of the phrases, 'Spirit-Rapping,' and 'Table-Turning,' has tended to give the general public a very low and inadequate idea of

* These phenomena occurred at one of the sittings at which Miss Lord was the medium, and to which we introduced Mr Coleman, with whom we were present.

the scope and object of this class of phenomena. According to their doctrine, these strange freaks which are played with material objects, are designed solely to arrest attention, and to convince the skeptical that unseen agencies are present, capable of holding communion with mortals; and that, this end having been attained, the real purpose of that which they regard as a beneficent dispensation, acquires its needful scope and comes into full play. This purpose they hold to be the communication from departed beings to their surviving relatives, of messages of solace, of warning, of encouragement, and of counsel, — conveyed occasionally by audible voices, but much more frequently in an alphabetic form.

"They believe that the ultimate end of these 'Spiritual Manifestations' is the advancement towards moral and religious perfection of the living through the loving ministrations of the dead; the proximate end being the counteraction of materialistic tendencies by the exhibition of cogent proofs of the reality of spiritual existence.

"If the extraordinary narratives were vouched for only by men utterly unknown, or of dubious credibility, they might scarcely be deemed worthy of serious attention. Even then we could scarcely avoid the reflection that the idea which constitutes the postulate of the Spiritualists, so far from being novel, has had adherents in every age and every nation. The belief in the possibility of intercourse betweeen spirits and mortals has found a place in almost every religious creed ever held by man; and pagan traditions and biblical records alike bear witness to supernatural communion. Nor can we entirely exclude the thought that these phenomena, if sufficiently attested to be accepted as real, would cast much light on many incidents in past secular history, which stand greatly in need of some rational elucidation, in place of the wholesale rejection of a mass of evidence which has hitherto been our desperate expedient. But are they so attested? This is the first point to be settled. The principal witnesses are literary men of note, merchants, lawyers, physicians, and divines; ministers of divers sects, men and women of unblemished repute, artists, poets, and statesmen. Of minor witnesses, the name is legion; but we have no personal knowledge of their claims to our belief. This much we know, that in America, and in our own country, there are many whose sanity no one doubts, whose general veracity no one would impeach, who aver that they have seen these strange things with their own eyes. It remains for us to say whether we will take their word.

"If we stamp all those who declare that they have witnessed these so-called 'Spiritual Manifestations,' as liars, of course the inquiry will be at an end. If, on the other hand, we are willing to believe that, in the narratives which they have given us, they have honestly recorded the impressions produced upon their eyes and ears, we shall next have to consider to what causes these phenomena may fairly be ascribed. Four hypotheses have been put forward: fraud, self-delusion, the operation of some hitherto undiscovered natural law, and spiritual agency. The idea of fraud, as a general explanation of the manifestations, may, we think, be fairly discarded. Imposture there may have been in cases where money was to be gained; but seeing that many of the most striking manifestations testified to, took place in private houses, where no paid medium was present, — this being especially true of the intellectual communications purporting to come from departed relatives, — it is difficult to believe that those who formed the circle could have been fools enough to practise a deliberate cheat upon themselves for no object whatever, to say nothing of the blasphemy against the holiest affections, which was involved in simulating a message from a deceased parent, wife, or child.

"It is not easy to understand what invisible mechanism would take a man out of his chair, float him round the ceiling, and then replace him in his seat; and that must be a very knowing apparatus for the production of raps which would spell out to an unknown foreigner the name of his step-son, who had been some years in the grave. But in purely private circles, — the vast majority of those which are held, — fraud is clearly out of the question. If self-delusion be the chosen explanation, then we ought to have it explained how it happens that the same delusion operates upon a dozen or more persons at the same time. If the operation of an unknown natural law be the solution adopted, it must be one law capable of producing all the phenomena recorded; for they appear to present themselves in very indiscriminate order at various *séances*.

"It is a current, but very grave error, to suppose that the most startling of these physical manifestations are opposed to known natural laws. It is generally said, for example, that the lifting of a table from the ground, — one of the commonest of the alleged phenomena, — is opposed to the laws of gravitation. Clearly it is not, if an unseen force be applied to it, powerful enough to counteract its attraction. An unseen force is no novelty in nature. Life is unseen, electricity is unseen: heat is unseen, until, by igniting matter it gives birth to flame. But this force must be one, capable of accounting for all the effects. It will not do to say that this phenomenon results from hysteria; that, from magnetism; the other, from thought-reading; a fourth, from the od force, whatever that may be.

"If the spiritual theory be resorted to, a vital point arises. Is it a good or an evil agency? The advocates of the Satanic theory have this great stumbling-block to get over, that the advice given in the messages communicated is said to be universally* good, the sentiments moral, and the doctrine piously Christian; and it can scarcely be supposed that the Author of Evil would labor for his own discomfiture. There may be a mixture of good and evil agencies; then we ought to discover how we are to discriminate between the two. For ourselves, we express no opinion on the subject; all we wish is to see the matter fairly investigated, with a total absence of that spirit of ridicule which is always offensive and proves nothing, and which is in the present case especially out of place. With the considera-

* Not universally. We have seen that we must still *try* the spirits; that they are as various as mortals in their moral and intellectual traits.

tion of '*Cui bono*' we have nothing whatever to do. The first question to be solved is, 'Is it true, or is it not?' The second, 'Whence is it?' If the first be answered in the affirmative, then, even should the second remain without reply, we may tranquilly leave the rest to the good providence of God."

"Alas for him who never sees
The stars shine through his cypress trees!
Who, hopeless, lays his dead away,
Nor looks to see the breaking day
Across the mournful marbles play;
Who hath not learned, in hours of faith,
The truth to flesh and sense unknown,
That life is ever lord of death,
And love can never lose its own!"

CHAPTER XI.

COMMON OBJECTIONS.—TEACHINGS OF SPIRITUALISM.

"I live; and this living, conscious being which I am to-day, is a greater wonder to me than it is that I should go on and on. How I *came* to be astonishes me far more than how I should *continue* to be." — *Rev. Orville Dewey.*

"I confess the awful mystery of life, and the perplexity which hangs around the question, *What it is, and what it all means.* Nevertheless, I am persuaded — as persuaded as I can be of any thing in this world — that the meaning is good and not evil, — good, I trust, to the individual as well as to the whole. There is a wondrous alchemy in time and the power of God to transmute our faults, errors, sorrows — nay, our sins themselves — into golden blessings." — *Rev. F. W. Robertson.*

IF we accept the fact that physical death does not affect the identity of the individual, it will be a necessary inference that there are as many intellectual and moral differences among spirits as among mortal men. In the spirit-world as well as in this, at each step of our progress, we can only take in the amount of truth we are organically fitted to receive and assimilate. There, as well as here, the saying of Locke holds good: "So much only as we ourselves consider and comprehend of truth, so much only do we possess of real and true knowledge. The floating of other men's opinions in our brains makes us not one jot the more knowing, though they happen to be true. Like fairy money, they turn to dust when they come to be used."

Our emancipation from this material husk does not alter those essentials of character which we have been born to here, or which we have failed to modify in the series of existences through which we may have passed. The moulding of our individuality must be our own work, mysterious as this may seem.

Beyond the mere fact, therefore, that spirits live and act (and what greater fact could we ask?), the teachings of spirits are to be received just as we receive those of fallible mortals, and to be subjected to the test of our own spiritual and rational powers. Pressed on by influences from all sides, we are yet to accept or reject them, according to the light which conscience may shed.

In regard to the vulgar notion that a spirit, on quitting his mundane tenement of flesh, parts with his identity, and attains at once to high spiritual knowledge and power, M. Jobard, of Brussels, remarks, "As well might a highway robber be looked upon as an honest man as soon as he is out of prison; or a madman, after clearing the walls of an asylum, be regarded as a sage! There is as much difference between spirits as between men. Every one takes with him into the next state of life his character and his moral and scientific acquirements. Fools here are fools there. Rogues, sensualists, tyrants, suffer from being deprived of their selfish stimuli. Hence we are instructed by the Holy Spirit to hold in low esteem those goods of earth, which we cannot assimilate to ourselves, nor take with us; but to attend rather to spiritual and moral goods, which do follow us, and which will serve eternally not only to delightfully occupy us, but as steps by which we shall rise higher and higher, on the great Jacob's ladder, into the boundless hierarchy of spirits."

It has been truly said, that the tendency of Spiritualism is to lift the phenomena of spiritual life out of the category of exceptional events into the region of divine law and order; and thus to promote our spiritual emancipation and development. By the light of this infant, or rather *adolescent* science, we now see clearly, that *the truth of a doctrine cannot be proved by a so-called miracle*. The meaning and worth of "a miracle" — *i.e.* the intervention of some intelligent, unseen agency — must rather be tested by the effect which it is calculated to produce upon the mind and heart; and this again can only be estimated by the devout and cultivated reason. The study of spiritual phenomena thus elevates the mind above a servile submission to mere dogmatic authority as well as above an ignorant resignation of its rights and faculties before a mere "sign and wonder."

"This emancipating tendency of the new science," says an English writer, "is quite sufficient to account for the opposition it has encountered at the hands of the *religious* world; while the innovating and revolutionary character of spiritual teaching induces a large section of the *irreligious* world to regard it with distrust and uneasiness. The weak and timid, and therefore false and unjust, conservatism of aristocratic England dreads each breath of free thought which tends to quicken the seeds of regeneration sleeping within her bosom. It makes many people uncomfortable to see old landmarks in religion, morals, or metaphysics threatened with annihilation. They regard the whole matter, much as the respectable country gentlemen in England fifty years ago regarded Methodism. If a man turned Methodist, it was equivalent to his becoming a radical, a blasphemer of social decorums and time-honored conventionalities. The case is much the same to-day; and, with a true instinct of self-preservation, the man of mere material, selfish aims, and hebdomadal religion, if he has any at all, recognizes in Spiritualism a disturber of his peace. This importunate proximity of unseen realities calls for a re-adjustment of his stagnant ideas; and it makes him tremble for the safety of the 'reserved seat,' to which he looked forward in the other world, and also of his reputation as an intellectual aristocrat in this.

"Such a fear is by no means a groundless one; for who can measure the influence which this despised Spiritualism is exercising on a score of worn-out *ologies* and *isms?* Its negative effects are those most obvious at present. It is a great truth, which has not yet woven a dress for itself, or elaborated appropriate organizations as outward and visible signs of its inward and spiritual grace. It wanders about in rags and tatters, and often in most disreputable company; so that *some moral courage is required even to acknowledge acquaintance, much more to associate with this truth, in the public roads of life.*

"We confess that we perfectly understand the aversion with which many earnest minds have been led to regard this subject. 'So far,' it is said, 'from these investigations having an elevating or emancipating effect upon the mind, so-called spiritual manifestations generally appeal to the lowest mental faculties, while pandering to idle curiosity and a thirst for sensational exhibitions.' There is much truth in this. And it is not enough to make the specious and oft-repeated reply to such taunts, that an evil and adulterous, or sense-bound generation needs a sign; and that the fittest for them are dancing-tables, knot-tying, and volant trumpets in dark closets, &c. A cultivated mind CANNOT look upon such things except as most disorderly and undivine; *although they may have a spiritual origin*, and, having such, are worthy the close examination of all who would acquaint themselves with the grounds of a spiritual belief.

"The higher manifestations of modern Spiritualism are not so obnoxious to contemptuous criticism. . . . But these and a hundred other objections, whether valid or not, do not disprove the fact, that Spiritualism is exercising a most beneficial influence on civilization, by leading to the discovery or illustration of spiritual laws. *Even supposing all these various manifestations to be disorderly and vicious, which I do not for a moment believe*, their illustrative value would be none the less. How much would the world know of physiology or the laws of health, if disease had not first necessitated the study of pathology?"

The motto on the cover of that excellent publication, the "London Spiritual Magazine," fairly expresses all that can be consistently claimed as comprehended by modern Spiritualism. For individual idiosyncrasies and speculations, whether of spirits or of earthly residents, Spiritualism is no further responsible than collective humanity is responsible for the vagaries of a drunken man or a lunatic. The following is the motto referred to:—

"Spiritualism is based on the cardinal fact of spirit communion and influx: it is the effort to discover all truth relating to man's spiritual nature, capacities, relations, duties, welfare, and destiny; and its application to a regenerate life. It recognizes a *continuous* divine inspiration in man: it aims through a careful, reverent study of facts, at a knowledge of the laws and principles which govern the occult forces of the universe; of the relations of spirit to matter, and of man to God and the spiritual world. It is thus catholic and progressive, leading to true religion as at one with the highest philosophy."

Expanding these views, the same magazine remarks: "Spiritualism is a science based solely on facts: it is neither speculative nor fanciful. On facts and facts alone, open to the whole world through an extensive and probably unlimited system of mediumship, it builds up a substantial psychology on the ground of strictest logical induction. Its cardinal truth, imperishably established on the experiments and experiences of millions of sane men and women, of all countries and creeds, is that of a world of spirits, and the continuity of the existence of individual spirit through the momentary eclipse of death; as it disappears on earth re-appearing in that spiritual world, and becoming an inhabitant amid the ever-augmenting population of the spiritual universe. Along with this primal truth comes the confirmation of the ancient truths of Deity, revelation from Deity to man, and the open communion of man in the body with man disembodied; with 'that great multitude which no man can number, of all nations and kindreds and people and tongues which stand before the throne.'

"That is the sum and substance of Spiritualism: it is the exponent and practical demonstrator of continuous spiritual being. Whatever truths independent of this assert themselves, must do so on the same substantial evidences, and must show their kinship to this grand central truth by their perfect harmony and oneness with it."

The general character of the higher spiritual communications of the present day is the absence of dogmatic teaching, and the assertion that it is only as we advance in virtue and in the deeper paths of knowledge that we can attain to further light in the science of things divine.

Whenever friend or foe, therefore, undertakes to commend or denounce any notion, good or bad, by the introductory claim that "Spiritualism teaches," &c., we may be pretty sure that the phrase ought to be so amended as to read, "A certain spirit teaches," &c.

If you ask why error and evil are allowed by Providence to exist in the spirit-world, you might, with equal propriety, ask why they are allowed to exist in this. We cannot reply better than in the words of Wollaston (1724), who says, "To ask why God permits evil, is to ask why he permits a material world, or such a being as man is; endowed, indeed, with some noble faculties, but incumbered at the same time with bodily passions and propensities. Nay: I know not whether it be not to ask why he permits any imperfect being; and that is, any being at all: which is a bold demand, and the answer to it lies perhaps too deep for us."

To ask why evil exists, may, to higher intelligences, be quite as irrational as it would be to ask why a triangle has three sides?

These remarks will render it superfluous, we hope, for us to refer further to that fighting with windmills which certain writers for both the religious and the secular press indulge in when they insist on charging any extravagant heresy or act upon the "teachings of Spiritualism."

"Spirit communications," says Mr. W. F. Jamieson, "partake more or less of the idiosyncrasies of the mediums through whom they are received. On the part of intelligent spirits, there is no claim to infallibility. They teach people to accept nothing without adequate proof. Seven-tenths of the alleged spiritual phenomena may be of mundane origin, though not impostures. But *one* incontrovertible fact proves spirit existence and communion as positively as a million facts can do. Years since I witnessed phenomena under circumstances that precluded imposition or trick of any kind. There may be ten thousand counterfeits, but they do not shake my confidence in that which is genuine."

Judge Carter, of Cincinnati, complains of the deceptive character of many of the communications. "I cannot," he writes, "now point to a single medium — and I have known many — and say that he or she is perfectly reliable."

To which we might reply, *perfect reliability implies perfect infallibility;* and the judge must seek for that, not among mediums, or spirits, or angels, or archangels, but of Omniscience alone.

The "committee on spiritual phenomena," who met at Cleveland, O., in 1867, report that "what at present passes for spirit communion among the people, is a mixed, and, for the most part, unanalyzed mass, rendering the identity of spirit presence very uncertain." And they add, "Many, if not all, of the disorderly manifestations, your committee deem wholly unspiritual, having their origin in half-controlled, diseased nerves, poor digestion, torpid liver, and general discord of mind and body." And they conclude, "We cannot suppose that a majority of the phenomena under consideration are projected and directed by spirits; but rather that while there is abundant evidence, direct and collateral, of spirit control, other causes enter largely into their production."

All which we can readily admit, and infer that it merely shows that we ought to discriminate between phenomena that may be explained by abnormal nervous action, and those which must be referred to some other cause; that it would seem to be not intended by the Author of our nature that we should surrender our individual reason to any authority, human or spiritual; that we should *try* the spirits, for they are a very mixed set, just like human beings; and that any implicit belief in the infallibility of spirit communications is likely to lead to a mental re-action quite as far from the truth as its opposite extreme. Perhaps the theory, held by certain Spiritualists, that there are *no evil spirits*, may have done something to color portions of the report of the Cleveland committee; and perhaps they have not yet fought their way out of the wilderness into a light which may one day be theirs. Their sweeping and indiscriminate condemnation of the "dark-circle" manifestations, and their loose, unqualified assertion, that "darkness is a condition assumed and insisted upon by tricksters," are so contrary to the experience of many investigators who have had as ample opportunities as they can have had to satisfy themselves on the subject, that their report does not carry the weight it deserves for much in it that is true. It is signed by F. L. Wadsworth, J. S. Loveland, E. C. Clark, and M. B. Dyott.

There is certainly nothing in the nature of the facts attested by Spiritualists (and by many who admit the facts without the hypothesis) to render it difficult to form a correct judgment as to the reality of occurrences whether in light or dark circles. Take the following instance quoted by Mr. Shorter: A distinguished London physician and physiologist, Dr. Wilkinson, in an account of a *séance* he attended, mentions among other phenomena witnessed by him, that a hand-bell, which had been brought by one of the party, was rung by an invisible agency; at the same time as it moved towards himself, he says, "I moved my fingers up its side to grasp it. When I came to the handle, I slid my fingers on rapidly; and now, *every hand but my own being on the table*, I distinctly felt the fingers, up to the palm, of a hand holding the bell. It was a soft, warm, fleshy, radiant, substantial hand, such as I should be glad to feel at the extremity of the friendship of my best friends. But I had no sooner grasped it momentarily, than it melted away, leaving me void, with the bell in my hand. I now held the bell tightly, with the clapper downwards; and while it remained perfectly still, I could plainly feel fingers ringing it by the clapper. As a point of observation, I will remark, that I should feel no more difficulty in swearing that the member I felt was a human hand of extraordinary life, and not Mr. Home's foot, than that the nose of the Apollo Belvidere is not a horse's ear. I dwell chiefly, because I can speak surely, on what happened to myself, though every one round the table had somewhat similar experiences. The bell was carried under the table to each, and rung in the hand of each. . . . They all felt the hand or hands, either upon their knees or other portions of their limbs. I put my hand down as previously, and was regularly stroked on the back of it by a soft, palpable hand as before. Nay: I distinctly felt the whole arm against mine, and once grasped the hand; but it melted, as on the first occasion. . . . While this was going on, and for about ten minutes, more or less, my wife felt the sleeves of her dress pulled frequently; and, as she was sitting with her finger-ends clasped and hands open, with palms semi-prone upon the table, she suddenly laughed involuntarily, and said, 'Oh! see, there is a little hand lying between mine; and now a larger hand has come beside it. The little hand is smaller than any baby's, and exquisitely perfect.'"

At a subsequent *séance* at Mr. Rymer's house, at Ealing, he describes a similar experience. The hand on this occasion purported (in a communication made) to be that of a deceased and intimate friend, "once a member of Parliament, and as much before the public as any man in his generation."—"I said," continues the narrator, "'If it is really you, will you shake hands with me?' and I put my hand under the table; and now the same soft and capacious hand was placed in mine, and gave it a cordial shaking. I could not help exclaiming, 'This hand is a portrait. I know it from five years' constant intercourse, and from the daily grasp and holding of the last several months.'" Others who were present at these *séances*—Mr. Rymer, Mr. Coleman, and Mrs. Trollope, in particular—have corroborated the testimony of this writer.

Commenting upon this testimony, Mr. Shorter remarks: "Whatever weight may justly attach to the testimony of men of known ability and attainments, any man of ordinary intelligence and powers of observation is generally able to judge, in an almost equal degree, of what Chalmers calls 'plain palpable facts' under his own observation. Any man, for instance, who can 'tell a hawk from a hand-saw,' can tell whether a table is resting on the floor, or is raised above it; whether a man is sitting in his chair, or is floating in the atmosphere of the room; whether sounds made by no visible agency, and which respond to his questions, mental or otherwise, are heard or not; whether a strong, heavy table, is, at his request, broken in fragments by no visible agency, 'in about half a minute,' or whether it remains whole. These things, and such as these, which rest on 'seeing and feeling and experimenting,' are so plain and palpable that the man who could not judge of their reality might conscientiously say with Dogberry, 'write me down an ass.' It is very easy to pronounce these things impossible; to say that they cannot be. But that which *does* happen *can* happen; and to tell people that an educated judgment would convince them that they did not see what they saw, and did not feel what they felt, can only furnish an illustration of that particular species of rhetoric called *bosh*. We are disciples of the Baconian philosophy, and cannot subscribe to that reasoning which denies facts when they do not square with our pre-judgments and accommodate themselves to our favorite theories."

Very frivolous and pointless are such objections as the following, brought by a critical American journal, called the "Nation," against the phenomena so pregnant with meaning to millions of investigators: —

"If the wonders of Spiritualism," says this authority, "are perfectly real, they are just as perfectly worthless. They prove nothing but the powerlessness of those who execute them, whether they be spirits or mortals."

"They move chairs, &c.; but it is nowhere asserted that the mediums move furniture half as well as day-laborers and porters."

"They are unable to tell any particular person what he did not know already."

This last objection has been so often disproved, and the narratives of this volume confute it so repeatedly, that we need not occupy any more space in considering it. Our own experience, given on page 112 of this volume, will serve as our answer.

With regard to the other objections, which are mere repetitions of those brought by unreflecting persons ever since the phenomena of 1848 were promulgated, a few considerations will show how much weight they carry.

Why, if these wonders, even though "real," are "worthless," have Faraday, Brewster, Babinet, the Cambridge professors, and many other eminent men of science, been so anxious to prove that they are *not* real?

If a phenomenon be worthless in itself, will it have no value if it carry evidence that it is the work of a spirit?

Yet the affirmative of this last question is the wholly heedless and irrational position taken by this editor. Every person of common sense will see its absurdity.

As for the objection that the mediums, or rather the forces operating in their presence, though they may move furniture, "cannot do it as well as day-laborers and porters," this would seem to be a very blind attempt at jocoseness, since the whole interest claimed for the movements referred to, lies in the inquiry, by what power are they done, not how skilfully are they done.

And yet objections thus slight and trivial are fair specimens of the kind of opposition which Spiritualism has encountered. Is it surprising that it has spread and grown as rapidly as it has?

The attempts to make Spiritualism responsible for the heresies and vagaries of certain persons calling themselves Spiritualists, are manifestly unjust. Accusations are often brought that Spiritualism teaches free love, pantheism, socialism, &c. As well say that the Newtonian philosophy teaches these things! Spiritualism is no more responsible for nominal Spiritualists, than Christianity is for nominal Christians, among which last may be counted free-love Anabaptists, Mormons, and the brigands of Italy.

"Pythonism" is the bad name which the "ministers of the Massachusetts Association of the New Jerusalem" (followers of Swedenborg) give to modern Spiritualism; the name being derived, they tell us, from the ***Python***, the mythological serpent, sprung from the mud after Deucalion's deluge, and which Apollo alone was able to destroy. "Hence the priestess of Apollo, at Delphi, was called a Pythoness." "Who cannot, in this mythological tradition," say the ministers, "see the serpent, engendered by the very lowest things of humanity," — Spiritualism, of course, being the chief of these "lowest things."

By ordinary persons, the supposed communications from the spirit-world will, we think, in this nineteenth century, be received as we receive communications through books, newspapers, and even weekly critical journals. Various and sometimes conflicting as these communications are, they merely show that spirits, like mortals, are very fallible, and often very conceited individuals, many of them it may be, groping in a moral and intellectual darkness denser than that which encompasses many souls yet fettered by the flesh. Spiritualism is merely an affirmation of the great fact of spiritual existence. It leaves us just where all codes and all revelations take us up; for the authority of a message, come whence it may, lies always in the completeness of its harmony, with the laws of our being as disclosed by the highest experiences of individuals and of the race.

Nor is Spiritualism any thing new, though never before in human history have men been so educated and prepared to receive its phenomena in a scientific spirit, and never before has priestcraft been so impotent to dictate terms, or to put its own convenient construction on facts appealing so directly to the common sense of mankind.

"The idea of the existence of spirits," says one of our French collaborators (Edward de Las Graves), "and of their intervention in human affairs, may be traced back to the most remote epochs of antiquity. We find it in all the philosophies: it forms the basis of all the religious systems of the ancients, and the Biblical narratives are full of it. The Greeks, the Romans, the Egyptians, the Druids, the Indians, and the Chinese had their oracles which they consulted. The Middle Ages could not bury the idea in the funeral piles which devoured their sorcerers and their witches. It has come down even to our own times, braving all persecutions, surviving all the revolutions, physical and moral, of humanity.

"Beyond a doubt this idea, imperishable because it is true, has often been associated with a thousand absurdities. Cupidity and the lust of domination have often made of it a powerful weapon, and have not feared even to disfigure, and pervert, and play false with it in order to subject it to their caprices, their ambitions, or their needs. But the time has come at length when the truth is destined to rise and glitter in all its splendor, chasing pitilessly the errors which ignorance and superstition have heaped up during the centuries."

All speech is spiritual. All communications addressed to our moral or intellectual faculties are of spiritual origin, whether they come from spirits incarnate or disincarnate.

Dr. William B. Potter, author of a pamphlet on the subject of modern Spiritualism, says of the communications which he himself has heard through mediums, that "endless contradictions and absurdities are mixed up with the most exalted truths, and the most profound philosophies. We are taught that God is a person, that he is impersonal; that he is omnipotent, that he is governed by nature's laws; that every thing is God, there is no God, that we are gods. We are taught that the soul is eternal; that it commences its existence at conception, at birth, at maturity, at old age. That all are immortal, that some are immortal, that none are immortal. That the soul is a winged monad in the centre of the brain; that it gets tired, and goes down into the stomach to rest; that it is material, that it is immaterial; that it is unchangeable; that it changes like the body, that it dies with the body, that it develops the body, that it is developed by the body, that it is human in form; that it is in but one place at a time, that it is in all places at the same time.

"We are taught that the spirit-world is on earth, — just above the air, — beyond the milky-way. That it has but one sphere, three spheres, six spheres, seven spheres, thirty-six spheres, an infinite number of spheres. That it is a real, tangible world; that it is all a creation of the mind of the beholder, and appears different to different spirits. That it is inhabited by animals, birds, &c.; that they do not inhabit it. That it is a sea of ether; that it is a plain, that it has mountains, lakes, and valleys; that it is a belt around the earth. We are taught that spirits eat food, — live by absorption; live on magnetism, thoughts, love. That they control media by will-power, by magnetism, by entering media, by standing by their side, by an influence beyond our atmosphere, by permission of the Lord.

"That spirits converse by thought-reading, by oral language. That their music is harmony of soul; that it is instrumental and vocal. That they live single; in groups of nine. That they marry without having offspring; that they have offspring by mortals; that they have offspring by each other. That their marriage is temporary; that it is eternal. That spirits never live again in the flesh; that they do return, and enter infant bodies, and live many lives in the flesh. That some are born first in the spheres, and afterwards are born on earth in the flesh. That the true affinity is born in the spirit-world at the same time that the counterpart is born on earth. That all spirits are good, that some are bad; that all progress, that some progress, that none progress. . . .

"That there is no high, no low, no good, no bad. That murder is right, lying is right, slavery is right, adultery is right. That whatever is, is right. That nothing we can know can injure the soul, or retard its progress. That it is wrong to blame any; that none should be punished; that man is a machine, and not to blame for his conduct. . . . That the spirit of the tree exists in perfect form after the tree is burnt. That monads are God's thoughts, and go through all forms of rocks, trees, animals, and at last become men. That spirit is substance, in absolute condensation; that matter is substance, whose particles never touch."

The reply to statements like these has been well made by Mr. A. E. Newton, who writes as follows: "To our view, the evidence of the basis-fact of modern Spiritualism — namely, 'the intelligent communication of spirits with minds in the flesh' — does not depend at all upon either the *truthfulness* or the *agreement* of their statements about any subject. Even should all who communicate, agree in denying that there is a spiritual world, or that any spirits exist at all, that denial would be no proof of such non-existence; on the contrary, it would be a very strong corroborative evidence in *favor* of spirit existence, for such testimony could not be supposed to originate in the minds of the mediums. *The testimony itself must come from mind*, and that mind must have existence. If not from the mind of the medium, or any one in the body acting through the medium, then it must be from a disembodied mind. The Cretans were once declared to be 'always liars;' and yet nobody doubts that the Cretans had existence, even though they themselves might affirm or deny the fact. *The proof of communication from the spirit-world depends on the evidence of mental action aside from and beyond that of the medium, or any mind in the flesh*, and not on the agreement, wisdom, or good sense manifested in such communications.

"But contradictions, even as to matters of fact, are often merely *apparent*, rather than *real*, arising from mutual misunderstandings as to the meaning of terms, and from too narrow and unphilosophical views of things.

"We would remind all who are perplexed with the statements of spirits in respect to the spirit-world, that it is doubtless vastly more extensive than earth, and hence may present a far greater variety of objective realities, and of modes of life and thought, than pertains to the earth-life. And, furthermore, since the spirit-world is the world of causes, *each external object must be to the beholder just what his perceptions make it;* that is, it appears *according to his power of insight as to its uses and relations.* Hence the same object may appear as one thing to one person, and as quite another thing to a person differently unfolded.

"This principle is exhibited to some extent in this rudimental sphere. For example, we have known two persons to attend the same concert of instrumental music, — one having little or no musical culture, the other possessing a very exquisite ear. To the first, some of the finest compositions were for the most part a mere jargon of inharmonious sounds which pained and tired the ear; while the other was by these same sounds transported to the seventh heaven of rapturous delight.

"So of objects seen: to the child or the uncultivated clown, that most gorgeous of spectacles, the evening sky, is a solid dome of comparatively limited dimensions, in which are hung up a multitude of little lamps for man's sole use; while the astronomer sees worlds on worlds filled with life and beauty, among which this earth is but a tiny speck floating in immensity."

Lord Lytton, whose abilities in many instances we have admired, has shown, in his novel of "A Strange Story," that he can both write a very stupid book, and venture to treat of things with which his acquaintance is superficial and inaccurate. He here gives no evidence that he has ever investigated the subject of the spiritual phenomena, ancient and modern, with any profundity of research or meditation. He so mixes up the crudest and most incongruous fancies of an imagination in search of the sensational with fragments of genuine truth, that his book has the effect on one of a wretched nightmare instead of a presentation of credible phenomena that can be reconciled with existing facts.

In a letter bearing date the latter part of the year 1867, Lord Lytton writes to Mr. Benjamin Coleman, "All the experiments I have witnessed, if severely probed, go against the notion that the phenomena are produced by the spirits of the dead; and I imagine that no man, who can take care of his pockets, would give up his property to a claimant, who could bear cross-examination as little as some alleged spirit, who declares he is your father or friend, and tells you where he died, and then proceeds to talk rubbish, of which he would have been incapable when he was alive. I can conceive no prospect of the future world more melancholy, than that in which Voltaires and Shakespeares are represented as having fallen into boobies, or, at best, of intellects below mediocrity."

See Kerner's answer, and the answers elsewhere in this volume, to obvious objections, like these, from persons who, like Lord Lytton, have gone but a little way in the path of investigation. All inquirers, like the friend who wrote the letter on the Rochester knockings (page 34 of this volume), have to pass through that phase of doubt at which the distinguished novelist seems to have arrived and stopped.

Because there may be mendacious, wanton, or frivolous spirits, who choose to assume great names, it does not follow that they represent the whole spirit-world, any more than Falstaff and Pecksniff represent all humanity.

Lord Lytton says that in all controversies on this question, he has found no clear definition of what is meant by *spirit*. And yet he talks very glibly of *matter*, and of agencies "operating upon or through matter," as if he well knew what *matter* is.

But we know just as much about spirit as we do about matter. It is true that we know nothing of the *essence* of spirit: it is equally true that we know nothing of the substance or essence of matter. But perhaps the reader will say, "We cannot *see* spirit, and, therefore, we know but little about it." "Did it ever occur to the reader," asks a scientific writer, "that we cannot see *matter* either? When we look at any object, it is not the object, after all, that we see, but merely the image of it formed on the retina of the eye. When I look at a house a mile distant, the object that I *really* see is not a mile distant, but within the eye. I do not see the house at all, but I see an image of light representing the house. Thus it appears that matter is just as invisible as spirit. We know some of the *properties* and *laws* of spirit, and this is precisely the extent of our knowledge of matter."

Newton was of opinion that if sufficient pressure were put upon the earth, it would be compressed to the size of a globe an inch in diameter. And if to that size, why not to that of a pea, and from that to a grain of mustard-seed, and from that to an invisible particle of dust. Newton virtually denied the existence of matter as substance. Nothing remains but a congeries of laws. "If the ultimate particles of matter are *mathematical points*,* as Newton assumed, it follows that if the particles of which the earth is composed were made to *touch* each other, the whole earth would be reduced to a mathematical point. And who can show that this hypothesis (that the laws of matter are, in fact, all there is of matter) is not scientifically correct?"

It would seem that Providence does not mean we shall be spared the trouble of thinking for ourselves. And so neither mortal nor spirit comes to us with the credentials of infallibility. "The commonplace character of a large portion of the spirit communications, the extravagant and turgid character of some, cease to perplex when we come to view them as proceeding from beings lately ordinary dwellers upon earth, and retaining still their earthly dispositions and ideas. True, the difficulty remains as to why some small portion at least of these communications should not bear the impress of transcendent wisdom and genius. The absence from them of any thing equalling, far less surpassing, the highest products of the human mind, argues, it must be admitted, some hinderance to intercourse with spiritual beings of an exalted order: may we not hope to overcome it?"†

In regard to the varied and contradictory character of the communications, "Honestas" remarks in "Human Nature," "As our childhood prepares us for maturer age, so our present life mediatorially renders us fit for the enjoyment of a future condition. But mediation implies that the characteristics of the former condition shall be preserved, and that they aid in bridging over the gulf that severs this life from the state hereafter. And with the preservation of our individuality, is it far-

* It was the conclusion of Faraday that matter, in its last analysis, is resolvable into *points of force*.

† Swedenborg tells us that the communications of angels and spirits are limited by the materials found in man's memory.

fetched to say, that the conditions which surround, sustain, and render its continuance possible, cannot be so world-wide different in the future life as to make intercommunication between the two states impossible; that the physical circumstances of spirits and of man have something akin, something in common, rendering superable that which was once believed insuperable? . . .

"Now, in the varied character of the communications, we have a standard given us to measure the actual amount of the change which that death, that transition, into a perhaps more subtle and elementary condition, effects. The change does not, however, carry with it complete severance; on the contrary, mediated by growth and development on earth, the soul is sustained by a condition of material laws, mediatorially rendered applicable by prior growth. In a word, the state hereafter cannot differ insuperably from that on earth.

"Spiritualism has often given offence because it has failed to satisfy the cravings of those who desire for perfection hereafter, — a perfection, it is unreasoning, illogical, to ask for. The varied character of the communications, so far from making me hesitate, strengthens my belief in the reality of Spiritualism; for it brings me back from an ideal to a reality; and in this reality I recognize the law of gradual step-by-step progress; no jump and bound into something uncongenial, but a progress into a mediatorially prepared and kindred state, in which the individuality of the soul is maintained. This individuality could not, however, be sustained, unless supported by the influence of great physical laws, which again co-operate and harmonize with those we recognize as operative in this, to us, natural world."

The "Banner of Light," the leading spiritual journal of America, introduces all its messages, purporting to come from the spirit-world, with these words of caution: "We ask the reader to receive no doctrine put forth by spirits in these columns that does not comport with his or her reason. All express as much of truth as they perceive, — no more."

Plutarch raises the same objections to the *style* of the oracles in his day that we raise to that of the spiritual communications of ours. "If the verses," asks one of his speakers, "are really bad, ought we to make Apollo their composer?" And the conclusion is, that "the first inspiration alone comes from him, which is, however, adapted to the nature of every prophetess."

An ingenious writer remarks, that to the reader familiar with spiritual phenomena, it is evident even from the sneering narrative of Gibbon, that the apostasy of Julian, and his intense enthusiasm in the cause of the fallen faith, was in truth due to communication with the invisible world. Spirits of departed pagans, still clinging to their earthly creed, seem to have impressed him powerfully, visiting him, and conversing with him in the forms of the Olympian gods. "We may learn," says Gibbon, "from his faithful friend the orator Libanius, that he lived in a perpetual intercourse with the gods and goddesses, that they descended upon earth to enjoy the conversation of their favorite hero, that they gently interrupted his slumbers by touching his hand or his hair, that they warned him of every impending danger, and conducted him by their infallible wisdom in every action of his life."

Still, notwithstanding the contradictory character of many of the communications purporting to come from spirits, it cannot be denied that among those that do not seem to be prompted by mere wantonness or an insane conceit, there is a wonderful agreement on certain important points. For example, this better class seem to be unanimous in rejecting the notions of the resurrection of the physical body; the eternity of hell torments; and the doctrine of vicarious atonement, in the sense in which the so-called evangelical theology teaches it.

They regard all punishment as remedial;* and the popular belief that the discipline of man and his moral responsibility cease at the period we call death, they denounce as a groundless assumption, in contradiction with the whole system on which

* See a communication in the "London Pall Mall Gazette," of Nov. 9, 1866, for much of this statement.

the moral and physical universe is conducted. Looking back myriads of ages, they bid us see always one slow, unbroken process of growth and development; and they point to the same in the history of man as a race, and of each man as an individual. They tell us there is no precedent in creation for any such dislocation of the action of organic law as would be involved in that sudden cessation of the operation of moral and intellectual discipline, which is popularly supposed to be the result of death; that to suppose that every person who has "faith" and some small "good works" is instantly elevated to an eternal happiness, is one of the most irrational of theories, as is likewise the doctrine that the good are subjected to a purifying process without any further moral responsibility; that is, without any further real moral discipline whatsoever.

If it be asked, How is this view to be applied to the millions in whom no moral discipline is commenced before death, this is the reply: We perceive that in the case of those in whom the moral discipline is really begun, and is carried on to the utmost perfection, a material portion of their existence is necessarily passed before the commencement of the discipline.

For years we all live a purely animal existence, and are apparently not a whit more like saints and sages in an embryonic stage than are the most degraded savages of Africa. How it is that an infancy and childhood of animalism and passion are an organic preparation for an intellectual and moral probation, we cannot tell; but there is the fact, not shocking, or distressing, or bewildering, because we are familiar with it, and we are cognizant of the subsequent development of the reasoning and moral character. Just such may be the whole terrestrial existence of the savages under the tropics, or in our own fields and cities. All analogy leads us to the supposition that it may be simply the infancy of an existence commenced here and developed hereafter. They live and die in ignorance of their nature and their coming destiny, like a babe that dies after a year of sickness and misery. This ignorance is, too, in harmony with that general law of ignorance slowly passing into knowledge, whose operation meets us wherever we turn our eyes. It is one of the great mysteries of our life. "If there is a God," we are tempted to ask, "why does he thus hide himself? And why cannot we speak with him as we speak with one another?" There is no answer. We do not know. But we do know that ignorance of things great and good and true is no proof that they do not exist. The ignorance of God in the savage and the pariah is no more a proof that he does not intend some day to make himself known to them, than their ignorance of the law of gravitation is a disproof of astronomical science.

Should this solution of the mystery of human existence be thought to indicate too near an approach to the ancient doctrine of pre-existence and metempsychosis, it can none the less be admitted as in harmony with much of the speculation that professes to come from spiritual sources.

As there are no acknowledged leaders of the present spiritual movement, it is of course impossible to lay down any statement of theological or religious doctrines in which Spiritualists agree. "Our cardinal rule of action," says Judge Edmonds, one of the best known of the American Spiritualists, "has been to build up no party, create no sect, cultivate no spirit of proselytism, make no parade of faith, but let it enter your soul and govern your life."

A convention of Spiritualists at Rochester, N.Y., September, 1868, adopted a series of resolutions which embody many of the conclusions at which a large number have arrived; but the convention wisely put forth these resolutions as presenting the opinions of *those persons only who voted in the affirmative.*

Spiritualism, they say, teaches, "That man has a spiritual nature as well as a corporeal; in other words, that the real man is a spirit, which spirit has an organized form, composed of sublimated material, with parts and organs corresponding to those of the corporeal body. That man, as a spirit, is immortal. Being found to survive that change called physical death, it may be reasonably supposed that he will survive all future vicissitudes. That there is a spiritual world, or state with its substantial realities, objective as well as subjective. That the process of physical death in no way essentially transforms the mental constitution or the moral character of those who experience it, else it would destroy their identity. That happiness or suffering in the spiritual state, as in this, depends not on arbitrary decree, or special provision, but on character, aspirations and degree of harmonization, or personal conformity to universal and divine law.

"Hence that the experience and attainments of this life lay the foundation on which the next commences. That since growth, in some degree, is the law of the human being in the present life; and since the process called death is, in fact, but a birth into another condition of life, retaining all the advantages gained in the experiences of this life, it may be inferred that growth, development, expansion, or progression is the endless destiny of the human spirit. That the spiritual world is not far off, but near around, or interblended with our present state of existence; and hence that we are constantly under the cognizance of spiritual beings. That as individuals are passing from the earthly to the spiritual state in all stages of mental and moral growth, that state includes all grades of character from the lowest to the highest.

"That, as heaven and hell, or happiness and misery, depend on internal states rather than external surroundings, there are as many gradations of each as there are shades of character,—each one gravitating to his own place by natural law of affinity. They may be divided into seven general degrees or spheres; but these must admit of indefinite diversifications, or 'many mansions,' corresponding to diversified individual character,—each individual being as happy as his character will allow him to be.

"That communications from the spirit-world, whether by mental impression, inspiration, or any other mode of transmission, are not necessarily infallible truth; but, on the contrary, partake unavoidably of the imperfections of the minds from which they emanate, and of the channels through which they come; and are, moreover, liable to misinterpretation by those to whom they are addressed. Hence, that no inspired communication, in this or any other age (whatever claims may be, or have been, set up as to its source), is authoritative any further than it expresses truth to the individual consciousness, — which last is the final standard to which all inspired or spiritual teachings must be brought for judgment. That inspiration, or the influx of ideas and promptings from the spiritual realm, is not a miracle of a past age, but a perpetual fact, the ceaseless method of the divine economy for human elevation. That all angelic and demoniac beings which have manifested themselves, or interposed in human affairs in the past, were simply disembodied human spirits, in different grades of advancement.

"That all authentic miracles (so called) in the past, such as the raising of the apparently dead; the healing of the sick, by laying on of hands or other simple means; unharmed contact with poisons; the movement of physical objects, without visible instrumentality, &c., — have been produced in harmony with universal laws, and hence may be repeated at any time under suitable conditions. That the causes of all phenomena, the sources of all life, intelligence, and love, are to be sought in the internal, the spiritual realm, not in the external or material. That the chain of causation leads inevitably upward or onward to an infinite spirit, who is not only a *forming principle* (wisdom), but an *affectional source* (love), — thus sustaining the dual, *parental* relations as father and mother to all finite intelligences, who, of course, are all brethren.

"That man, as the offspring of this infinite parent, is his highest representative on this plane of being, — the perfect man being the most complete embodiment of the 'Father's fulness' which we can contemplate; and that each man is, or has, by virtue of this parentage in his inmost, a germ of divine essence, which is ever prompting to the right, and which in time will free itself from all imperfections incident to the rudimental or earthly condition, and will triumph over all evil. That all evil is disharmony, greater or less with this inmost or divine principle; and hence whatever aids man to bring his more external nature into subjection to, and harmony with, his interiors, — whether it be called 'Christianity,' 'Spiritualism,' or the 'Harmonial Philosophy;' whether it recognize the 'Holy Ghost,' the Bible, or a present spiritual and celestial influx, — is a 'means of salvation' from evil."

Dr. Henry T. Childs, of Philadelphia, a medium, writes as follows: "Having seen and conversed with many spirits in different conditions, the facts are as clear to me as they can be, that the after-life and this are subject to progression, and that not from a pure stand-point of excellence in which there are no evil tendencies, but from whatever condition the spirit may be in. I have yet to find a spirit who does not feel that progression and growth are synonymous, and they ever mean a reaching forward to something better and leaving something that is *evil.* Death is nothing more than an incident in the continuous life-line of humanity, changing the surroundings, but leaving the interior just as it *was.*

"With this experience, I *know* there are *evil spirits;* spirits who, like men, may delight in mischief and perverseness; who have not realized their own rights sufficiently to respect the rights of others; and my reason teaches me just what these *facts* have demonstrated to me. Seeing evil all around us, I see also the beautiful spiral pathway of progress, which is ever leading us up out of these conditions."

The following lines by Daniel Norton embody, we believe, the views of great numbers of Spiritualists on the subject of the ministry of evil: —

"Sin leads to pain, and pain repentance brings:
Thus sin, though evil, is a savior;
For in its train comes knowledge of those things
To soul and body hurtful; and the stings
Of conscience bring us wisdom. Wisdom brings
The pledge of future good behavior.
Blessed be Darkness, then! It bringeth light
From out the darkness, brighter glowing.

Blessed be Evil; for it bringeth Right,
As day is more effulgent after night;
Blessed be Sorrow; it begets the might,
To set life's truer current flowing."

To this doctrine it is objected by the theologians, that, as it makes an entrance into evil necessary in order to serve as a self-conscious return to good, it exalts evil itself into goodness; the idea of sin and responsibility is destroyed, and views are introduced that would prove fatal to all true morality, as they would imply that no being can be properly educated, except through a process of sinning.

Perhaps this is an extreme construction to put upon the doctrine. If life be disciplinary, all our experiences must be designed for our information and our good. Then comes up the terrible problem of free agency; and from this, we shall see, as we proceed, many great and learned men can find no escape except in the hypothesis of a pre-existent state of the soul.

As a specimen of the expositions of one of the most eloquent of American mediums, Mr. Selden J. Finney, we give the following abridgment of one of his discourses. Of course it merely expresses the views of an individual in regard to the philosophy of modern Spiritualism; but we quote it as manifesting no ordinary degree of philosophical culture and insight. "The great, distinguishing feature of this philosophy," says Mr. Finney, "is that it begins with the demonstration of a transcendent spiritual nature, called the soul, within the body of man. This it defines as an organic, spiritual entity; and proves that it lives on, after the physical body is dead, in higher spheres, subject to the same laws of intellectual, social, and moral being that rule us here; though, in those higher spheres, it has been translated into more refined conditions and relations. . . .

"It demonstrates that all angels are planet-born men and women. It proves the unity of nature, and so shows that our hells are kindled here by our own hands in our own breasts.

It shows that when every physical sense is paralyzed, the mind and soul may be all the more untrammelled, — as in trance and clairvoyance, — and can soar afar into the deeps of external nature or hold blessed communion with spiritual intelligences. The wonders of clairvoyance, of trance-mediumship, of inspirational speaking, of table-moving, of impersonation, — in fact, of all the great classes of mediumship, — are the external proofs of the reality of our philosophy; while the vast revelations that constitute the contents of the best communications are the ideal elements. . . .

"Even a brute can be surprised by the movement of a table without contact of visible power; while under the inspiration of the gifted seer and poet, the great fields of eternal day break on our rapt vision. This philosophy opens, on the one hand, the grand questions of physiological psychology; and, on the other, the profound questions of transcendental theology. . . .

"In demonstrating the independent entity of the soul, which can, even while in the body, transcend the limits of sensation and hold converse with immortals, Spiritualism destroys the sensationalism of English philosophers, the subjective atheism of Spencer, and the materialism of the French encyclopedists; while, on the other hand, it corrects the too ideal tendencies of Hegelianism in Germany, and holds it to account on that middle ground of philosophy where sense and soul touch and unite.

"The idealism of Berkeley, which reduced all the external world to a mere phantasm of sensation; to a mere picture on the nerves of the body, whose cause was for ever shut away from our reach; and the pantheism of Spinoza, or more especially of his one-sided disciples, — here find their grave, in common with that subjective idealism of Spencer, Sir William Hamilton, and Mr. Mansel, which is of late so much in vogue. Sensationalism has a half-truth; idealism and pantheism have a half-truth; but so long as each claimed to be the only truth, all were false in a double sense, and blind.

"The truth in each of these schools is revived, emancipated, and united in the spiritual philosophy. Idealism would re-create the external world from the depths of unaided consciousness. Sensationalism would create consciousness from the external world as a mere material force, which goes out like any other fire in the ashes of its body. But Spiritualism, in demonstrating the dual nature of man, in showing that we live in two worlds at once, and are vitally related to each, having powers that lay hold on the forces and verities of both at once, unites in itself the truth of each, unmixed with the errors of either.

"Does Mr. Spencer tell us that spirit is 'utterly inscrutable'? The spiritual philosophy answers, 'Man is a spirit *per se*, and can cognize spiritual beings of the immortal life; has done so; has identified the persons of the departed; your theory must be false.' Does Mr. Mansel set 'limits to thought'? The spiritual philosophy pulls them down, and opens again the fair fields of spiritual naturalism to the contemplation of thinkers. Does Sir William Hamilton call the idea of God a 'revelation'? The spiritual philosophy answers, 'Yes;' but a 'revelation made through those natural powers and faculties of the soul, which connect us with the soul of the world, and which transcend the physical senses, as the immortal transcends the mortal life of man, and not by any means a supernatural revelation, made in a book.'

"The great contest in philosophy has been and is waged over 'method.' The sensational philosophy reasons only inductively; from external facts toward their causes. Idealism reasons only deductively from ideas which it finds in the reason, toward their effects. But neither method can *give* any facts or ideas to begin with. Both facts and ideas are assumed in the outset by both methods. Hence it is evident that neither method is alone, or often together, full and complete. How do we find the facts and ideas to start with, if, after all, we cannot *get* our facts by induction; for induction begins with facts as given, and cannot proceed one inch, except on the assumption of facts from which to reason and infer? Induction cannot set out from zero and reason to entity. It must begin with some previously known and acknowledged facts or principles. It cannot discover by induction the original facts from which induction can alone set out.

"So with deduction: it sets out with ideas which it finds in the mind. It cannot descend to effects from zero, any more than induction can rise from zero to causes. Neither can *originate* its facts or its principles. Both are dependent for their respective data on some power superior to either method of reasoning. These methods are both second-hand processes; neither is aboriginal, primary. Now, what is that power which gives us the facts on the one side, and the principles on the other from which to set out? Whatever it be, it is self-evidently superior to either induction or deduction; *for on its directly given data both methods proceed.* Both methods are then secondary; both are the mere mechanics of that power which gives the data to begin with. Hence reasoning is only that process by which things and principles are *accounted* for and related, but never authorized.

"There is, hence, the necessity for some power that is aboriginal, direct, authoritative, and supreme, implied by both methods of reasoning. This power must, therefore, be in direct contact with both the facts and the ideas with which these two methods begin, and on which they depend. This power can be nothing less than intuition. Intuition is the direct and immediate perception of facts on the one side, and of principles on the other. No reasoning can begin upon any other ground. The data of all reasoning is given at first hand in intuition alone. Hence intuition is the only power of discovery. When it reveals the external facts, it acts through the external senses; when it reveals ideas, principles, laws, it acts through the soul.

"And here comes to view the spiritual method of philosophy. It is direct, intuitive, aboriginal, authoritative, and supreme. All possible speculation rests at last on its revelations. I say 'revelations.' When we see the external forms of the outward world, a revelation is made; when we discover an idea, another revelation is made. 'Revelation' is the great aboriginal fact in all mentality. We no more *will* to see the world than we will to *be*. We do not come to know that we *are*, or that any thing else is by induction any more than we will to *be*, by induction. The consciousness of the existence of the me, and of the *not* me, is as direct a revelation as it is possible to conceive. These are the great aboriginal intuitions of all souls, and form the ground of all possible reasoning. Now, if it be possible to get the greater, it is possible to get the lesser facts of existence by such aboriginal intuition, — direct 'revelation.' Indeed, all the *contents* of existence are included in this primal intuition of existence itself. And if the *existence* itself can be thus given intuitively, directly, and with supreme authority, so can all the *contents* of existence be so given.

"Hence the spiritual method of philosophy. All perceptions by the senses are direct intuitions of all that sensation reveals or perceives. Sensation may be, and doubtless is, limited to the phenomenal alone; but if so, its intuition of phenomena is direct and authoritative. So spiritual intuition perceives directly and at first hand the eternal laws and ideas which rule the whole phenomenal empire of the world. Hence all reasoning is dependent on intuition as the great revelator of all things and principles. It is the supreme voice of the absolute in the soul of man; or, rather, it is the world, the universe, of both phenomena and power arisen into self-cognition. The consciousness of man is the self-cognition of the universe.

"Axioms of mathematics are self-revelations of eternal ideas, — 'self-evident truths.' They are eternal. Axioms are given as eternal, and as absolute. They admit of no contradiction, no limitation, and no suspension. They are absolute authority. Other axioms have the same character. Axioms are not inferences, not deductions. They do not depend upon logic; logic depends upon them. All reasoning *derives from*, not gives authority *to*, them. Hence these are intuitions of eternal principles. Now, if the greater can be given by intuition, so can the less. And hence the spiritual method opens anew the royal road to knowledge. Clairvoyance is a practical proof of the feasibility and utility of the intuitive method. If the uneducated shoemaker's apprentice, blindfolded and paralyzed, can, through supersenuous channels, inact the great facts of science (as has been proved and tested in this country often), then we have a practical and experimental proof and exhibition of the reality and truth of the spiritual method of philosophy.

"Mere metaphysical argument alone is inadequate to reach the masses. But when to spiritual metaphysics we add the experimental illustration of the transcendent nature and relations of the soul, we secure both sides of the required demonstration. And when on the top of all this, we place the wonderful facts of spiritual intercourse, our philosophy becomes irresistibly demonstrative. It recognizes the intuitive method as authority in *revelation*, and the inductive and deductive methods as the two wings of *demonstration*. The first reveals ideas and facts,— the original *data* of all philosophy. The last two show the logic and relations of those data.

"Hence the completeness of the spiritual philosophy. Does sensationalism ask for 'facts'? The experimental branch of our philosophy gives them in abundance. Does idealism demand ideas and deductions? The ideal side gives them at first hand. Does pantheism demand recognition of the Infinite Presence and Power? Intuition gives us the direct revelation thereof in the very substance of the soul and its relations.

"It is vain for Mr. Spencer, Mr. Mansel, and others, to deny to us any absolute knowledge, or any knowledge of the absolute. The 'absolute' of Spencer, Mansel, and others, is nonentity defined as Being. This is evident from Mr. Spencer's summary of the argument for the 'relativity of all knowledge.' He says, 'We have seen how, from the very necessity of thinking in relations, it follows that the relative itself is inconceivable except as related to a real non-relative.' We reply, A '*non-relative*' related to the '*relative*,' is a contradiction in terms, and an impossible conception. Mr. Spencer's 'non-relative,' is used to mean the 'absolute,' 'the infinite, — the *real reality underlying all appearances.*' And yet it is said to be out of all relations, — 'non-relative.' And yet the relative itself is conceived as dependent for its conception on its relations to this 'non-relative.'

"If this is not self-contradiction with a vengeance, what can be? Mr. Spencer's 'non-relative' is nonentity defined as the 'absolute,' 'the infinite,' — a 'real reality underlying all appearances.' Can the 'infinite,' 'real reality,' be destitute of all relations? It is absurd. The very argument for the 'relativity of all knowledge,' destroys itself; for the very idea 'relative,' is acknowledged to be dependent on its relation to the 'absolute.' The characteristics of Mr. Spencer's 'non-relative' are those of zero. The 'infinite' of Nature and of the soul are not identical with this 'absolute' of Spencer. He is therefore wrong. An 'infinite reality underlying all' things must be the aboriginal *esse* of the entire universe, the one indivisible substance and power of all forms and all force. Hence it is in contact with the soul, with the mind. Nay: it is the substance of both body and soul.

"And who shall then attempt to set limits to our knowledge? No man can do it, until he can comprehend the infinite possibilities of eternal progress; until he can take the latitude and longitude of all possible truth; until he can measure all the possible developments of immortal ages; until he can rise out of his own limitations to a realm where he can embrace and outline the whole future career of the immortal intellect of man. And this is self-evidently impossible.

"The very ground on which Mr. Spencer plants himself to prove the 'relativity of all knowledge,' is, by his own claims, and in his own words, '*the ever-present sense of real existence.*' He confounds the idea of some knowledge of the 'infinite,' with infinite knowledge. His whole system is that of subjective atheism; or, if you choose, of objective idealism. He plants us in an ontological vacuum between the objective world and the 'absolute' Nature; and after granting the clear conception of the one, and the 'ever-present sense' of the other, denies us any absolute knowledge of either.

"He attempts, it is true, to save religion; but he saves it to us as the pursuit of an 'utterly inscrutable power,' of whose nature and character, whether divine or devilish, we can never have any knowledge whatsoever. And yet he bids us worship this 'utterly unknowable power.' What is that religion good for that bids us worship 'we know not what'? It may be deity, it may be devil. And are we to be told that, though religion can never rise to the idea of divinity, can never know there is a God, — in other words, can never have a philosophy of religion, — we must still push on after both deity and a religious philosophy? Is this the way religion, the grandest pursuit of man, is to be saved to the nineteenth century? What is this but atheism under another name? What is the difference to me, whether it be proved that I can never *know* God, or that there *is* no knowable God? Is it not all one as to worship? Can we be rationally called upon to worship utter inscrutability under pretence that it may be divine for aught we know? To such absurdities has modern sensationalism and inductive philosophy driven itself.

"But Spiritualism relegates man to the aboriginal sources of all inspiration and all revelation. It plants itself on the demonstration of the spiritual entity and supersensuous relations of the soul. It illustrates its philosophy in its experiments. It rises inductively from this demonstration to the divine idea, — to God; or, starting with this divine idea, reasons deductively down to the idea of the soul and its immortality. Starting with the fact that man is a spirit *per se*, it rises to the inference that all aboriginal substance may be spirit, *per se*. Or, starting with the idea of God as infinite spirit, shows that there is no room for 'matter' as aboriginal substance in the universe. If one admits the idea of infinite spirit, — God, — he cannot escape the great spiritual idea that there is but one substance in the universe; viz., Spirit. If one start with the idea of the spiritual entity of the soul, he lands in the same conclusion. Both paths lead to the same great idea. And when we perceive the unity of nature; when we regard the mutual transformability of bodies, and of all forces; when we discover in the analyzed sunbeam and starbeam the elements which have been precipitated and hardened into rocks and coal and iron and other metals; when we behold everywhere the reign of the same invisible power, ever changing in form, but ever the same in *esse*, — the soul is carried, as on the tide of inspiration, up to the same great idea that spirit 'is all, and in all.'

"Our philosophy shows that man is made of the same stuff as the universe is. Hence, his fraternity with all things. For how could man receive life, power, substance, light, heat, gravitation, electricity, beauty, and wisdom, if he were not composed at bottom of substance and power and law, one and identical with these? All substance and power is one, or no universe could arise out of them. Hence man is the autocrat of creation. He carries sheathed within his flesh the potent secrets of all things.

"And here, it will be seen, is a religious philosophy, which carries with it all the causes of ultimate success. In its view, all creation is trembling with the tides of divine life. Hence its high estimate of true science. Can science discover a truth our philosophy will not consecrate and use? No; for science is only the study of modes and symbols of divine life and action. Spiritualism is the only religion on earth, that can 'have science for symbol and illustration.' It is the mathematics and ethics of eternal law. It is true it makes religion natural; but then it makes nature spiritual and divine. It does not degrade God to 'matter;' it elevates 'matter' to spirit. It does not reduce religion to 'material' science; it elevates science to the divine business of justifying, explaining, and demonstrating religion. . . . It is spiritual power alone that thus renews the world. The meaning of spiritual is real, in our philosophy.

"Hence the spiritual idea of man: man is nature, physical and spiritual, essential and phenomenal, gone up into organic, self-conscious moral unity and volition. He has a sense for each external phenomenon, and a spiritual faculty for all eternal verities. It was spiritual inspiration which moved the poet to write, —

> 'Even here I feel
> Among these mighty things, that as I am,
> I am akin to God; that I am part
> Of the use universal, and can grasp
> Some portion of that reason in the which
> The whole is ruled and founded; that I have
> A spirit nobler in its cause and end,
> Lovelier in order, greater in its powers,
> Than all these bright and swift immensities!'

"As the solid earth is but precipitated sunbeams, so the nature of man is organized spirit. The body is but the secreted shell of the soul. . . . A day will come to every soul, when into the channels of its purified being will pour the love, the truth, and the beauty of the world. To be passive to the spirit of nature, is the secret of genius, and the path of salvation. Thus does the spiritual philosophy revive the hopes, and strengthen the soul of man."

We translate the following from the "Livre des Esprits" of Allan Kardec: —

"The morality of the superior order of spirits is substantially that of Christ in this evangelical maxim: Do unto others as you would have them do unto you; that is to say, do good, and not evil. In this principle, man may find the universal rule of conduct for his slightest acts.

"They teach us that egoism, pride, sensuality, are passions which bring us nearer to the animal nature in attaching us to matter; that he who here below detaches himself from matter by contempt of merely worldly futilities, and by love of the neighbor, draws nearer to a spiritual nature; that each one of us ought to render himself useful according to the faculties and the means which God has put into our hands for proving us; that the strong and the influential owe support and protection to the feeble, for he who abuses his power to oppress his fellow-creatures, violates the law of God.

"They teach, finally, that in the world of spirits, as nothing can be concealed, the hypocrite will be unmasked, and all his turpitudes exposed; that to the state of inferiority and of superiority of spirits are attached pains and enjoyments which are unknown upon earth. But they also teach that there are no faults that are irremissible, and that cannot be effaced by expiation. Man finds the means of remission and improvement in the different existences that afford him an opportunity to advance according to his desire and his efforts, in the way of progress, and towards the perfection which is his ever-receding goal."

In an able series of essays, entitled "What is Religion?" Mr. Shorter shows the fallacy of the notion that Spiritualism is a new religion; or, indeed, a religion at all. He shows that religion is not the mere acceptance of other people's beliefs; that belief, simply as such, and separate from the moral element of faith, or, in other words, mere opinion about religion, no more makes a man religious, than his opinion about shoemaking makes him a shoemaker; that history is not religion; that literature is not religion; and that morality simply is not religion; for though religion comprehends morality, as the larger comprehends the less, yet morality may be practised from such motives as not to include religion.

But, contends Mr. Shorter, if Spiritualism be not religion, it leads up thereto; it evidences, illustrates, confirms, enforces it; and gives certainty to what in many minds had become doubtful. It brings heaven and hell sensibly nearer and more real to us as states of being, the necessary consequence of what we have been and are, and so opens out to us broader, grander, nobler views of man's nature and destiny than is possible to those to whom nature and the present life are all; or, than is common when religion consists mainly in the acceptance of tradition and dogma, which are held but as the accident of education and geographical position.

It shows men, to use the words of Henry More, the Platonist (1659), that "no other Nemesis should follow them than what they themselves lay the train of."

While Spiritualism corroborates and elucidates the genuine

truths of religion, it also exposes and corrects many of the delusions and mistakes into which men have blundered in their speculations on matters associated with it. Mr. Shorter instances, as an illustration, the old controversy on which theologians are still divided, and, so long as they move only in the old ruts, are likely to remain so, — the question, whether at death the soul retains its active, conscious powers, and at once enters on its new life; or, whether it only wakes to consciousness to be re-united to its resuscitated body at some period unknown, when the great assize of all humanity is to be held, and the affairs of the world finally wound up.

Now it needs no argument to show, that, if there be any truth in Spiritualism, there can be none in the latter of these two views. If the departed still perceive, remember, think, love, suffer, and enjoy, and communicate with us, it must be evident that they are neither in their graves, nor are they like an antediluvian toad imbedded in a coal seam, in a state of torpor or suspended animation; but that, on the contrary, they are in the present plenitude of their life, with all that appertains to it.

Religion, then, is something to be experienced and lived: it is not now to be discovered or invented. If modern Spiritualism dates from 1847, and constitutes a new religion, *wherein* is it *new?* What is there in religion since 1847 that there was not in it in 1846? The immortality of the soul; the existence of a spirit-world; the manifestations and ministry of spirits, and communion with them; the assurance that Divine mercy and spiritual progression are not limited to the natural world and the present life; that the future retribution is not arbitrary, penal, and vindictive, but the inevitable consequence of the acts here done and the character here formed, — these are all ideas of the old world and of the old faith.

"Our place and state, our condition and surroundings in the spirit-world are determined by a law of moral gravitation, — the attraction of spiritual affinity. Prudence, even at its best, is not religion. To set down that feeling as religious, which springs from a fear of hell, or a desire to secure the good things of heaven, is to degrade the ministry of religion. To inspire men with dread,—to supplement the jail and the gallows,—this, indeed, were hangman's work for religion; and scarce less degrading were it to religion to employ her to coax men (as children are coaxed with sugar-plums) by the promise, that if they will but be good, they shall certainly hereafter be made very comfortable and be well paid for it.

"It is a terrible and mischievous burlesque of religion that would thus make it the minister to human selfishness, provided only that it be a little more subtle, enlightened, and far-sighted than ordinary, and coated with a thin varnish of sentiment. The aim of religion is not to cultivate selfishness of any kind, not to disguise it under fine names and fair pretences; but to deliver men from selfishness of every sort and degree, here and everywhere, now and at all times, in time and in eternity. Especially is this so of the religion of Christ."

By self-denial, by unceasing combat against evil, by prayer for strength to do and to suffer, we prepare ourselves for heaven, — not to sit down there in idle beatitudes, but to carry out more fully the ends of our being in works of good uses towards all creation. We go there to work, not to sit supinely and enjoy a flow of pleasure. And the cost we are paying here daily is to fit us for the work.

"We are immortal," says Andrew Jackson Davis, "because, 1. *Nature was made to develop the human body;* 2. *The human body was made to develop the human spirit;* and, 3. *Every spirit is developed and organized sufficiently unlike any other spirit, or substance in the universe, to maintain its individuality throughout eternal spheres.*

"Each human spirit *possesses within itself an eternal affinity of parts and powers;* which affinity there exists nothing sufficiently superior in power and attraction to disturb, disorganize, and annihilate.

"Death is but the local or final development of a succession of specific changes in the corporeal organism of man. As the death of the *germ* is necessary to the birth or development of the *flower*, so is the *death* of man's physical body an indispensable precedent and indication of his spiritual *birth* or resurrection. That semi-unconscious slumber into which the soul and body mutually and irresistibly glide, when darkness pervades the earth, is typical of death. Sleep is but death undeveloped; or, in other words, sleep is the incipient manifestation of that thorough and delightful *change*, which is the glorious result of our present rudimental existence. Night and sleep correspond to physical death; but the brilliant day and human wakefulness correspond to spiritual birth and individual elevation.

"There is every reason why man should rest, with regard to life and death, and be happy; for the laws of nature are unchangeable and complete in their operations. If we understand these laws, and obey them on the earth, it is positively certain that our *passage* from this sphere, and our *emergement* into the spirit-country, will be like rolling into the blissful depths of natural sleep, and awakening from it, to gaze upon, and to dwell in, a more congenial and harmonious world."

"As to the immortality of the soul," writes Bianca Mojon (who died in Paris in 1849), "I maintain that we all feel it independently of revelation. It is not a mathematical certainty, (?) — that does not belong to moral questions; but it is precisely a moral certainty. Without this belief, there is neither religion, charity, nor possible virtue. Not that I believe those who deny immortality to be incapable of virtue; but I maintain that they are actuated by a confused feeling of immortality, which, in spite of every thing, works in them. Their opinion is but a negative doubt, and the want of an intellectual sense. I conjure you, then, with tears in my eyes, not to withhold this support and consolation from your children. Do not throw them into the void and desolation of metaphysical doubt."

The credibility of a future state of existence is fully sufficient to become a practical principle, however low the evidence may appear; for, at the very lowest, *we cannot prove the negative*. Death, even to our senses, is not an annihilation, but only a new combination of matter.

At a convention of Spiritualists in Cleveland in 1867, Mr. Burtis, of Rochester, an aged man, is reported to have said, "I am hardly nineteen years old. It is about that time since these tiny raps came to my house, and awakened me to a consciousness not only of the life beyond, but of this life also. I had been here many years, but it was only from that time I began to live."

How well does this saying of the old man confirm these words of Jean Paul Richter: "A man may, for twenty years, believe the immortality of the soul; in the one-and-twentieth, in some great moment, he for the first time discovers with amazement the rich meaning of this belief, the warmth of this naphtha-well!"

"Christianity, when rightly understood," says Mr. H. J. Slack, "presents itself as the synthesis of all that heathen times endeavored to reach. It is a purification and completion of the wisdom of the past, not an antagonism, as some would teach."

In the narrative by Plato of the last days of Socrates, his friend Crito is represented as asking him the question, "How and where shall we bury you?" Socrates rebukes the phrase instantly: "Bury me," he answers, "in any way you please, if you can catch me to bury. . . . Say rather, Crito, say, if you love me, where shall you bury my body; and I will answer you, Bury it in any manner and in any place you please."

But how can a soul issue from the lifeless body without our seeing it? asks the skeptic. To which it may be replied, that to appeal to sense to prove the non-agency, and therefore the non-existence of an object not perceptible by sense, is hardly sound logic.

"When the materialist argues," says Sir Benjamin Brodie, "that we know nothing of mind except as being dependent on material organization, I turn his argument against him, and say that the existence of my own mind is the only thing of which I have any positive and actual knowledge."

The Catholic Church admits the facts of Spiritualism. Recently in New York, Father Hecker, an eminent Catholic divine, declared that "with the truth underlying Spiritualism there is no issue, so far as the Catholic Church is concerned. It has ever been a household affair in the Church." As to the *character* of the manifestations, he believed they were demoniacal rather than angelical. "The Church," he says, "has an order of exorcists to combat the demoniac influence, and provides for the use of the exorcistic ritual whenever the signs establishing demoniac 'possession' are clearly proved; and, singular to say, those signs are the identical ones now used by Spiritualists to prove their doctrine; namely, speaking in tongues with which the medium is not, in the natural state, conversant; disclosing a knowledge of things hidden; showing strength above that appertaining to the years and constitution of the medium. Spiritism, doubtless, has intercourse with the other world, but not with the heavenly portion of it.

"No one who reads can doubt that there has ever been an intercourse between the human race and the spirits of the other world. This is the most deep, the most mysterious instinct of the human soul. And there is nothing connected with this that shocks us. Shakespeare, the great poet of the heart, introduces the ghost in 'Hamlet' in order to corroborate, as it were, the theory that the spirits of the departed are our familiars still. Socrates believed that he saw and conversed with his familiar spirit. So strong is the belief on these points, that the great Dr. Johnson avowed that he would not maintain that the dead were seen no more, against the concurrent testimony of the world."

The theory that the modern spiritual phenomena proceed, without exception, from fallen spirits or devils, has been urged by several learned writers, both Catholic and Protestant. The "Dublin Review," the great organ of Catholic theology in Ireland, has an important article in its issue for October, 1867, in which it admits the phenomena, but pronounces them infernal. It quotes from a work by Father Perrone, published at Ratisbon in 1866, in which that learned ecclesiastic gives a selection from the names of several eminent persons, lay and clerical; among the latter, Lacordaire, Sibour (the late archbishop of Paris), the late Cardinal Gousset, and others, on whose minds a full conviction of the genuineness of the spiritual phenomena had been wrought. The writer in the "Dublin Review" remarks as follows in regard to these phenomena: —

"Among men of keen and cultivated minds they were at first received, not only with disbelief, but with laughter and derision: they were rejected as untrue, not because not proven, but because incapable of proof, because they were impossible; and, indeed, impossible they are, as we shall see, to mere human power and skill. Among the characteristics of the world in modern times, a tendency to believe in the preternatural most certainly can *not* be reckoned. The phenomena of magnetism and spiritism at least *appear* preternatural. The predisposition was dead against accepting them: it was predicted that, before the generation that witnessed their rise had died out, they would have disappeared and been forgotten. Well, years have rolled on; and men who formerly would not without impatience read or listen to the accounts of these phenomena (the present writer was one of these), had at length been led to examine what was making such a noise in the world, and from mature, and for a time prejudiced, examination, have been led to conviction. In this way have been brought round several of the ablest and most learned men in Europe, Catholic theologians, physicians, and philosophers; and others, Catholic, Protestant, and free-thinking. Authority does not necessarily, nor even generally, prove an opinion; in a matter of mere opinion, the most inquiring and cautious men may be greatly deceived, and have been so deceived. But here there is question of facts and of the testimony of the senses; of facts sensible to the sight, the hearing, the touch; of facts and testimonies repeated over and over again, beyond the possibility of calculation, in the greater part of Europe and America, and recorded year after year down to the present day. It is quite impossible that about such facts such a cloud of such witnesses should be all deceived."

In regard to the question, By what agency are these phenomena produced? our reviewer condenses very closely the author whom he follows. The various hypotheses put forward are examined *seriatim*, until certain conclusions (given in the form of propositions) are reached. "His first proposition is, that though some of the *physiological* phenomena of animal magnetism, somnambulism and Spiritualism, viewed in themselves and apart from accompanying adjuncts, may be ascribed to material natural causes, most of them, or the whole taken in the aggregate, can by no means be referred to such a source;" while to refer the *psychological* phenomena to unknown laws of nature, as some do, "is extremely unphilosophical and absurd; for they contradict laws of nature that are certain and universally known. For example, it is a law of our nature that we cannot read with our eyes closed and bandaged, that we cannot speak a language we never learned," &c.

The closing propositions affirm that good angels cannot be the cause of these phenomena; of which no other cause can be admitted, save bad angels or devils. On these propositions, Spiritualists, according to their peculiar experiences, are likely to join issue, and to contend that the existence of bad spirits, able to manifest their power on this material plane, implies the existence of good.

But it is something to have proved a spirit, whether good or bad. If the modern phenomena have done this, they have done what many minds will accept as the evangel that will lift them from utter darkness into the light of immortality. Spiritualism will not be disturbed by this cry of diabolical agency. It was raised centuries ago against one of whom it was said, he "spake as never man spake." To the learned Sadducees of the nineteenth century, it will be a great step out of the fog if they can be made to believe even in a devil.

"If," says an eloquent writer, "we can make no other use of these lower spirits, — these stragglers on the outer boundaries of the spirit-world, — we may at least accept them as adventurous travellers seeking a new world; accept the floating weeds on the heaving waves — *as signs that land is near*. But how any one, with God's spirit to whisper to him, and nature to smile upon him, and angels in the flesh to love him, and the Bible open before him, can talk in this way, *and seek to frighten us from the bright path now opening before us, with the smell of fire and brimstone*, and horrible phantasmagoria of nothing but evil, I cannot tell. For myself, I am resolved to go on; for, at present, I have seen nothing of all this. The fiends have not mocked me, but the angels have whispered to me; and if I am told that they are only the children of falsehood in disguise, still I will go on. Surely, I shall come up with the outposts of the Great King before long; for surely God and the angels are not altogether banished from a world where, I am told, the spirits of evil are allowed to lurk for prey!"

Cudworth, one of the noblest of the English Platonists and Spiritualists, writing nearly two centuries ago, dismisses, with something like scorn, the notion that evil powers can ever establish any evil creed; for, whatever is evil or immoral, is in itself a standing proof against itself, though it came with all the power of miracles. "Though all miracles promiscuously," he says, "do not immediately prove the existence of a God, nor confirm a prophet, nor whatever doctrine, yet they do all of them evince that there is a rank of invisible, understanding beings, superior to men, which the atheist cannot deny."

We have already seen that the followers of Swedenborg give a bad name to modern Spiritualism. This is sometimes instanced in opposition, inasmuch as Spiritualists do not deny the limited seership of Swedenborg. But since the majority of skeptics in regard to a future existence would be relieved of their unhappy doubts, if it could be proved to them that there is any thing like spiritual agency, good or bad, going on in the world, would it not be more generous in our Swedenborgian friends to brave the perils of an investigation, and do what they can to place these phenomena, significant of spirit, on an impregnable scientific basis? For our own part, we are quite willing to run all the risk which those who cry "*Pythonism!*" and "*Diablerie!*" brandish to deter us, if we can be the means of conveying light and hope to one poor human mind, groping amid the mists of unbelief on the great question of the ages, — Does the conscious individualism of man terminate with the phenomenon called *death?*

"The relation of Swedenborgianism to Spiritualism," says William White, "is a story for a humorist. Years ago, when familiarity with spirits was rare, Swedenborgians used to snap up and treasure every scrap of supernatural intelligence. The grand common objection to Swedenborg was his asserted acquaintance with angels and devils: it seemed an insuperable obstacle to faith. Many of the early Swedenborgians had wonderful private experiences to relate. Spirits rapped in Noble's study. Clowes professed himself an amanuensis of angels. Swedenborgians, it might be supposed, were ready to run wild after spirit manifestations. But it so happened that clairvoyants and mediums, while they confirmed *in general* Swedenborg's other-world revelations, *contradicted him in many particulars.* This was intolerable, — contradict our heavenly messenger! At once the old line of argument was abandoned. Nothing was now wickeder than converse with spirits. Intercourse with them is dangerous, disorderly, and forbidden by the Word! True, Swedenborg did talk with spirits, but he held a special license from the Lord; he warned us of its perils; and his example is no pretext for all, and sundry. . . . In return, the Spiritualists rank Swedenborg among their chief mediums, and question and adopt his testimony *at discretion;* but this liberal indifference only adds fire to the jealousy of the Swedenborgians, and fiercer and thicker fall their blows."

With regard to this alleged danger of spiritual intercourse, so much insisted on by the followers of Swedenborg, it has been well replied that, were the danger to the full as great as represented, the objection would still be insufficient. We have only to take up a newspaper to be convinced that it is very dangerous to hold intercourse with men in the natural world; that there are here plenty of spirits who lie, and cheat, and rob, and murder. Even in "respectable society," in "the Church," and among its ministers, there are many who pretend to be what they are not; with whom, for instance, charity is often on the lip while bitterness is in the heart. Are we, therefore, to abandon society, to abandon religion, to shut out all human intercourse? God forbid! The prosecution of natural science is, we know, attended with danger, sometimes with destruction: are we therefore to abandon it? Is the knowledge of spiritual things less important, less noble than of material things? And is the fear of danger the most noble and heroic virtue that Christianity has enshrined?

If, as Swedenborg says, spirits associate only with their like, then to false and malignant men alone is there danger from false and malignant spirits: those who earnestly seek truth and goodness do not incur it.

The literal sense of the teachings of the Old Testament is generally rejected by the Swedenborgians; but it is now thought convenient to adhere to it, so far as it prohibits spiritual intercourse. That must be permitted to Swedenborg only. But will it be contended that all which was prohibited to Jews under the old dispensation, is prohibited to Christians under the new?

There have been many seers who, like Swedenborg, Thomas L. Harris, and others, have claimed infallibility; but in reverent minds this very claim must be conclusive against them. Swedenborg, it is well known, relates that, while he sat eating in a tavern in London, the Infinite One appeared to him in the form of a man, and talked quite familiarly with him, upbraiding him for eating so much, &c.

Probably no medium, while subjected to the limitations of our earthly condition, can be implicitly trusted in what he may affirm as to the *identity* of a spirit. "It must not be supposed," says Delachambre, author of a "System of the Soul," published in 1665, "that the form of the soul and of angels is fixed and determinate, like that of solid bodies: it is vague and changeable like that of the air and the liquids, which assume the form of all the solid bodies surrounding them: and the difference is this that the vivacity of the forms that upervene to the latter is of necessity, and that which is found in spiritual substances depends on their will; for, as they move as they please all their parts, they also assume whatever form they desire."

If spirits have this plastic power (and all the modern phenomena go to prove it), their capacities of deception as to identity may be far greater than we imagine. Perhaps our own spiritual insight, purity, and elevation must be the measure of our ability to detect spiritual impostors. Perhaps Supreme Wisdom intentionally keeps us unrelieved from the necessity of exerting our own faculties for the prosecution of truth. "If God," says Lessing, "should hold all truth inclosed in his right hand, and in his left only the ever-active impulse to the pursuit of truth, although with the condition that I should always and for ever err, and should say to me, Choose! I should fall with submission upon his left hand, and say, 'Father, give! Pure truth is for thee alone!'"

A noble saying; offspring, we believe, of a profound insight into spiritual laws; signifying that our own individuality and the great ends of our being are best promoted by that discipline which compels us to think for ourselves, do for ourselves, and seek light for ourselves; seek it not only from the exercise of our meditative powers, but from communion with all good influences and spirits and men. But if we think to find spirits who will relieve us of the trouble of exercising our own mental and moral faculties, we must not complain should we become the dupes of such as are unscrupulous, false, or fanciful. Good spirits, we may be sure, will not try to violate the laws of our being by making us mere passive instruments under their control, thus taking from us all spiritual dignity and freedom.

CHAPTER XII.

SPIRITISM. — PRE-EXISTENCE. — METEMPSYCHOSIS.

"Moreover, something is or seems,
That teaches me with mystic gleams,
Like glimpses of forgotten dreams,
Of something felt, like something here;
Of something done, I know not where, —
Such as no language may declare." — *Tennyson.*

WE have said that the modern spiritual movement is leaderless. This fact is one of its remarkable features. All attempts to identify any one man, or set of men, with any exposition of its principles, claiming to be authoritative, and accepted as such, have proved utter failures. Even organizations and conventions seem, by the necessity of their nature, too narrow for its ever-widening circles. The unprecedented progress of Spiritualism, numbering as it does its recipients by millions in the United States alone, is a success that is to no extent due to the concentration and exclusiveness of a creed, or to the machinery of a sect. It has baffled all the calculations of those who believe there can be no efficient propagandism without a creed and an organization.

All the arguments that could disaffect the worldly and alarm the timid, have been freely used to check its advance. With some few honorable exceptions, the learned and the influential have not only stood aloof, but have denounced it either as an imposture, or as a diabolical snare. Many persons, calling themselves Spiritualists or mediums, have done all they could to give it a bad name by laying at its door their own offences against good morals, good English, or good taste. It has been stabbed in the house of its friends, as well as mobbed and maligned by its enemies; and yet never did its great truths shine for so many eyes with an immortal lustre, as at this day. Never was it so secure in the triumphs of the future; and never did it see Science herself so thoroughly rebuked and confounded by the overwhelming testimony by which it is upheld above the sneers of the unthinking, and the arrogant hostility of those who decide without examining, and make their own experience the measure of God's truth.

In France, when the news of the Rochester knockings began to make a noise, M. Latour, editor of the "Medical Union," who had experimented sufficiently to satisfy himself that they were not all fraud or delusion, wrote (May, 1853) as follows: "What, then, is this phenomenon? Grand Dieu! I cannot hazard the least opinion on the subject. But this I will say, that science should seriously seek to produce these singular facts, to study and determine their laws, and divine their nature, if it is possible. What is there hidden in the discovery of these phenomena, and what is reserved for the future? *If the scientific world neglect and disdainfully deny them without experimenting, they will fall into unworthy hands*, and be obscured by exaggeration and enthusiasm, and serve for the propagation of mystical practices: they will serve to nourish folly and credulity and charlatanry; but if the *savans* wish it, they will perhaps find the germ of some great discovery. To animate inert bodies, and make them obey the will, is, no doubt, repugnant to human reason. Two thousand years ago it was observed by hazard that a piece of amber, on being rubbed, had the property of attracting light bodies. This phenomenon, passed over by the science of the ancients, has now become the pivot on which turns modern science. Can human reason fully explain why the needle always turns towards the north? Does it know the intimate nature of magnetism, electricity, of caloric, of light? And, while regarding this table on which I write, I cannot help crying, as Galileo of old did of the universe, '*And yet you turn.*'"

The warning of this man of science to his brethren does not seem to have been heeded; and much that he predicted has undoubtedly come to pass. Spiritualism has been tended and nursed by the lowly and unpretending; and the stubborn facts that have been elicited are not now to be set aside.

The man who has perhaps approached nearest to the character of a leader of the spiritual movement in France, is M. Hippolyte-Léon-Denizard Rivail, who, under the pseudonym of Allan Kardec, has had remarkable success, by his writings and teachings, in making a belief in the spiritual solution of the modern phenomena carry with it the Platonic doctrine, somewhat modified, of pre-existence and re-incarnation. Before Kardec, this ancient doctrine had been advocated by several of the French philosophers and mystics; by St. Simon, Prosper Enfantin, St. Martin, Fourier, Pierre Leroux; and, lastly, by Jean Reynaud, the literary associate of Leroux, and author of a remarkable work entitled "Terre et Ciel," of which we shall have more to say.

Kardec was born in Lyons, in 1804. A pupil of Pestalozzi, he became one of the propagators of the educational system of that distinguished reformer. Born of Catholic parents, Kardec gave much thought from an early age to the subject of religious reform. In 1850, he began the examination of the American spiritual phenomena. Becoming convinced of their genuineness, he applied himself to the deduction of their philosophical consequences. In this task, he employed both the inductive and the deductive processes; investigating phenomena, interrogating the supposed spirits through a great variety of mediums, and then examining the results by the light of his own reason and spiritual intuitions. The information thus procured he has condensed and arranged methodically, adding his own remarks and explanations in a manner to distinguish them from the rest of the text.

To his system he gives the name of *Spiritism*. "The words *Spiritualism*, *Spiritualist*," he says, "have a well-defined acceptation: to give them a new one by applying them to the doctrine of spirits, would be to multiply the causes, already so numerous, of amphibology. Properly speaking, Spiritualism is the opposite of materialism: whoever believes he has within him something distinguished from matter, is a Spiritualist; but it may not follow that he believes in the existence of spirits, or in their communications with the visible world. To designate this latter belief, we employ, in place of the words *Spiritualism*, *Spiritualist*, the words *Spiritism*, *Spiritist*."

In 1858, Kardec established "La Revue Spirite," a monthly magazine, which is still published. He is the author of several volumes, in which his peculiar doctrines are set forth in a remarkably clear, matter-of-fact style, methodical and precise, free from all mysticism and prolixity. One great cause of his success is perhaps his lucid, intelligible way of treating the profoundest questions relating to the dual nature of man.

According to the doctrine* of Spiritism, the soul is the intelligent principle which animates human beings, and gives them thought, will, and liberty of action. It is immaterial, individual, and immortal; but its intimate essence is unknown: we cannot conceive it as absolutely isolated from matter, except as an abstraction. United to an ethereal or a fluid envelope or *perisprit*, it constitutes the concrete spiritual being, determinate and circumscriptive, called *spirit*. By metonymy, we often employ the words *soul* and *spirit*, the one for the other, speaking of happy souls or happy spirits, &c.; but the word *soul*, according to Spiritism, suggests the idea rather of an abstract principle, and the word *spirit* that of an individuality.

The spirit, united to the material body by incarnation, constitutes the *man*; so that in man there are three things: the *soul*, properly so-called, or intelligent principle; the *perisprit*, or fluid envelope of the soul; the *body*, or material envelope. The soul is thus a simple being; the spirit, a double being, composed of the soul and the perisprit; the man, a triple being, composed of the soul, the perisprit, and the body. The body, separated from the spirit, is inert matter; the perisprit, separated from the

* We are indebted to "Le Dictionnaire Universel" of Maurice Lachâtre for the substance of this statement.

soul, is a fluid matter without life or intelligence. The soul is the principle of life and of intelligence.

It is not true, therefore, as certain critics have pretended, that in giving to the soul a fluid, semi-material envelope, Spiritism has made of it a material being.

The first origin of the soul is unknown, because the principle of things is one of the secrets of Omnipotence; and it is not given to man, in his actual state of inferiority, to comprehend all. On this point, one can only formulate systems. According to some, the soul is a spontaneous creation of Divinity; according to others, it is a very emanation, a portion, a spark of the divine essence. The problem is one on which we can only establish hypotheses, inasmuch as there are reasons for and against.

Against the second of these opinions, the following objection is brought: God being perfect, if souls are portions of Divinity, they ought to be perfect, by reason of the axiom that the part is of the same nature as the whole; whence the question would arise, Why are souls imperfect and in need of further improvement?

Without stopping at the different systems touching the intimate nature and the origin of the soul, Spiritism considers it as it is manifested in the human race: it ascertains, by the proofs, of its isolation and of its action, independent of matter during life and after death, the great facts of its existence, its attributes, its survivance, and its individuality. Its individuality is shown in the diversity which exists in the ideas and qualities of each in the phenomenon of the manifestations; a diversity which implies for each a proper existence.

A fact not less important is proved by observation: it is that the soul is essentially progressive, and that it makes acquisitions unceasingly in knowledge and in moral wisdom, since we find it at all stages of development. The almost unanimous teaching of spirits tells us it is created *simple and ignorant;* that is to say, without knowledge, without consciousness of good and of evil, with an equal aptitude for either, and for acquiring all.

Creation being incessant, and from all eternity, there are souls arrived at the summit of the ladder when others are arriving at the consciousness of life; but, all having the same point of departure, God creates no one of them better endowed than another, and this is in conformity with his sovereign justice: a perfect equality presiding at their formation, they advance more or less rapidly, by virtue of their free will, and according to the pains they take. God thus leaves to each the merit or demerit of his acts, and the responsibility increases as the moral sense develops. So that of two souls created at the same time, the one may arrive at a certain height more quickly than the other, if it labors more actively for its amelioration; but those who lag behind have it equally in their power to reach that height, although not so soon, and after many rude experiences, for God does not shut the future to any of his children.

The incarnation of the soul in a material body is, according to Spiritism, necessary to its improvement, by the labor which the corporeal existence demands, and the intelligence it develops. Not being able, in a single life, to acquire all the moral and intellectual qualities which are needed to conduct it to its goal, it arrives there *in passing through an unlimited series of existences, whether upon this earth or in other worlds, in each of which it takes a step in the way of progress, and gets rid of some of its imperfections.* Into every existence the soul brings what it has acquired in its preceding existences. And thus is explained the difference which exists in the innate aptitudes, and in the degree of advancement of races and people.

According to Spiritism, the universe is an immense laboratory, where humanity, "emanating from an ethereal fluid, becomes elaborated, individualizing itself by incarnation, purifying itself in bodies as in so many crucibles; and, through a progressive advancement, by virtue of its inherent perfectibility, arriving finally at the state where it is the crowning work of creation."*

* "La Raison du Spiritisme, par Michel Bonnamy. Paris: 1868." See, also, "Du Spiritualisme Rationnel, par G. H. Love. Paris: 1862." A work by a man of science, in which the doctrine of pre-existence and re-incarnation is supported by arguments drawn from physical facts.

Such is the system of Spiritism in regard to the soul's re-incarnation. It is claimed for it, that it reconciles to our notions of divine justice the fact of those striking differences, moral and intellectual, in human beings, from the very moment of birth.

To its doctrine of pre-existence, it is objected, that it shifts the difficulty, but does not remove it; for, go back as far as we may, we come at last to a point where we have two souls, supposed to have been created equal. Now, if in virtue of their free will, one of these souls takes a bias to good, and the other to evil, how can there have been perfect equality in their conditions, or in the temptations to which they were exposed? Perfect equality throughout ought to lead to an equality of results. Why, then, should one soul get the start of another in goodness?

To this objection the Rev. Dr. Edward Beecher, the principal American advocate of the doctrine of pre-existence, replies, in his able "Conflict of Ages," as follows: "The real and great difficulty lies, not in the idea that free agents should sin, but in the idea that God should bring man into being with a nature morally depraved, anterior to any will, wish, desire, or knowledge of his own, or with a constitution so deranged and corrupt, as to tend to sin with a power that no man can overcome in himself or in others; and that, in addition to this, He should place him in a state of so great social disadvantage, and, as the climax, expose him, so weak, to the fearful wiles of powerful and malignant spirits. This difficulty, pre-existence does touch and entirely remove, by referring the origin of his depravity to his own action in another state, and showing that the system of this world is a system of sovereignty established over beings who have lost their original claims on the justice of God.*

"If now a difficulty is alleged still to exist as to their first

* Here Dr. Beecher diverges from the teachings of Spiritism, in his attempt to accommodate the theory of pre-existence to the demands of the Calvinistic theology. It will be seen that Origen believed that all our punishment here is remedial, and that there is no spirit so evil that he may not ultimately reform.

sinning in a previous state, it is enough to say that this is not the same difficulty that existed before, but altogether a different one; that is, how beings, created with an uncorrupt moral constitution, and in a spiritual system arranged in the best possible manner to favor their perseverance in right, could possibly sin. Suppose, then, that this question is not answered, and cannot be (although I do not concede that it cannot); but suppose it. What then? It merely leaves a mysterious fact; but it does not, as in the former case, present an alleged fact, which the human mind can see to be within the range of its faculties, and to be positively unjust. It therefore removes a dispensation positively unjust, and, in place of it, presents one that is simply mysterious.

"But it resorts to mystery in a proper place; for, since the past history of the universe is not revealed in detail, nothing exists to forbid the idea that, whatever were the circumstances in which men sinned, and whatever were the reasons of their sinning, still they were such as in the highest degree to show forth the honor, justice, and love of God, and to throw the whole blame on man. What, then, if we cannot state exactly these circumstances and reasons? What if we cannot reconstruct the past history of each man? Still we know nothing, and we see nothing, to forbid a full belief, based on confidence in God, that, in all his dealings with them, he was honorable and just. . . . These disclosures of the Bible settle the question as to the origin of evil. They no less clearly prove that the origin of the sin of man is not to be looked for in this world."

Another reply to the objection may be, that we know not through what equalities of struggle and temptation all souls may have passed. The saint of to-day may have been a direful sinner in some previous state; and many of the differences among souls may be simply the result of the different number of disciplinary existences through which they have passed, some having entered on conscious life in advance of others. Should this view strike us as a humbling one, it may be none the less salutary on that account; for why, when we think of it, should we merit it of Providence that we should be born with better propensities, or under more auspicious circumstances, than our neighbor?

The theological dogma that all the numerous millions of Adam's posterity deserve the ineffable and endless torments of hell for a single act of his, before any one of them existed, is admitted by all to be repugnant to that reason which God has given us ; subversive of all possible conceptions of justice. Even Pascal, Calvin, Mansel, and other orthodox writers, while accepting the terrible doctrine, admit thus much, substantially, as to its character, humanly considered; but they escape from the difficulty, by assuming that we must not measure divine by human notions of justice; that morality may be one thing on earth, and another in the heaven of heavens!

With irresistible force has Mr. John Stuart Mill replied to this attempt to pacify faith at the expense of reason and the moral sense. He says, "Mr. Mansel combats as a heresy of his opponents the opinion that infinite goodness differs only in degree from finite goodness. Here, then, I take my stand upon the acknowledged principle of logic and morality, *that when we mean different things, we have no right to call them by the same name and to apply to them the same predicates, moral and intellectual.* If, instead of the glad tidings that there exists a Being in whom all the excellences which the highest mind can conceive, exist in a degree inconceivable to us, I am informed that the world is ruled by a Being whose attributes are infinite, but what they are we cannot learn, except that the highest human morality does not sanction them, — convince me of this, *and I will bear my fate as I may.* But when I am told that I must believe this, and at the same time call this Being by the names which express and affirm the highest human morality, *I say in plain terms that I will not.* Whatever power such a Being may have over me, he shall not compel me to worship him. I call no being good who is not what I mean when I apply that epithet to my fellow-creatures; *and if such a Being can sentence me to hell for not so calling him, to hell I will go.*"

Whittier has expressed similar sentiments, though in a tenderer strain: —

> "Not mine to look when cherubim
>   And seraphs may not see;
> But nothing can be good in Him,
>   Which evil is in me.
> The wrong that pains my soul below,
>   I dare not throne above:
> I know not of His hate, — I know
>   His goodness and His love."

Kardec's doctrine of pre-existence differs, it will have been seen, from that of those seers who proclaim the eternity as well as the immortality of the soul. Kardec believes in the unremitting exercise of the creative power of Deity.

To many Spiritualists the doctrine of re-incarnation seems to be quite repulsive; though it is largely taught in communications claiming to be from spirits. Mr. Shorter asks, What must be the thoughts of the pious mother who imagines the possibility that the infant at her breast is a graduate from the galleys or the prison! To this objection it may be retorted, What must be the thoughts of those pure-minded and faithful parents, who, in spite of all their parental care and affection; see their children turn out criminals or sensualists? Would a theory that exempted those parents from the dread that something for which they themselves were responsible — something vicious in their own souls — had dragged those children down, be wholly unacceptable?

"The prevalent idea with regard to spirits," says Kardec, "renders the phenomenon of their manifestation, at first sight, incomprehensible. These manifestations can only take place through the action of spirit upon matter, and therefore those who believe that spirit is the absence of all matter, ask, with some appearance of reason, how it can act materially. Now, here is precisely their error; for spirit is *not* an abstraction: it is a defined being, limited and circumscribed. The spirit, clothed in the body, constitutes the soul; but when, at the hour of death, it quits the body, it is not divested of all envelopment. All spirits tell us that they preserve the human form; and, indeed, when they appear to us, it is in such forms as we can recognize.

"Let us observe them attentively at the moment that they have quitted this life. They are in a state of perplexity; all seems confused around them. On the one hand, they behold their body, whole or mutilated, according to the manner of death; on the other hand, they see and feel themselves alive. Something tells them that this body belongs to them, and they cannot understand being separated from it. They continue to see themselves in their original form; and this sight produces amongst some of them, for a short time, a most singular illusion, — that of believing themselves still in the flesh. They require to become accustomed to their new condition before they can be convinced of its reality. This first uncertainty being dispelled, the earthly body becomes to them an old garment which they have thrown off for ever, and which they no longer regret. They feel lighter, and as if relieved of a burden. They experience no longer physical pain, and rejoice in being able to rise and pass through space, as they sometimes fancied they did in their earthly dreams.

"Nevertheless, notwithstanding the absence of their earthly body, they retain their personality. They possess a form, but one which neither impedes nor embarrasses them. In fact, they have still their individuality and consciousness of being. What, then, must we conclude? Briefly, that the soul leaves not all in the grave, but that she carries something away with her to her new home."

According to Kardec, death is the disintegration of the material body which the soul abandons. The spiritual body, or *perisprit*, now disengages itself, and accompanies the soul which thus still finds itself "clothed upon." This new body, though fluid, ethereal, vaporous, and invisible to us in its normal state, is as real as matter itself, though up to the present time we have been unable to seize and analyze it.

This second envelope of the soul exists, then, during the corporeal life. It is the medium of all the sensations of which the spirit is conscious, and through which the spirit conveys its will to its exterior body, and acts upon the various organs. To use a material comparison, it is the electric wire which serves to receive and transmit the thought; it is, in short, that mysterious, imperceptible agent, spoken of as nervous fluid. To recognize this spiritual body, is to obtain the key to a multitude of problems hitherto unexplained.

The spiritual body is not one of those hypotheses to which science sometimes has recourse to explain a fact. Its existence is not only revealed by spirits themselves, it is the result of observation. Whether in the earthly body or out of it, the soul is never separated from its spiritual encasement.

The spirit body, then, is an integral part of the man; but this encasement alone is no more the spirit than the body alone is the man; for the spiritual body cannot think: it is to the spirit what the body is to the man, the agent or instrument of his action. The human form and that of the spirit body are identical; and when the latter appears to us, it is generally with that particular exterior with which we were formerly familiar. We might think from this that the spiritual body, though separate from all parts of the outer body, moulds itself in some way upon it, and preserves the impress of it; but it appears that this is not the case. Making allowance for the organic modifications necessitated by the surroundings in which men are placed, with the exception of some details, the human form (says Kardec) is to be found in the inhabitants of all worlds; at least, so say the spirits. It is, moreover, equally the form of all non-incarnate spirits, and those who have only the spirit body.

It is the form in which, through all ages, angels and purified spirits have been represented, from which we may conclude that the human shape is the type of all human beings, in whatever state or worlds they may be found. But the subtle substance of the spirit body has not the tenacity nor rigidity of the material body. It is, so to speak, flexible and expansive, and therefore the form it takes, though traced or copied from that of the body, is not absolute: it bends itself to the will of the spirit. Freed from these fetters which confined it, the spirit body can extend, contract, or transform itself; in a word, can lend itself to any metamorphosis, according to the will which acts upon it. It is through this property of its fluid encasement that the spirit which desires to make itself known can take, when necessary, the exact appearance it had when living, even to the bodily peculiarities, and the very style of dress, by which it can be recognized. We see, then, that spirits are beings like ourselves, forming around us a population, invisible to us in the normal state.

But the spirit body, though fluid, is, nevertheless, a kind of matter, and this results in the facts of tangible apparitions. Under the influence of certain mediums, there have been seen hands, possessing all the properties and appearances of living hands, warm and palpable, which offer the resistance of a solid body, which will seize and hold you, and in a moment vanish again like a shadow. The definite action of these hands — which evidently obey a will in executing their movements, and playing even on a musical instrument — prove that they are the visible parts of an invisible intelligence. Their tangibility, their temperature, and, in short, the impression they make on the senses, — for they have been known to leave an impress on the skin, to give blows so hard as to be painful, or caress most delicately, — prove that they are of some species of matter. Their instantaneous disappearance proves, moreover, that this matter is eminently subtle, and is of the nature of those substances which can alternately pass from the solid to the fluid condition, and *vice-versâ*.

As we have already seen, Spiritism teaches that the *essential* nature of the spirit proper, that is, the *thinking* being, is entirely unknown to us. It reveals itself to us only by its acts, and its acts can affect our material senses but through some intermediate substance. Thus the spirit requires matter to act upon matter. It has, for its direct instrument, the spirit-form, just as man has the body; hence the spirit-form is matter, as we have seen. It has, further, the universal ether, a sort of vehicle on which it can act, as we act on the air, to produce the effects of dilation, compression, propulsion, and vibration.

Looked at in this manner, the action of spirit on matter is, in Kardec's philosophy, easily conceived; and hence it is to be understood that all the effects which result from it enter into the class of natural facts, and have in them nothing miraculous. They have appeared supernatural simply because their cause was unknown. This once known, the marvellousness disappears, and this cause is entirely in the semi-material properties of the spirit body. It is a new order of facts, which will find their explanation in a newly discovered law, and which will very shortly astonish us no more than does the intercourse now made possible through electricity.

It may be asked, perhaps, how the spirit, with the help of so subtle a substance, can act upon heavy and compact bodies, lift tables, &c. Surely no man of science, says Kardec, would raise such an objection; for, not to mention unknown properties which this new agent may possess, have we not under our own eyes analogous examples? Is it not in the most rarified gases and the imponderable fluids that industry has found its most potent motive powers? When we see the air overturn whole edifices, steam propel enormous masses, gaseous powder burst asunder mighty rocks, and electricity tear up trees and pierce the solid walls, what is there strange in allowing that a spirit, with the aid of its spirit body, can lift a table, especially when it is known that this spirit body can become visible, tangible, and exhibit the attributes of a solid body?

On the subject of pre-existence, Henry More, the Platonist, and the friend and correspondent of Descartes, writes (1659) as follows: "This consequence of our soul's pre-existence is more agreeable to reason than any other hypothesis whatever; has been received by the most learned philosophers of all ages, there being scarce any of them that held the soul of man immortal upon the mere light of nature and reason, but asserted also her pre-existence; that memory is no fit judge to appeal to in this controversy; and, lastly, that traduction * and creation are as intricate and inconceivable as this opposed opinion."

Among the advocates of pre-existence, More enumerates Zoroaster, Pythagoras, Epicharmus, Empedocles, Cebes, Euripides, Plato, Virgil, Cicero, Hippocrates, Galen, Plotinus, Jamblichus, Proclus, Boethius, Psellus, Synesius, Origen, Marsilius Ficinus, Cardan; and, lastly, the great authority of the scholastic period, Aristotle, who, in his treatise on the Soul, speaks of the body that the soul is to actuate; and, blaming those who omit that consideration, says, "that they are as careless of that matter as if it were possible that, according to the Pythagoric fables, *any soul might enter into any body;* whereas, every animal, as it has its proper species, so it is to have its peculiar form. 'But those that define otherwise,' saith Aristotle, 'speak as if one should affirm that the skill of a carpenter did enter into a flute or pipe; for every art must use its proper instrument, and every soul its proper body.' Whereas, as Cardan also has observed, Aristotle does not find fault with the opinion of the soul's going out of one body into another (which implies its pre-existence); but that the soul of a beast should go into the body of a man, and the soul of a man into a beast's body, — this is the absurdity that Aristotle justly rejects, while the other opinion he seems tacitly to allow of."

Of Marsilius Ficinus, whom More reckons among the advocates of pre-existence, he relates that Marsilius had made a vow with his fellow-Platonist, Michael Mercatus, that the one who might die first should appear to his friend and confirm the truth of what they had often made a subject of discussion; namely,

* Tertullian taught what was called "*material traducianism*," according to which, life, having its source in the blood, was naturally transmitted from parent to offspring. There was a "*spiritual traducianism*" taught by another Christian school, according to which the soul of the offspring was engendered by the soul of the parent. Besides the Platonic doctrine of *pre-existence*, held by Origen, there was a fourth doctrine called "*creationism*," according to which the Deity creates the soul for each individual body at or shortly after the moment when the body itself begins to exist.

the soul's immortality. Michael, being intent on his studies on a certain morning, heard a horse approaching with all speed, and observed that he stopped at the window; and therewith heard the voice of his friend, Marsilius, crying out aloud, "O Michael, Michael! vera, vera sunt illa!" — *O Michael, Michael! those things are true, are true!* Whereupon Michael suddenly opened the window, and, espying Marsilius on a white steed, called after him; but, as he looked, the apparition vanished. Michael sent presently to Florence to inquire how Marsilius was, and learned that he died about the hour he had appeared at the window.

Of Sir Henry Vane, the younger, Burnet says, "His friends told me, he leaned to Origen's notion of a universal salvation of all, both of devils and the damned, and to the doctrine of pre-existence. When he saw his death was designed, he composed himself to it with a resolution that surprised all who knew how little of that was natural to him."

"It is not without reason," says Saint Martin (1743–1803), "that we look upon the duration of this corporeal life as a time of chastisement and expiation; but we cannot look upon it as such, without forthwith thinking that there must have been for man *a state anterior and preferable to the one wherein he now finds himself*. . . . Each of his sufferings is an index of the happiness wanting in him; each of his privations proves that he was made for enjoyment; and his present subjection announces an ancient authority; in one word, to feel now that he has nothing, is a secret proof that once he had all. . . . As our material existence is not life, our material destruction is not death."

Joseph Glanvil, to whose investigations into the facts of witchcraft and other spiritual phenomena we have already referred, published in 1662, but without his name, a treatise to prove the reasonableness of the doctrine of the pre-existence of souls. He was also the author of a letter, still in existence among the Baxter manuscripts, full of curious learning in defence of the doctrine.

In Germany, Kant, Schelling, and Jul. Müller, used the doctrine of the metempsychosis to explain the beginning and root of sin in humanity. Herder, Lessing, Schubert, and Lichtenberg seem to have favored it. Van Helmont, the younger, who died in 1699, taught it in Holland.

Herder has some remarkable dialogues on the subject of metempsychosis ; and, though the vein of them is tentative rather than dogmatic, it is easy to see the drift of his meditations. We quote a few passages: —

"In nature every thing is related; morals and physics, like body and spirit. Morality is only a more beautiful *physique* of the spirit. Our future destination is a new link in the chain of our being, which connects itself with the present link most minutely and by the most subtile progression, as our earth is connected with the sun, and the moon with our earth. . . .

"Perhaps there are appointed for us places of rest, regions of preparation, other worlds in which, — as on a golden heaven-ladder, — ever lighter, more active and blest, we may climb upward to the fountain of all light, ever seeking, never reaching the centre of our pilgrimage, the bosom of the Godhead. For we are and must ever be limited, imperfect, finite beings. But wherever I may be, through whatever worlds I may be led, I shall remain for ever in the hands of the Father who hath brought me hither and who calls me further; for ever in the infinite bosom of God. . . .

"Hereafter, when death shall burst these bonds, when God shall transplant us like flowers into quite other fields, and surround us with entirely different circumstances, then —— Have you never experienced, my friend, what new faculty a new situation gives to the soul? A faculty, which, in our old corner, in the stifling atmosphere of old circumstances and occupations, we had never imagined, had never supposed ourselves capable of? . . .

"The younger Van Helmont, in his 'De Revolutione Animarum,' has adduced, in two hundred problems, all the sayings and all the arguments which can possibly be urged in favor of the return of souls into human bodies according to Jewish ideas. . . . These assert that the soul returns into life on this planet twice or thrice, — in extraordinary cases oftener, — and accomplishes what it had left unfinished. . . . And is there no weight in the arguments from reason in support of these ideas? Shall not the Long-suffering and the Just give every one space and time for repentance? Has not the fruition of life been to many imbittered and abridged without any fault of their own?

"Look at the thing humanly: consider the fate of the *mis-born*, the deformed, the poor, the stupid, the crippled, the fearfully degraded and ill-treated; of young children, who had scarce seen the light and were forced to depart. Take all this to heart, and you must either have weak conceptions of the progress of such people in the world to come, or they must first have wings made for them here, that they may learn to soar, even at a distance, after others; that they may be, in some measure, indemnified for their unhappy, or unhappily abbreviated existence in this world. Promotion to a higher, human existence is scarcely to be thought of in their case.

"None can give as God gives, and no one can indemnify and compensate like God. To all beings he gave their existence of his own free love. If some appear to have been more neglected than others, has he not places, contrivances, worlds enough, where, by a single transplantation, he can indemnify and compensate a thousand-fold? A child prematurely removed, — a youth whose nature was too delicate, as it were, for the rude climate of this world, — all nations have felt that such are loved by the gods, and that they have transferred the treasured plant into a fairer garden. . . . How many may have been made happy in another world through having been unhappy here!"

With regard to the Pythagorean notion of the transmigration of souls into the bodies of brute animals, Herder says, "With some nations, — for example, the Egyptians and Hindoos; and perhaps, too, with Pythagoras, — it was designed as a moral fable, representing the doctrine of ecclesiastical penance in a sensuous and comprehensible form: "You who are cruel shall be changed into tigers, as even now you manifest a tiger-soul. You who are impure, shall be swine," &c. These representations, addressed to the senses, and clothed with the authority of religion, would, undoubtedly, have a greater effect than metaphysical subtleties.

"Our language," says Herder, "all communication of thought, what bungling work it is! Hovering on the tip of our tongues, between lip and palate, in a few syllabled tones, our heart, our innermost soul would communicate itself to another, so that he shall comprehend us, shall feel the ground of our innermost being. Vain endeavor! Wretched pantomime with a few gestures and vibrations of air! The soul lies captive in its dungeon, bound as with a seven-fold chain; and only through a strong grating, and only through a pair of light and air-holes, can it breathe and see. And always it sees the world on one side, while there are a million other sides before us and in us, had we but more and other senses, and could we but exchange this narrow hut of our body for a freer prospect. . . .

"Sacred to me is the saying, 'Blessed are the poor in spirit, for theirs is the kingdom of heaven. Blessed are they that mourn, for they shall be comforted. Blessed are the pure in heart, for they shall see God.' Purification of the heart, the ennobling of the soul, with all its propensities and cravings; this, it seems to me, is the true palingenesis of this life, *after which, I doubt not, a happy, more exalted, but yet unknown metempsychosis awaits us.* Herewith I am content."

Schubert, a devoted Spiritualist, and who satisfied himself of the genuineness of the phenomena produced in the presence of Mrs. Hauffé, the Seeress of Prevorst, wrote the "History of the Soul." His psychological views are not unlike those we have given in our abstract of Mr. Wake's recent work. Schubert shows, first, how the soul is, as it were, reflected in and by the body; how it gives form and perfection to our material organization. Next, entering upon the analysis of mind, he sets forth the distinction between the soul and the spirit; a distinction which St. Paul seems to have recognized. Origen also regarded the soul as intermediate between body and spirit. In the Alexandrian philosophy, we have the *pneuma* denominated as the rational soul, and the *psyche* as the sensitive soul, that which desires or lusts. Irenæus says, "There are three of which the perfect man consists: flesh, soul, spirit; the one, the spirit, giving figure; the other, flesh, being formed. That, indeed, which is between these two is the soul, which sometimes following the spirit is raised by it; and sometimes consenting to the flesh, falls into earthly lusts." Dr. George Bush held that the *pneuma* is to the *psyche* what the soul is to the body. The *psyche* is the spiritual body or body of the spirit. This view does not differ much from the teachings of Kardec.

Schubert regarded the soul as the inferior part of our intellectual nature, — that which shows itself most distinctly in the phenomena of our dreams, — the power of which also is situated in the material constitution of the brain. The spirit, on the contrary, is that part of our nature which tends to the purely rational, the lofty, the divine!

That profound and intrepid thinker, Lessing, who anticipated by a century much of the advanced thought of the present, remarks as follows on this subject of pre-existence: "Why may not each individual man have existed more than once in this world? Is this hypothesis, therefore, so ridiculous because it is the oldest?* because it is the one which the human understanding immediately hit upon before it was distracted and weakened by the sophistry of the schools? . . . Why should I not return, as often as I am able, to acquire new knowledges, new talents? Is it because I carry away so much at one time as to make it not worth the while to return? Or, because I forget that I have been here before? It is well for me that I forget it. The remembrance of my former states would allow me to make but a poor use of the present. Besides, what I am *necessitated* to for-

* Delitzch pronounces this statement incorrect. Franck, on the other hand, says that metempsychosis was the earliest form in which the dogma of immortality presented itself to the human mind. But again we find immortality, but no metempsychosis, in the most ancient poems of India, the "Rig-Veda," for example.

get now, have I forgotten it for ever? Or because, on this supposition, too much time would be lost to me? Lost? What have I then to fear from delay? *Is not the whole eternity mine?*"

According to Herodotus, the Egyptians were the first to entertain the doctrine of metempsychosis. They believed that the soul was clothed successively with the forms of all the animals that live on the earth, and that it then returned, after a cycle of three thousand years, into the body of a man to recommence its eternal pilgrimage. From them the Greeks may have received the idea, which was a leading feature of the doctrine of Pythagoras, who claimed to recollect his former self in the person of a herald named Æthalides; Euphorbus, the Trojan; and others; and he even pointed out, in the temple of Juno, at Argos, the shield he used when he attacked Patroclus. He taught that after the rational mind of man is freed from the chains of the body, it assumes an ethereal vehicle, and passes into the regions of the dead, where it remains till it is sent back to this world to be the inhabitant of some other body, human or brutal; and that after suffering successive purgations, when it is sufficiently purified, it is received among the gods, and returns to the eternal source from which it first proceeded.

Ritter says that the sum of the Pythagorean doctrine of immortality was this, — *that condition would accurately follow character.* Pletho states that the Pythagoreans, as the Platonists after them, conceived the soul to be a substance not wholly separate from all body, nor wholly inseparate; but partly separate, partly inseparate, separable potentially, but ever inseparate actually.

The later Pythagoreans maintained that the soul has a life peculiar to itself, which it enjoyed in common with demons or spirits before its descent to the earth, and that there must be a degree of harmony between the faculties of the soul and the form which it assumes: this last is also the idea of Swedenborg.

Plato, in his "Phædo," maintains pre-existence of the soul before it appears in man; and of this pre-existent condition it retains dim reminiscences; and after death, according to its peculiar qualities, it seeks and chooses another body. Every soul, according to him, returns to its original source in ten thousand years. After completing each life, it spends a thousand years in the spiritual world in a condition corresponding to that life, after which it passes into a new body corresponding to its ethical quality.

"Plato's conception of the immortality of the soul," says Grote, "includes pre-existence as well as post-existence; a perpetual succession of temporary lives, each in a distinct body, each terminated by death, and each followed by renewed life for a time in another body. In fact, the pre-existence of the mind formed the most important part of Plato's theory about immortality; for he employed it as the means of explaining how the mind became possessed of general notions. Not *all* learning, but an important part of learning, consists in reminiscence; not indeed of acquisitions made in an antecedent life, but of past experience and judgments in this life. Of such experience and judgments every one has travelled through a large course; which has disappeared from his memory, yet not irrecoverably. Portions of it may be revived, if new matter be presented to the mind, fitted to excite the recollection of them, by the laws of association."

According to Socrates, priests, priestesses, and poets (Pindar among them) tell us that the mind of man is immortal, and has existed through all past time in conjunction with successive bodies.

The idea of metempsychosis re-appears in the speculations of the Neo-Platonists, in the cabala of the Jews, and in the teachings of some of the Church Fathers. Porphyry conceives that it is to expiate sins committed in a pre-existent state that we are now clothed with a body, and that as our conduct was more or less culpable we assume more or less material bodies. By fulfilling exactly and with resignation the duties imposed upon us, we return by degrees through the state of heroes, angels, archangels, &c., to the Supreme Being. There is also a descending scale of diabolical life.

The Cabalists thought that the destiny of every soul was to return into mystical union with the divine substance, but that in order to do this it must first develop all the perfections of which it has the germ within itself. It is sent through life after life till it acquires all the virtues possible to it.

Origen, born at Alexandria, A.D. 185, distinctly maintains the doctrine of metempsychosis, and finds in it the final cause of creation. In his view, according to Gieseler, the Godhead can never be idle. Before the present world there was an endless series of worlds, and an infinite succession of them will follow. All intellectual beings were originally created alike; but they were never without bodies, since incorporeality is a peculiar prerogative of Deity. After a great moral inequality had arisen among them by their difference of conduct, God created the present world, which affords a dwelling-place to all classes in proportion as they answer their moral condition. The fallen intellectual beings he put into bodies more or less gross, according to the measure of their sinfulness. Still they all retain their moral freedom, so that they may rise again from the degraded circumstances in which they exist. Even the punishments of the condemned are not eternal, but only remedial; the devil himself being capable of amelioration and pardon. When the world shall have answered its purpose as the abode of fallen spirits, it will then be destroyed by fire; and by this very fire souls will be completely purified from all stains contracted by intimate union with the body. But as spirits always retain their freedom, they may also sin again, in which case a new world like this will be again necessary; the earths to which incarnated spirits are sent corresponding to their moral condition.

Evil, according to Origen, is the only thing which has the foundation of its being in itself and not in God, and which is, therefore, founded in no being, but is nothing else than an estrangement from the true Being, and has only a subjective and no objective existence at all, and is in itself nothing. Therefore, he says, "The proposition of the Gnostic, that Satan is no creature of God, has some truth for its foundation; namely, this, that Satan, in respect to his nature, is a creature of God, but not as Satan."

Origen set his theory of the pre-existence of souls in opposition to creationism, which supposed individual souls to arise from the immediate act of creation on the part of God; for this theory appeared to him irreconcilable with the love and justice of God, which maintains itself equally towards all his creatures; and also in opposition to the traducianism of Tertullian, for this theory appeared to him too sensuous.

He infers a moral destiny of the embryo, originating in a pre-existent state, from this fact, among others; namely, that Jacob and Esau, while yet unborn, and prior to all earthly agency, are objects respectively of divine love and hate.

The doctrine of the transmigration of souls as a means of penance was held by the Manicheans. It also existed among the ancient Italians, the Celtic Druids, the Scythians and Hyperboreans, and is still entertained by the heathen nations of Eastern Asia, the Caucasian, and other tribes. Coupled with the notion of transmigration into the bodies of brutes, among the ancient Egyptians it led, as it still does with the Hindoos, to the veneration of certain animals, and the fear of eating their flesh, since their bodies may be the abode of departed ancestors and friends. The Pythagoreans would not kill animals for the same reason.

Origen's notion of earths being created to correspond with the moral *status* of the spirits sent to be re-incarnated upon them, has been recently reproduced in the speculations of an English clergyman, the Rev. William Hume-Rothery, who furnishes the following statement of his views on the subject: —

"From the all-good and all-wise Being, who is the only Creator, nothing of evil and misery can possibly proceed. Yet in this world, to go no further, we do find an awful amount of wickedness and wretchedness; and in nature itself, which none but God can produce and preserve, there are deadly poisons, savage and disgusting animals, famines and pestilences, &c., which certainly, according to the judgment that God has given us, are evils and blemishes in his creation. Now, as there is but one Creator, a Being of spotless purity and absolute wisdom, how can evils and malformations and embodied savagery and consuming maladies and the entire family of wrongs come into existence?

"This is the explanation: God, in making man, endows him with free will, which is essential to manhood. By virtue of free will, man can live either according to the will of the Creator, or he can disobey this ever-righteous will. So far as he obeys the Creator's will, to that extent he is orderly and happy. But so far as he opposes the divine will, in that same degree are confusion and misery introduced into his life and world. The soul, consisting of the will and understanding, is the primary creation, being that which is usually denominated *spirit;* the body, which is the soul itself developed into a bodily form, is the next proceeding creation; and the world, comprising the three kingdoms of nature, with all objects of the senses, is the ultimate ground of creation, which usually goes by the name of *matter*, in which the states of the soul are brought down, spread out, and revealed in a region of space and time. Thus the soul, the body, and their world are a great unit of life, which assumes form in three different degrees or planes, but is *distinctly* one. Thus, too, spirit and matter, or life and its embodiments, or, which is exactly the same, life and its phenomena, are the beginning and the ending of a human being; and all the evils and disfigurements in nature, and all its blessings and beauties, are the embodiments and revealments of blessings and curses in the soul of man; good, both vital and phenomenal, flowing from harmonious co-operation with the Lord; and evil, both spiritual and natural, being produced by man's violation of the inflowing creative life of God.

"Such is the universal order of creation. Every natural world in the universe is the effectuated life and outward revelation of a world of created beings. Dream-land is thus created. All poetical imagery is brought forth after this manner. The wild fancies of the drunkard, which are called *delirium tremens*, burst into existence in this way. The phenomena of death owe their birth to corresponding changes of mental state. The human soul — willing and thinking here on the lowest platform of life, viz., that of *effects* — when indrawn by the Lord into a deeper ground of affection and thought, viz., that of causes, is evolved into a corresponding body and a corresponding world, in which latter its inmost states are represented in detail, as in the body they are represented in the sum."

On the subject of pre-existence, Cudworth, in his "Intellectual System" (1678), remarks: "It is well known that, according to the sense of antiquity, these two considerations were always included in that one opinion of the soul's immortality; namely, its pre-existence as well as its post-existence. Neither were there ever any of the ancients before Christianity, that held the soul's future permanency after death, who did not likewise assert its pre-existence, — they clearly perceiving that if it was once granted that the soul was generated, it could never be proved but that it might be corrupted. And therefore the assertors of its immortality commonly began here, — first, to prove its pre-existence, proceeding thence afterwards to establish its permanency after death."

"To admit," says Schopenhauer (1820), "that that which has not existed during an infinite period, must yet continue to exist during all eternity, is certainly a very bold hypothesis. That only which has had no commencement, or is truly eternal, can alone be indestructible. The Hindoos are more consistent. While they admit a continuation of existence after death, they also believe in a life anterior to our birth in this world, and declare that all which is, is eternal."

On the contrary, Irenæus dogmatically remarks: "If any person maintain that those souls which only began a little while ago to exist cannot endure for any length of time, but that they must on the one hand either be unborn in order that they may be immortal; or, if they have had a beginning in the way of generation, that they should die with the body itself, — let them learn that God alone, who is Lord of all, is without beginning and without end."

"Christianity," says Mr. J. W. Jackson, "notwithstanding the large infusion of Hellenism by which it is characterized, is still so essentially Judaic in some of its aspects, that it has never yet dared to promulgate the great Platonic veracity of pre-existence (the necessary correlate of post-existence), save in connection with its founder, the presumable incarnation of the eternal Logos. . . . Will a logically and metaphysically trained people be satisfied with the absurd assurance that an everlasting existence can have had a beginning in time? . . . The Christian will have to learn that his boasted doctrine of immortality is but a half-truth, the mere hemisphere of the sublime veracity that man, like his divine Father, is not only *immortal* but also *eternal*."

With regard to the objection that so much must be obliterated from man's memory before he can be born with his anterior experiences all a blank, it may be answered that the facts of somnambulism and double consciousness offer numerous analogies with such a dispensation. Every experienced physiologist must know instances wherein whole tracts of memory, extending through periods of many years, have, by physical accident or disease, been suddenly obliterated, and, after a long suspense, been as suddenly restored. The cases are numerous of aged persons, who, in their dying moments, have been able to converse in languages which they had utterly forgotten since their early childhood.

It may be interesting to note what the great Swedish seer has to say on this subject of pre-existence. Swedenborg's explanation is as follows: "It is not allowed that any angel or spirit should speak with man from his own memory, but only from the man's. If a spirit were to speak with a man from his own memory, the man would appropriate the spirit's memory as his own, and his mind would become confused with the recollection of things he had never experienced. In consequence of the memories of spirits getting muddled with men's, *some of the ancients conceived the idea that they had existed in another realm previous to their birth on earth*. Thus they accounted for the possession of memories which they were sure had not originated in ordinary experience."

This law, it is suggested, will serve as a reply to the frequent complaint that in spiritual communications we have nothing new or extra-human.

It is the theory of Andrew Jackson Davis, the highly esteemed American seer, that memory is something more than a mental faculty of registration; that the mind is a compound of eternal principles, each of which being from God, and of God, is self-intelligent, from which intelligence memory is inseparable. Thus, Davis holds the doctrine of the pre-existence of the psychical principle, though not of that of the individualized spirit. He teaches that "all souls existed from the beginning in the divine soul; all individuality which is, has been, or will be, had its pre-existence, has its present existence in creative being." Thus our "soul-matter" had an existence, but not a conscious existence, before we came on this earth. But may we not say the same of all "soul-matter," whether of men or brutes?

M. Cahagnet, author of "Arcanes de la Vie Future Dévoilés," bases his pneumatology on communications supposed to be from spirits, and obtained through clairvoyants and somnambulists. He admits pre-existence, but no re-incarnation. It is his theory that all the souls in the universe were created by God at once, and are eternal, as well as immortal; that they were all placed in worlds of perfect happiness, but yet not with all their affections and faculties called forth; and that they are sent down to worlds of material life for discipline, and to make them the better appreciate the heaven which they will soon regain; for without experience of evil they cannot properly estimate the good.

He does not admit the notion of the non-existence of space and time in the spirit-world; says that if spirits occupied no space, they would be nothing; and if there were no time, there could be no succession of events. These errors, he says, arise from the fact that the rapid action of spirits is incalculable by our time. It is also an error that spirits can be in several places at the same time; but they can transfer themselves from place to place with such speed, and can communicate with other spirits in such rapid succession, that it seems to take place at once. A spirit can see the whole of his existence in a moment, as has been experienced repeatedly by drowning persons. Sir Humphrey Davy, while under the effect of the nitric-oxide gas, exclaimed, "The whole human organism is an assemblage of thoughts."

In our remarks on *psychometry*, in a future chapter, this subject will be considered further.

In his "Conflict of Ages," Dr. Edward Beecher advocates the doctrine of pre-existence in the interests of the "Orthodox" theology, and in the hope of removing "the causes of paralysis and division from our common Christianity." We have seen (page 303) that in many of the communications supposed to come from spirits, moral evil is regarded as a means of education. In the philosophy of Hegel (where "each his dogma finds"), a similar view is taken of sin, as being not only incident to human nature, but one of the appointed means of its development and advancement. Origen, too, regarded all God's penalties as simply remedial, and believed in the ultimate restoration of all souls. But to these views the large majority of Christian theologians are, as we have seen, opposed. Dr. Beecher himself says, "The multitudes who are saved owe eternal life to the free grace of God. All who are *lost* perish entirely by their own original revolt from God, *persisted in during this life*."

Again he remarks, "It has been conceded repeatedly, that the acts ascribed to God, in his dealings with the human race through Adam, do appear dishonorable and unjust, according to any principles of equity and honor which God has made the mind of man to form. And yet, simply on the basis of Rom. v. 12–19, and without any adequate search for a more legitimate mode of interpretation, good men have for ages gone on to ascribe these acts to God. . . . Notice, then, the full confession of the great body of the Church, that the only defence against the charge of doing this has been the theory that all men had forfeited their rights as new-created beings, by 'an act over which they had not the slightest control, and in which they had no agency,' and which took place before they existed; and also the confession of Calvin, that nothing is so remote from common sense as this defence; and of Pascal, that nothing appears so revolting to our reason. . . . And, now, *is it nothing practical that pre-existence can deliver the Church at once from such a state of things?*"

Dr. Beecher is solicitous for the deliverance of the Church from a dilemma that outrages his reason. But a deliverance that would still require the hypothesis of an eternity of hell torments for nine-tenths of the human race, is hardly an improvement on the theology that would make us all subject to damnation because of the sin of Adam. Reason asks for a more "practical" exhibition than that which Dr. Beecher's plan would supply of the resources of Infinite Wisdom and Love.

It is with a sense of relief, therefore, that we turn from his ingenious attempt to extenuate the theological notion of eternal damnation by grafting on it the doctrine of pre-existence, to the celebrated "Terre et Ciel" (Earth and Heaven) of Jean-Ernest Reynaud (born 1806). This eloquent writer goes far to exhaust both the theological and the philosophical argument in behalf of pre-existence. He believes in the continuity of human life through successive incarnations, with the perpetual progress of nature and of man towards God, always infinitely removed.

In no other work on the subject are the objections to pre-existence so ably met, or the theory itself made so attractive by the charms of a persuasive style and by appropriate expositions, scientific, historical, and religious.

"Terre et Ciel" has been ably reviewed, if not answered, by M. Caro in his "Études Morales sur le Temps Présent;" and has called forth the denunciations of some of the doctors and bishops of the Church. To these Reynaud has replied in a manner to indicate that, in theological discussion, he is entirely at home. He shows that the Church has left this subject of pre-existence an open question; and that St. Augustine himself, who is sometimes quoted against it, had not, in his old age, made up his mind in regard to it.

Reynaud utterly rejects the theological notion of hell. This earth he regards as a specimen of the only kind of hell to which God will subject his children, and the object of his placing us here is not penal, but disciplinary and with a view to progress. Reynaud does not agree with Origen that we are here in the way of a descent from what we have been in an anterior life. On the contrary, we are here in an ascending passage.

"We are not," says Reynaud, "sinners because we are the sons of Adam: we are the sons of Adam because we are sinners." It may be more gratifying to our self-love to think that we are suffering here through the fault of Adam rather than through our own; but by such a sentiment we derogate from divine justice.

From the infinity of the universe, Reynaud argues in favor of his system. The infinity of creation is an earnest of the immortality of intellectual beings. The very great and the very little are both conditioned alike in view of infinity. Strong in the consciousness of our spiritual dignity, we may feel ourselves superior to all merely material grandeurs, however stupendous, and we may look upon the vortices of the firmament, with its systems behind systems, with the same regard that we look on whirlwinds of dust.

Let us not suppose that these immense separations between planetary worlds and systems, which, in view of the velocity of our freed spirits, have hardly the thickness of partitions, are insuperable abysses. It is not to the soul that they are barriers, but only to those organs to which our souls are temporarily united. All these worlds are but one for the immortal soul. Thanks to that infinity in which mere plurality is lost, the principle of unity, overshadowed for an instant by that of number, re-assumes the plenitude of its empire; and, as there is but one God, there is but one heaven. The fixity of this heaven is in the unalterable order of its changes; its incorruptibility is in its permanence; its immateriality is in the immensity of its extent. And this earth which we tread under our feet; where we come, turn by turn, to accomplish our task, in company with our kind; upon which we appear, without remembering whence we come; from which we disappear, without knowing whither we go; where we live, without being able to say with certainty who or what we are, — this earth rolls through the heavens, is one of the elements of the heavens, and constitutes us residents of the heavenly expanse. Let us give back to religion the eloquent words which Kepler, breaking through the vaults of the antique firmament, traced as a line of light in his "Harmonies," to illumine astronomical science for ever: *Hoc enim cœlum est, in quo vivimus et movemur et sumus, nos et omnia mundana corpora,* — "For this is heaven, in which we live and move and are, we and all mundane bodies."

The lot assigned to us on earth, Reynaud tells us, is far more tolerable than that which would be ours in the theological heaven, were it out of our power to aid still in mitigating suffering and evil; for here, in spite of all the obstacles that impede us, we are free at least to yield to the noble instinct which bids us help all suffering creatures; free to expect confidently from the bounty of God the end of all that evil, the view of which afflicts us. The thought of relatives, friends, fellow-creatures, suffering in hell while we, on our heavenly heights, had no power to help them, would be like that paralysis we have in nightmare, when we cannot move to avert some terrible danger, and when we strive to cry out in our despair. Such a state would itself be the most frightful of punishments; and so Reynaud repudiates the common theological notion of hell as blasphemous, revolting, unsound.

"While waiting," says Reynaud, "the illuminations of the higher life, I content myself with concluding, from the ordinary simplicity of Providence in the execution of his designs, that the souls of the departed will find themselves carried where their merits or demerits may make it fitting, by means as easy and spontaneous as those which govern matter; mounting of their own accord to a higher condition or descending to a lower, conformably to the rules of justice, in the same manner as bodies, by reason of their variations in weight, mount or descend in our atmosphere. . . .

"If, in the succession of the various phases of our immortality, repose may sometimes become the recompense of the just, it must be on condition of its being but a transient alternative; a refreshment, as it were, after fatigue, and serving to repair the strength for new and nobler efforts."

In opposition to the general theological notion, he maintains that the superior life, instead of being one of passive beatitude, will be one of sovereign activity; and the more active, the more elevated it is; that is to say, the more nearly akin to that divine model whose life overflows with an indefatigable activity through all the worlds.*

"And so heaven is not a permanent dwelling: it is a road; and the celestial hierarchy which fills it, ascends unceasingly, like a column of incense. But what fate awaits us at the extremity of this road? And what is the end of all this movement? Is it God, within whose abysses souls go successively to merge themselves, as the theologians of Bouddha, in their insensate mysticism, have dreamed; and not only they, but many others, who, even under the discipline of the Church, misled by an imaginary love of God, have fallen into a like spiritual suicide?

"It is just here that Christianity triumphs; for on this capital question it is Christianity alone that gives us the true lesson. No, says this superior religion: it is not God who occupies this mysterious summit; it is God and man together; it is the simultaneous type of the two natures; it is the God-Man; and, if the theologian will have his own expression, it is the divine exemplar, Jesus Christ. And so, even at this inaccessible sum-

* Reynaud's language here reminds us of a remarkable passage in the writings of Origen, where he attributes all existence, whether of men or of the lower animals, to "the exuberant fulness of life in the Deity, which, through the blessed necessity of his communicative nature, *empties itself into all possibilities of being*, as into so many receptacles."

mit, it is always man; man conceived by faith in the double perfection of his personal development and of his personal union with the second hypostasis; man, finally, such as he is when in perfect accord with and well-pleasing to God. Man is therefore the master of his own endless elevation. At any degree in his sublime ascension, neither does his personality, nor his activity, nor his perfectibility have a tendency to be engulfed and lost; for always, high above him, he sees the ideal of man, the ineffable archetype of creation, the common model of all the free beings of the universe. . . .

"But shall friends and relatives be re-united? Nothing can prevent our so ordering our existences as to travel for ever in company, through the abysses of the universe, with all those we love.

"Friends, relatives, parents, if you have profoundly at heart the wish not to lose one another by death, bind yourselves together in the same life, the same morality, and the same hopes; and you will rejoin one another there above, even as you were associated here. . . .

"Even in our birth we are free, for it is we who determine the conditions. If we resemble our parents, it is because we resembled them virtually before we were born. If we find ourselves in prosperous or adverse circumstances, it is because these are such as are best adapted to our progress and our needs. Let us be consoled therefore in the thought that there is no fatality weighing upon us; that there are no evils to which we are now subjected, from which we may not, by the good government of our actions, deliver ourselves radically at death." . . .

But what shall we say of our *ignorance?* "Not only is memory powerless in regard to the times that preceded our birth into this world, but it is not capable of representing to us even the times that followed that event. It tells us nothing of the period passed at the maternal bosom. It fails us in a multitude of instances. Beyond the cradle all is as dark as beyond the tomb. . . .

"And yet who shall venture to say that our being may not

contain within its profundities the wherewith to illumine some day all those spaces successively traversed by us since our first hour? Do we not find, even in the experiences of this present life, that certain recollections, which seemed absolutely extinguished, revive all at once, and render back to us a past which we had supposed lost for ever?"

The facts of Spiritualism, as we shall see in a succeeding chapter, abundantly confirm the hypothesis here suggested. And, were it not so, is it so certain that our *identity* is absolutely dependent on the formal presentations of our *memory?* "Do we," says Pierre Leroux, "in any phenomenon whatsoever of our lives, have at the same time memory of all preceding phenomena?" No, he replies: we are then occupied with a certain object, and our anterior life escapes us; and memory is but a past fact of our life perceived by us as present. Therefore our identity, our personality, our *ego*, is not a product of memory. *To remember* is but an accidental phenomenon of this *ego*, the same as to perceive, to see, to judge, &c. . . . "But why," asks Chaseray, "may not the memory, like the attention, the meditative faculty, the judgment, like all the intellectual faculties we possess, and like all the new ones we may acquire, follow the law of progress, and, from life to life, go on improving and gaining in extent?" Analogy shows that this may well be.

We continue our quotations from Jean Reynaud: —

"That astonishing faculty then which we call memory, is of a nature to preserve for us in the depths of our being, and unknown to ourselves, impressions which, from the fact of their having momentarily ceased to be disposed in a manner to come up at our appeal, continue none the less to make part of our domain, where they abide, as it were, dormant; and hence, why should it not be the same with the action of this faculty in regard to events which have preceded the actual period of our existence here? . . .

"The body, through the senses with which it is furnished, is needed for conveying impressions to the soul; but as to the *preservation* of those impressions, that is no longer the body's affair. It is the soul which has received; it is the soul which keeps them.

"Our own experience offers confirmation of the fact. Is there in the organs, by means of which we are to-day in communication with the universe, I will not say simply a single molecule, but a single form, which belonged to the organs which served us in infancy? Since that period, how many bodies has not our vital faculty taken to itself, used, dissipated! And yet, in spite of all these mutations, does not the soul preserve its memory? How many things there are on which I had not thought for years, which I had let fall completely from my remembrance, but which, all at once, in association with places or with persons, or roused by an effort of attention, start up and re-appear to me! Is there not here an indication of what may be produced hereafter in sublime proportions?

"Notwithstanding those apparent interruptions of such moment to us, and which the vulgar in trembling call *death*, our life, considered not in its earth-bound span of a day, to which the prejudices of our education reduce it, but *in its infinite line*, is in reality as continuous in all its development as in the short period laid bare to us between the cradle and the tomb. . . .

"In admitting even what our present experience may lead us to conclude in regard to the suspension of all remembrance of anterior existences, — this, namely, that death must produce on unprepared natures the effect of a heavy blow, and that, in striking, it stuns the memory, —*yet to stun it, is not to annihilate it.* After a suspension of days, months, or years, the memory may recover itself. The fact that we may have no reminiscence of anterior existences *now* is no reason why we may not have it at some future time. . . .

"Each one of us carries in his actual form and organism the secret history of his anterior emotions; so accurately, that spiritual eyes, penetrating to the depths of our being, see at a glance all that we *have* been in all that we *are*.

"Our history therefore is not only in that Book of Life which theologians put in the hands of God; it is inscribed in our very substance; our being itself is the unfailing record we carry with us, from stage to stage, through the worlds. . . .

"It is wholly arbitrary to suppose that immortality preserves life without preserving at the same time the faculty of repentance equally with all others. The quibbles by which we may try to justify the hypothesis of the abandonment of the damned, may be employed with equal force to support the idea that God ought to abandon, without remission, every culpable soul, even in this life. *The culpable soul is not blinded more irremediably after having passed through death than it was before;* for death is but an accident, as incapable of changing the nature of the soul as of changing the disposition of God. That which the soul was on the eve of death, it will be the next day. . . .

"In reflecting on the spectacle of the universe, such as it presents itself to us from the point of view of modern times, it seems to me that the mind is naturally disposed to conclude that there must exist *a first series of worlds* more or less analogous to this earth, in which the souls of men, at their entrance on the limitless career which opens before them, still frail and not attaching themselves firmly enough to the laws of duty, find themselves exposed to the discipline of temptation, succumb to it or else triumph over it; little by little advance, in the way of amelioration, from one world to another, in the midst of trials always proportioned to the degree of feebleness and culpability, and arrive at last, after labors more or less prolonged, at the merit of being admitted into *the worlds of the higher series.* There shall be accomplished the definitive deliverance from all evil: the love of the good shall henceforth be so paramount that no one shall lapse from it; but all, on the contrary, animated by the desire of elevating themselves, and seconded in their efforts by the incessant grace of God and the co-operation of the blissful societies in the bosom of which they live amid all the splendors of nature, shall display to this end the activity of all their virtues, and draw nearer by a continual progress, more or less rapid, according to the energy of each individual, to the infinite type of perfection. . . .

"It is impossible not to recognize that there is no tradition which throws a light so clear as this on the ideas of liberty, of personality, of immortality. Delivered from all arbitrary constraint, man presents himself as the direct author of his destiny: not the sport of fatality, not the victim of original sin, it is *the individual himself who has determined in an anterior life the initial conditions of his present life, even as he is to determine in this the conditions of his life to come;* and the terrestrial world, with its diversities of good and of evil, gives us the image of that world we run the risk of entering to-morrow, unless we have known how to qualify ourselves for something higher. Above the region of troubled and confused existences in which forgetfulness takes place, from re-birth to re-birth, expands the region of luminous existences, in which memory, acquiring all its force, renders to each, with the full possession of his past, the full identity of his person, his completed individuality."

Such is the theodicy of "Terre et Ciel." If the author is not always successful in his endeavor to make his liberal philosophy harmonize with the theology of the Church by compelling old dogmas to assume a new and spiritual aspect, we cannot deny to him the merit of investing this ancient theory of the pre-existence and transmigration of souls with a fresh and abiding interest. His work has passed through six editions in France, and is deserving of an English version. In our own renderings of detached passages, we have done it but slender justice.

Pierre Leroux, the associate of Reynaud in editing a philosophical dictionary, was an eloquent advocate of the doctrine of the transmigration of souls. Of recent French works, in which the same doctrine is proclaimed, we have "Du Spiritualisme Rationnel, par G. H. Love (1862);" "La Raison du Spiritisme, par Michel Bonnamy (1868);" "Conferences sur l'Ame, par Alexandre Chaseray (1868)." The author of the last-named work does not appear to either admit or deny the recent phenomena. From his two concluding pages, we quote the following *résumé* of his views: —

"All discussion relative to Deity, to the government of the

world, to the origin and end of things, can result in nothing conclusive, for its object surpasses the reach of human intelligence. Besides, every proposition of this nature is of a secondary interest for man. Upon all these points, I declare myself a positivist. In my opinion, the only question veritably important in philosophy, *consists in knowing whether we have an immortal soul.* I am still so far a positivist in this, that I dismiss the Spiritualism of St. Thomas and of Descartes as undemonstrated and undemonstrable, and that I recognize *the method of physical observation as alone capable of conducting to certitude.* But science is as yet very uncertain; and I cannot resign myself to wait, when the question for us is between nothingness and eternal life. I have then, provisionally, recourse to those metaphysical reasonings which render as very probable the continuance of the soul at death. I attach myself with ardor to that verity which physiology, I do not doubt, will one day make clear to all.

"This capital point, immortality, being then sufficiently established by metaphysics, we remain in the presence of two principal systems in respect to the destiny of the soul: that of one single life followed by an eternity of recompenses or of chastisements; and that of an indefinite series of re-births, permitting a progress slow and continuous.

"Justice, reason, good sense, militate in favor of this latter system, which I adopt, in company with a crowd of emerited thinkers. And so the immortality of the soul, the universal solidarity of intelligent beings, their free will, and their successive transformations, according to the good or bad exercise of this free will, — such is my philosophical programme. My thought, firmly established on this basis, remains calm and serene, and has so remained through long years, catching glimpses of a future without end, and of felicities less sudden, less abruptly marvellous than those of the Christian heaven, but which satisfy both my reason and my taste more fully, and of which the hope is not counterbalanced by the frightful risk of a fall to the bottom of an infernal abyss."

In his way to these conclusions, Chaseray says,—

"I shall not invoke the reasoning which consists in saying, 'The soul is a substance simple, immaterial, and indivisible; consequently it cannot perish, since death is a decomposition, a disjunction of parts.' This reasoning is good only for Spiritualists,* for it accepts as admitted a contested point; this, namely, that there exist immaterial substances, and that the soul is an immaterial substance.

"Here, you see the weak side of metaphysics. It reasons; but as it is compelled to take its stand on an hypothesis, a supposition, we ruin its reasoning in contesting the truth of its premises. . . .

"A proof of the duration of the soul, which is more generally admitted, is that which springs from the aspiration of man towards the Infinite. The aspiration of a being, it is said, is the measure of its destiny. Now, the aspirations of man being without bounds, it follows that his destiny, too, is limitless.

"M. Flourens draws a similar conclusion from the infinite problems which the human mind strives in vain to solve. 'These problems,' he says, 'which we cannot solve in this world, must have their solution in another; and here, if I mistake not, we have one of the surest signs that there *is* another.'"

"Where, in the plan of creation," asks Reimar, "do we find instincts falsified? Where do we see an instance of a creature instinctively craving a certain kind of food, in a place where no such food can be found? Are the swallows deceived by their instinct when they fly away from clouds and storms to seek a warmer country? . . . Yes: the voice of Nature does not utter false prophecies. . . . And if this be true with regard to the impulses of physical life, why should it not be true with regard to the superior instincts of the soul?"

* By *Spiritualists*, Chaseray here means those who are such through metaphysics; not Spiritualists through the phenomena of somnambulism, and the powers manifested by mediums, seers, &c.

Chaseray objects to that refined Spiritualism which would reject all notion of actual space-filling substance. Because anatomists can detect nothing that leaves the body after death, he does not admit that nothing may actually be disengaged. To combat the notions of the extreme materialists, who deny the existence of a soul, *he becomes himself a materialist, to a certain extent;* that is to say, he considers man as composed of two principal elements, a body and a soul, of which *the one*, after death, remains, by its very grossness, perceptible to the eye, and which follows the dissolution of its parts and their transformation into new principles, animal or vegetable, and of which the other escapes the view and the touch by its subtilty. "Let us distrust," he says, "our imperfect senses: there are so many substances which we can neither see nor feel! Let us not go so far as to deny the duality of the human being, because the scalpel of the anatomist is impotent to make a principle eminently subtile reveal itself to our eyes.

"If this principle really exists, and if the soul is material, does it necessarily follow that it is mortal? No: man is not necessarily pushed into nothingness in the hypothesis of materiality. 'Though man should be wholly nothing but matter,' says Charles Bonnet, in his 'Essai Analytique,' 'he would be none the less perfect, none the less a candidate for immortality.' This was also the opinion of George Salzer, who sought to prove that the immortality of the soul does not depend exclusively on its simplicity; that a materialist, who admits a soul distinct from the body, might attribute to this soul another life after our terrestrial death."

It is, then, through mistake, that some writers maintain that if the soul survives the body, *it must be because it is immaterial.* The question of immortality is not necessarily bound up in that of spirituality. The philosophers of antiquity, and all the early Fathers of the Church, believed in a material or corporeal soul; and, among the moderns who believed the same, we may cite Averroès, Politian, Pomponatius, Cardan, Viviani, Hobbes.

"Our soul," says Irenæus, "is not incorporeal except in comparison with gross bodies." "The matter of the soul consists in heat," says Lactantius. "The soul is nothing, if it is not a body," says Tertullian. *Nihil, si non corpus!* In the third century, Roger Bacon recognized a spiritual matter and a corporeal matter, a spiritual form and a corporeal form. It was St. Thomas Aquinas who first introduced into the Church the doctrine of pure Spiritualism; and, according to him, millions of spirits might find room to dance on the point of a needle. Descartes did the same for philosophy.

From the writings of Cabanis, Broussais, and Azaïs, the leading materialists of the early part of this century, Chaseray says that passages favorable to the idea of an immortal soul may be produced. Cabanis, for example, who sees nothing but organism, who explains all by organism, who regards the brain "as an organ designed specially to produce thought, as the stomach and intestines are to accomplish digestion, the liver to filtrate bile," &c., cannot at last avoid the declaration, that to bring all this about, to give life to this carcass, there must be "a principle or vivifying faculty which nature fixes in the germs or spreads in the seminal fluids." In a posthumous letter, he is still more explicit. "It is impossible for us to affirm," he says, "that the dissolution of the organs involves that of the moral system, and, above all, of the cause which renders us susceptible of perception; since we do not know it in any manner, and in all probability are interdicted from ever knowing it. Now, it suffices for those who would establish the persistence of this cause after the destruction of the living body, that *the contrary opinion cannot be demonstrated by positive arguments.*"

Rather more modest was the profound Cabanis than the impetuous Messrs. Vogt, Büchner, and Moleschott, who cannot repress their indignation because people will still heed this old wives' story of a future state.

M. Azaïs is a materialist of a still more decided type; and he says, "The soul which resides in the central part of the brain is formed by the agency of the organs of sense, and by the magnetic commerce of these organs with external beings." And yet this philosopher does not draw from his premises the conclusion which we might expect; namely, that the destruction of the organs involves the dispersion of ideas and the complete annihilation of the intelligent being. On the contrary, he says, "While time enfeebles, alters, destroys the body of the sage, it perfects his soul; and this progress indicates a high destiny. It is, in reality, the soul of the sage which is the ultimate object of the composition of the world; it is the soul of the sage which ought to be strengthened and preserved by the laws which govern the universe. No other result would be worthy of this sublime work."

The physiologists, too, are claimed by Chaseray in support of his views.

Charles Bonnet, the great naturalist and physiologist (1720–1793), believed, like most modern Spiritualists, that to maintain the human personality, which must consist, above all, in the *memory*, and to maintain a link between the present and the future state of man, there must be an ethereal and indestructible body to which the soul remains united after death, and which is the germ of the new body destined to perfect the faculties of man in another life.

M. Flourens has declared that the "bad philosophies must not pretend to find their support in physiology."

Alfred Maury, after allowing these words to escape him, namely, "The intelligence is, after all, a function of the brain," hastens to add, in a note, "I do not pretend to deny the action of the soul, but I would remark that this action is always closely connected with the play of the organism."

Milne Edwards says, that he does not regard the organization as being *all* in the economy of living bodies.

M. Gratiolet writes, "The system which best satisfies common sense, is that which admits the individuality of souls and their existence independently of a *certain* body."

Finally, the celebrated professor, Rodolph Wagner, says at Göttingen, in the midst of an assembly of *savants*, "The moral which flows from materialism is this: Let us eat and drink; to-morrow we shall be no more."

Other *savants*, somewhat numerous, it must be admitted, the Vogts, Rostans, Büchners, Robins, Moleschotts, persist in seeing in the phenomenon of thought nothing more than a cerebral function, a pure effect of organism.

"To think without a brain," objects M. Etienne Vacherot, "seems to me as great a miracle as to feel without a nervous system, and to perceive without organs."

"And yet," retorts Chaseray, "this sort of miracle is accomplished every day; and the least contested phenomena of somnambulism and animal magnetism subvert the biological notions based on the ordinary state of man. The organism would seem to be the instrument indispensable for the exercise and development of the faculties of the soul, rather than the condition necessary for the existence of these same faculties. Without organism, the soul is in repose; it may be compared to the engineer without a locomotive to conduct, or the musician without an instrument on which to perform.

"This hesitation of science, *this division of the doctors into two camps*, denotes the absence of positive proofs on either side, and consequently leaves subsisting in their entirety the metaphysical and moral proofs which make the immortality of the soul so probable. We must not allow ourselves to be imposed on by the audacity of these *organizationists*."

If M. Chaseray would acquaint himself with the great facts recorded in this volume, he would see that the "audacity" of these *organizationists* is, as we have shown from their own admissions, wholly based on their denial, or, as we contend, their *ignorance*, of phenomena known at this time to several million intelligent contemporaries.

But M. Chaseray is not without hope of a scientific proof of the fact of the soul's immortality. He says, "The day when physiology shall have proved the existence of the soul, shall have made it appear that an incorruptible substance separates itself at death from the discarded organism, this proposition will pass from the domain of metaphysics into that of the positive sciences; from probable it will become certain. I do not despair of this success."

Christian Garve, a German writer (1742-1798), seems to have entertained a belief not unfavorable to the theory of progressive existences. He writes, "The greatest encouragement to intellectual progress arises from our belief in one supreme Fountain of Wisdom, toward which we may continually advance; while, as we reverently approach that Source of mental light, the obscurities hanging about our present defective vision will gradually pass away. Without such a faith, I must look upon the world from a melancholy point of view.

"I behold around me a vast universe crowded with innumerable objects of interest, all possessing powers and qualities of which myself and my fellow-creatures can only understand a minute part.

"Is there not a Supreme Mind which comprehends the whole more perfectly than we understand the minutest portions of it? If I doubt this, how hopeless must appear my efforts toward intellectual satisfaction! For how can I, in my short life, hope to gain, by the slow process of experimental inquiry, a knowledge of this vast world around me, or to answer the deepest questions which my own rational nature suggests? If myself, and other finite creatures like myself, are the only intellectual beings, how little can we ever know of ourselves and of the universe!

"Is it not more cheering to believe that the rays of light in our own mind descend from one central Sun, than to imagine that our finite minds are the only illumined spots amid a wide creation left in darkness? . . . If this picture of the world were true, what proportion would there be between the massive and innumerable objects of material nature, and the few intellectual beings called mankind! . . . Let us believe that as our feeble corporeal frames are surrounded and supported by a vast material world, so our finite minds are under the sway of an infinite intellectual Power. We shall now see a just proportion between mind and matter. The world now becomes a noble object of unceasing study. The attainment of truth appears at least possible."

Among those American Spiritualists who accept the pure Platonic doctrine of pre-existence, if not of metempsychosis, we may mention A. Bronson Alcott, a rare example, in these times, of the veritable sage. He writes: "Every creature assists in its own formation, souls being essentially creative, and craving form. . . .

"Throughout the domain of spirit, desire creates substance, wherein all creatures seek conjunction, lodging, and nurture. Nor is there any thing in nature, save desire, holding substances together; all things being dissolvable and recombinable in this spiritual menstruum. . . .

"Under the sway of occult forces we partake of preternatural insights, having access to sources of information unopened to us in our wakeful hours. Vast systems of sympathies, antedating and extending beyond our mundane experiences, absorb us within their sphere, relating us to other worlds of life and light.

"For never is the sleep so profound, the dream so distracting, as to obliterate all sense of the personality; despite these vagaries of the night, these opiates of the senses, memory sometimes dispels the oblivious slumbers, and recovers for the mind recollections of its descent and destiny. Some reliques of the ancient consciousness survive, recalling our previous history and experiences. . . .

"Ancient of days, we hardly are persuaded to believe that our souls are no older than our bodies, and to date our nativity from our family registers, as if time and space could chronicle the periods of the immortal mind by its advent into the flesh and decease out of it. . . .

"None of us remember when we did not remember, when memory was nought and ourselves were unborn. Memory is the premise of our sensations: it dates our immortality. . . .

"Moreover, the insatiableness of our desires asserts our personal imperishableness. Yearning for full satisfactions, while balked of these perpetually, we still prosecute our search for them, our faith in their attainment remaining unshaken under every disappointment. Our hope is eternal as ourselves, — a never-ending, still-beginning quest of our divinity. Infinite in essence, we crave it in potence. The boundlessness and elasticity of the mind, its power of self-recovery, uprise from temporary obstructions, self-imposed or from temperament, are assurances made doubly sure of our soul's infinitude and longevity. . . .

"'Every thing aspires to its own perfection, and is restless till it attain it, — as the trembling needle till it find its beloved north. And the knowledge of this is innate as is the desire, else the last had been a torment and needless importunity. Nature shoots not at rovers. Even inanimate things, while ignorant of their perfection, are carried toward it by a blind impulse. But that which conducts them knows. The next order of beings have some sight of it, and man most perfectly till he touch the apple.' Our delights suckle us life-long, our desires being memories of past satisfactions; and we here but sip pleasures once tasted to satiety. . . .

"Still heaven is, our hearts affirm against every disappointment; and whether behind or before us, as memory or as hope, 'tis to be ours; our port and resting-place sometime in the stream of ages."

The poets have often availed themselves of the Platonic theory of pre-existence. Virgil, in the Sixth Book of his Æneid, teaches very distinctly the doctrine of the transmigration of souls. "These souls," says Anchises, "destined for other bodies, drink, in the waters of Lethe, a long oblivion of things past." Robert Southey, in one of his published letters, remarks, "I have a strong and lively faith in a state of continued consciousness from this stage of existence; and that we shall recover the consciousness of other stages through which we previously may have passed, seems to me not improbable." And again he writes, "The system of progressive existences seems, of all others, the most benevolent; and all that we do understand is so wise and so good, and all we do, or do not, so perfectly and overwhelmingly wonderful, that the most benevolent system is the most probable."

In his novel of "Lucretia," Lord Lytton observes: "What we call eternity may be but an endless series of those transitions which men call *deaths;* abandonments of home after home, ever to fairer scenes and loftier heights. Age after age, the spirit, *that glorious nomad*, may shift its tent, fated not to rest in the dull Elysium of the heathen, but carrying with it evermore its elements, — activity and desire. Why should the soul ever repose? . . . Labor is the purgatory of the erring; and it is none the less the heaven of the good."

Walter Scott, in his diary, under the date of Feb. 17, 1828, remarks, "I cannot, I am sure, tell if it is worth marking down, that yesterday, at dinner-time, I was strongly haunted by what I would call the sense of pre-existence, in a confirmed idea that nothing which passed was said for the first time."

Tennyson repeatedly refers to this mood; and in "The Prelude," by Wordsworth, we find the following passage: —

"Our childhood sits,
Our simple childhood, sits upon a throne
That hath more power than all the elements.
I guess not what this tells *of Being past*,
Nor what it augurs of the life to come."

In his "Intimations of Immortality, from Recollections of Early Childhood," Wordsworth is still more direct in his reference to that key to many mysteries, the doctrine of pre-existence: —

"Our birth is but a sleep and a forgetting:
The soul that rises with us, our life's star,
Hath had elsewhere its setting,
And cometh from afar:
Not in entire forgetfulness,
And not in utter nakedness,
But trailing clouds of glory do we come
From God who is our home."

CHAPTER XIII.

PSYCHOMETRY.

"'Tis immortality deciphers man,
And opens all the mysteries of his make.
Without it, half his instincts are a riddle;
Without it, all his virtues are a dream." — *Young.*

EVERY step taken in advance by science is in harmony with the great facts which Spiritualism reveals. We think that our popular friend, Agassiz, notwithstanding his exhibition of vexation when he found he could not manipulate the spiritual phenomena as easily as he could some rare specimens of cod and haddock, was nearly right when he said, "We trust that the time is not distant when it will be universally understood that the battle of the evidences will have to be fought on the field of physical science, and not on that of the metaphysical."

The further science carries its analysis, the more does the material world lose that character of rigidity which our external senses attach to it; and the more does it seem plastic under spiritual laws. Modern chemistry has shown us that all solid bodies may exist as aeriform; that even iron may be converted into an invisible gas; and the diamond which to our senses is inert, ponderable matter, may be volatilized in the fire of the burning mirror so as to develop neither smoke nor cinders. On the other hand, fire, essentially volatile, can be condensed in the calcination of metals, so as to become ponderable.

From these facts, De Montlosier deduces the interesting conclusion, that all the bodies of the universe might be volatilized and made to disappear in those spaces which our ignorance calls ***the void***; and that, in its turn, what we call ***the void*** might be condensed, so that the number of the celestial bodies might be multiplied a hundred-fold; and, through all this, the universe would not have changed in its nature and essence, though it would be changed in its phenomenal aspect.

One of the most eminent physiologists of the day, Schroeder van der Kolk, remarks: "In my opinion, this untoward distinction between the *material* and the *immaterial* has singularly contributed to confuse our ideas. Should we not proceed more surely in distinguishing in nature *that which it is possible for us to perceive by the senses, and that which escapes their scrutiny? Who gives us the right to admit that the limits of nature do not go beyond those of our organs?*"

Everywhere, under the appearance of concretion and hardness, living elemental forces are latent, and the slightest variation in the equilibrium and correlation of these might alter the face of the universe, and the most solid substances might vanish like a dream.

In the remote distances between the planets, there is no inactive void. "The space," says Oersted, "is filled by ether and penetrated by the attractive forces by which the whole universe is held together. The ether itself is an ocean, whose waves form light, that great connecting link which conveys messages from globe to globe and from system to system. The wonders unravelled by science prove that we are not isolated beings, but that we are related to the whole universe."

Thus science comes in to confirm that great deduction of Spiritualism, which assures us of the solidarity of all life and intelligence in whatever world or system they may be developed, — that we none of us are aliens in God's universe, but cosmopolitans, entitled to the freedom of the whole of it; ay, born to make all the past and all the future our heritage; our earnestness and our efforts being always the measure of our acquisitions in goodness and in knowledge. And for this infinite work we have an eternity before us.

Besides the assurances of immortality which Spiritualism gives, in revealing to us the phenomena of spiritual action and intelligence, it offers an added and more important confirmation in the revelation it makes of powers in the human soul, which proclaim that it is not merely the creature of space and time, but that in eternity and infinity is to be found its native atmosphere; and that such are its capacities of clairvoyance, that not only the remotest planet, but the past eternities, may be hereafter scanned by its unconditioned vision. The argument for immortality, drawn from these capacities, has already been presented, and it seems to us unanswerable.

Another stupendous fact, which the phenomena we have been dealing with disclose, is this,—and it is one which, more than all other considerations, except the consciousness that God sees us, ought to keep us from defiling the soul by any act which in our better moments we may deplore,—*Memory is imperishable;* and all thought and all action leave their eternal record in the organic structure of our very souls. Nothing happens, not the most fleeting and seemingly trivial occurrence of our lives, that may not be, ages and æons hence, reproduced to our own consciousness, as well as to that of others, independently of our own will or co-operation.

"There is a power," says Voltaire, "that acts within us without consulting us."

Much goes on in the soul, of which consciousness takes no note *at the time;* but all mental processes, conscious or unconscious, leave their record, and that record is ineffaceable. Modern physiologists tell us of "latent thought," of "unconscious cerebration," of the "automatic action of the mind," &c.; and Dr. Maudsley says, that "consciousness is not co-extensive with mind;" that "mental power is being organized before the supervention of consciousness;" and that "the preconscious action of the mind, and the unconscious, are facts of which self-consciousness can give us no account."—"The brain not only receives impressions unconsciously, registers impressions without the co-operation of consciousness, elaborates materials unconsciously, calls latent residua again into activity without consciousness, but it responds also as an organ of organic life to the internal stimuli which it receives unconsciously from other organs of the body."

All this is true, but not the whole truth. It is but a partial view of the facts. Consciousness may not take note at the moment of all this unconscious or automatic action, but it is inscribed where consciousness can read it in some supreme moment, God's moment, perchance, when long latent memories start up with a vividness that commands the concentration of all our faculties in one effort of attention. We have heard how, when persons are drowning, the incidents of a lifetime pass in a few seconds before the mental ken. We ourselves experienced the sensation once, when we anticipated instant death from an accident in a carriage.

In his "Biographia Literaria," Coleridge mentions a case, also authenticated by Abercrombie, of a young and ignorant woman who, during a fever, talked incessantly in Latin, Greek, and Hebrew; and who, as it was afterwards discovered, had lived with a learned man, who was a great Hebraist.

"This authenticated case," says Coleridge, "furnishes both proof and instance, that reliques of sensation may exist for an indefinite time, in a latent state, in the very same order in which they were originally impressed; and as we cannot rationally suppose the feverish state of the brain to act in any other way than as a stimulus, this fact (and it would not be difficult to adduce several of the same kind) contributes to make it even probable that all thoughts *are in themselves imperishable;* and that, if the intelligent faculty should be rendered more comprehensive, it would require only a different and apportioned organization, the body celestial instead of the body terrestrial, to bring before every human soul the collective experience of its whole past existence. *And this, perchance, is the dread book of judgment, in the mysterious hieroglyphics of which every idle word is recorded.*

"Yes, in the very nature of a living spirit, it may be more possible that heaven and earth should pass away, than that a single act, a single thought, should be loosened or lost from that living chain of causes, with all the links of which, conscious or unconscious, the free will, our only absolute self, is co-extensive and co-present."

"All mental activities, all acts of knowledge," says Sir William Hamilton, "which have been once excited, persist; *we never wholly lose them*, but they become obscure. This obscuration can be conceived in every infinite degree, between incipient latescence and irrecoverable latency. The obscure cognition may exist simply out of consciousness, so that it can be recalled by a common act of reminiscence. Again, it may be impossible to recover it by an act of voluntary recollection; but some association may revivify it enough to make it flash after a long oblivion into consciousness. Further, it may be obscured so far that it can only be resuscitated by some morbid affection of the system; or, finally, it may be absolutely lost for us in this life, and destined only for our reminiscence in the life to come."

The facts of clairvoyance go to prove that the "absolute loss," of which Sir William speaks, does not take place even in this life. Sir William was somewhat in advance of our Cambridge professors. He was well persuaded of the essential facts of clairvoyance. "However astonishing," he says, "*it is now proved beyond all rational doubt* that, in certain abnormal states of the nervous organism, perceptions are possible through other than the ordinary channels of the sense."

In his essay on the "Philosophical Teaching of Magnetism," M. Dupotet says, "Let thy actions be virtuous; for know that thy soul will remember them all thy after life on earth, and the remembrance of them will be ineffaceable. Not on sand are human actions engraven, but in the conscience. *Whatsoever thou shalt have thought, shall be known by all who wish to know it.* For thee no more dissimulation is possible; no longer any mask. As thou wilt be able to read in others, so they in thee; and thy most trifling actions will appear like a cloud under a serene sky."

The phenomena of clairvoyance show that there is not the slightest exaggeration in all this. The remembrance of our slightest acts and thoughts may be suspended; but it is *eternally reproducible.*

We have seen that, all unconsciously, science, at every step of its progress, is revealing analogies with spiritual facts. A lecture was recently delivered at the Royal Institution by Professor Tyndall, in which he demonstrated, that a ray of light was allowed to traverse a strip of glass every time he caused it to set up a musical sound; the glass being held in a vice, and the light from an electric lamp polarized upon it. The same learned professor delivered a lecture on "The Rhythm of Flames," or "On Sounding and Sensible Flames," when he exhibited a flame some twenty inches in height, which fell down to eight on the slightest tap on an anvil. It responded to the tinkle of a bunch of keys or a few pence shaken together, the creaking of boots, the rustling of a silk dress or a piece of paper; while certain intonations of the voice threw it into violent commotion.

Grove says, in his "Correlation and Continuity," p. 161, "Myriads of organized beings may exist imperceptible to our vision, even if we were amongst them; and we might be equally imperceptible to them."

The universe is a vast whispering gallery, a boundless system of correlated influences; and the soul of man has the eternal freedom of the infinite "mansions."

The faculty which some *sensitives* have, like Mr. J. V. Mansfield, of learning the contents of a letter, or the mood of the writer, by simply feeling of the paper, has been so repeatedly tested as to be placed beyond a doubt. Others, as we have seen in the case of Home, by simply being brought in contact with a person, or by touching a lock of his hair, will have revealed to them incidents in his past life to an extent wholly inexplicable.

Heinrich Zschokke, the celebrated German writer, was one of these. He was instinctively a Spiritualist from his youth up, was well acquainted with the phenomena of rhabdomancy (divination by a rod or wand), which, he says, presented him with a new phase of nature, and which was, moreover, of considerable use to him in his mining operations. From personal experience, he believed in spiritual impressions and presentiments, especially as conveyed in dreams. But his most remarkable faculty was what he describes as a singular kind of prophetic gift he called his inward sight, but which was always an enigma to him. The following is his detailed account of it:—

"It is well known that the judgment we not seldom form at the first glance of persons hitherto unknown, is more correct than that which is the result of longer acquaintance. The first impression, that through some instinct of the soul attracts or repels us with strangers, is afterwards weakened or destroyed by custom, or by different appearances. We speak in such cases of sympathies or antipathies, and perceive these effects frequently among children, to whom experience in human character is wholly wanting. Others are incredulous on this point, and have recourse rather to the art of physiognomy. Now, for my own case: It has happened to me sometimes on my first meeting with strangers, as I listened silently to their discourse, that their former life, with many trifling circumstances therewith connected, or frequently some particular scene in that life, has passed quite involuntarily, and, as it were, dream-like, yet perfectly distinct before me. During this time, I usually feel so entirely absorbed in the contemplation of the stranger life, that at last I no longer see clearly the face of the unknown, wherein I undesignedly read, nor distinctly hear the voices of the speakers, which before served in some measure as a commentary to the text of their features.

"For a long time, I held such visions as delusions of the fancy, and the more so, as they showed me even the dress and motions of the actors, rooms, furniture, and other accessories. By way of jest, I once in a familiar family circle at Kirchberg related the secret history of a seamstress who had just left the room and the house. I had never seen her before in my life: people were astonished and laughed, but were not to be persuaded that I did not previously know the relations of which I spoke; for what I had uttered was the *literal* truth. I, on my part, was no less astonished that my dream-pictures were confirmed by the reality. I became more attentive to the subject, and, when propriety admitted it, I would relate to those whose life thus passed before me the subject of my vision, that I might thereby obtain confirmation or refutation of it. It was invariably ratified, not without consternation on their part. 'What demon inspires you? Must I again believe in possession?' exclaimed the *spirituel* Johann von Riga, when, in the first hour of our acquaintance, I related his past life to him, with the avowed object of learning whether or no I deceived myself. We speculated long on the enigma, but even his penetration could not solve it. I myself had less confidence than any one in this mental jugglery. Sò often as I revealed my visionary gifts to any new person, I regularly expected to hear the answer, 'It was not so.' I felt a secret shudder when my auditors replied that it was true, or when their astonishment betrayed my accuracy before they spoke. Instead of many, I will mention one example, which pre-eminently astounded me.

"One fair day in the city of Waldshut, I entered an inn (the Vine), in company with two young student-foresters; we were tired with rambling through the woods. We supped with a numerous society at the *table d'hôte*, where the guests were making very merry with the peculiarities and eccentricities of the Swiss, with Mesmer's magnetism, Lavater's physiognomy, &c. One of my companions, whose national pride was wounded by their mockery, begged me to make some reply, particularly to a handsome young man who sat opposite us, and who had allowed himself extraordinary license. This man's former life was at that moment presented to my mind. I turned to him and asked whether he would answer me candidly if I related to him some of the most secret passages of his life, I knowing as little of him personally as he did of me? That would be going a little further, I thought, than Lavater did with his physiognomy. He promised, if I were correct in my information, to admit it frankly. I then related what my vision had shown me, and the whole company were made acquainted with the private history of the young merchant; his school years, his youthful errors, and, lastly, with a fault committed in reference to the strong box of his principal.

I described to him the uninhabited room with whitened walls, where, to the right of the brown door, on a table, stood a black money-box, &c. A dead silence prevailed during the whole narration, which I alone occasionally interrupted by inquiring whether I spoke the truth. The startled young man confirmed every particular, and even, what I had scarcely expected, the last mentioned. Touched by his candor, I shook hands with him over the table, and said no more. He asked my name, which I gave him, and we remained together talking till past midnight. He is probably still living!

"I can well explain to myself how a person of lively imagination may form, as in a romance, a correct picture of the actions and passions of another person, of a certain character, under certain circumstances. But whence came those trifling accessories which *nowise concerned me*, and in relation to people for the most part indifferent to me, with whom I neither had, nor desired to have, any connection? Or, was the whole matter a constantly recurring *accident?* Or, had my auditor, perhaps, when I related the particulars of his former life, very different views to give of the whole, although in his first surprise, and misled by some resemblances, he had mistaken them for the same? And yet, impelled by this very doubt, I had several times given myself trouble to speak of the most insignificant things which my waking dream had revealed to me. I shall not say another word on this singular gift of vision, of which I cannot say it was ever of the slightest service; it manifested itself rarely, quite independently of my will, and several times in reference to persons whom I cared little to look through. Neither am I the only person in possession of this power. On an excursion I once made with two of my sons, I met with an old Tyrolese, who carried oranges and lemons about the country, in a house of public entertainment, in Lower Hanenstein, one of the passes of the Jura. He fixed his eyes on me for some time, then mingled in the conversation, and said that he knew me, although he knew me not, and went to relate what I had done and striven to do in former time, to the consternation of the country people present, and the great admiration of my children, who were diverted to find another person gifted like their father. How the old lemon merchant came by his knowledge, he could not explain, neither to me nor to himself; he seemed, nevertheless, to value himself somewhat upon his mysterious wisdom."

Emile Deschamps communicates to "Le Monde Musical," of Brussels (1868), the following account of his own experience in psychometry: "If a man believed only what he could comprehend, he would believe neither in God, in himself, in the stars which roll above his head, nor in the herbage which is crushed beneath his feet. . . .

"In the month of February, 1846, I travelled in France. I arrived in a rich and great city; and I took a walk in front of the beautiful shops which abound in it. The rain began to fall; I entered an elegant gallery. All at once I stood motionless; I could not withdraw my eyes from the figure of a lovely young woman, who was all alone behind an array of articles of ornament for sale. This young woman was very handsome; but it was not at all her beauty which enchained me. I know not what mysterious interest, what inexplicable bond, held and mastered my whole being. It was a sympathy subtle and profound, free from any sensual alloy, but of irresistible force, as the unknown is in all things. I was pushed forward into the shop by a supernatural power. I purchased several little things, and, as I paid for them, said, 'Thank you, Mademoiselle Sara.' The young girl looked at me with an air of surprise. 'It astonishes you,' I continued, 'that a stranger knows your name, and one of your baptismal names; but, if you will think for a moment of all your names, I will repeat them all to you. Do you think of them?' 'Yes, monsieur,' she replied, half-smiling and half-trembling. 'Very well,' I added, looking fixedly in her face, 'You are called Sara Adele Benjamine N——.' 'It is true,' she replied; and after some minutes of surprise she began all at once to laugh; and I saw that she thought that I had obtained this information in the neighborhood, in order to amuse myself with it. But I knew very well that I had not till this moment known a word of it, and I was terrified at my own instantaneous divination.

"The next and the next day I hastened to the handsome shop; my divination was renewed at every instant. I begged of Sara to think of something, without letting me know what it was; and, immediately, I read on her countenance her thought not yet expressed. I requested her to write with a pencil some words, which she should keep carefully concealed from me; and, after having looked at her for a minute, I, on my part, wrote down the same words in the same order. I had her thoughts as in an open book; but she could not in the slightest degree read mine, such was my superiority; but at the same time she imposed on me her ideas and her emotions. Let her think seriously on any subject, or let her repeat in her own mind the words of any writing, and instantly I was aware of the whole. The mystery lay betwixt her brain and mine, not betwixt my faculties of intuition and things material. Whatever it might be, there existed a *rapport* between us as intimate as it was pure.

"One night I heard in my ear a loud voice crying to me, 'Sara is very ill, very ill!' I hastened to her: a medical man was watching over her and expecting a crisis. That evening Sara had entered her lodgings in a burning fever; she continued in delirium all night. The doctor took me aside, and told me that he feared the worst result. From that apartment I saw the countenance of Sara clearly, and, my intuition rising above my distress, I said in a low voice, 'Doctor, do you know with what images her fevered sleep is occupied? She believes that she is at this moment at the grand opera at Paris, where she indeed has never been, and a *danseuse* gathers, amongst other buds, some hemlock, and, throwing it to her, cries, "*That is for you.*"'

"The physician thought I was delirious too; but some minutes afterwards the patient awoke heavily, and her first words were, 'Oh! how beautiful is the opera! but why did that handsome girl throw to me that hemlock?' The doctor was stupefied with astonishment. A medicine containing hemlock was administered, and in some days Sara was well."

According to Goethe, this same faculty of psychometry or inward sight was possessed by Lavater, the celebrated physiognomist (1741–1801). Goethe tells us that Lavater's insight into the characters of individuals "surpassed all conception;" and he speaks of it as one of those gifts which "seem to have something of magic in it." However this may be, we have his authority for asserting that Lavater believed in special providences, especially in answer to prayer, and that he had "a perfect conviction that miracles can be wrought to-day as well as heretofore." He tells us, too, that "his [Lavater's] system of physiognomy rests on the conviction that the sensible corresponds throughout with the spiritual, and is not only an evidence of it, but, indeed, its representative;" and, like Swedenborg and Spiritualists in general, he held that the future life was a continuation of the present, though under different conditions.

"Whatever may be conjectured or inferred," says Lavater, "in regard to the state of the soul after death, may be stated in the following thesis or axiom: 'MAN SHALL REAP AS HE HAS SOWN.' It is impossible to discover a more lucid or simple principle, or one capable of a wider application.

"There exists a general, natural law which governs every world, and every department of the physical, moral, intelligent, visible, and invisible worlds. It is this: 'Whatever is susceptible of affinity, attracts; the same species are mutually drawn to each other, unless thwarted by obstacles fortuitously interposed.'

"Every soul freed from matter *not only knows itself;* not only do the errors, distractions, and blindness which opposed it in the contemplation of itself, and in the knowledge of its powers, weakness, and shortcomings, cease, but it feels itself attracted toward every thing which has affinity for it, by an interior, irresistible force; while it feels repulsion for whatever is alien to it.

"Its moral or religious character gives it a determinate direction. Whoso is good goes toward the good. Its needs, its attractions for the good, give it this direction. The impure soul is repelled among the impure. Just as a heavy weight, tossed into open space, would fall swiftly into the abyss, so impure, immoral, and irreligious souls will inevitably go to join their like."

Lavater, in this, merely sums up what is highest and most uniform in the teachings of Spiritualism.

Among American poets of promise was Forceythe Willson. Born in Indiana in 1837, he died in 1867. He, too, was a psychometrist. He would take a letter, and, pressing it to his forehead, announce accurately the character and personal appearance of the writer. He, too, like Oberlin, professed to have interviews with his departed wife. There is a remarkable poem from his pen, entitled "The Voice," which seems to have reference to the fact. We quote the following passages: —

"My soul to ecstasy was stirred;
It was a Voice that I had heard
A thousand blissful times before,
But deemèd that I should hear no more
'Till I should have a spirit's ear
And breathe another atmosphere. . . .

"'Where art thou, blessed spirit, where,
Whose voice is dew upon the air?'
I looked around me and above,
And cried aloud, 'Where art thou, Love?
Oh let me see thy living eye
And clasp thy living hand, or die!'
Again upon the atmosphere
The self-same words fell, '*I am here!*'

"'Here? Thou art here, Love?'—'*I am here!*'
The echo died upon my ear!
I looked around me everywhere,
But ah! there was no mortal there!
The moonlight was upon the mart,
And awe and wonder in my heart.

I saw no form! — I only felt
Heaven's peace upon me as I knelt,
And knew a Soul Beatified
Was at that moment at my side."

Between Willson and a neighbor a coolness had arisen. But as Willson was about to leave town, the neighbor met him at the cars, and, holding out his hand, said, "We must not part with a cloud between us." Willson grasped the proffered hand with emotion, and replied, "*The good man within me* told me to say to you just what you have said to me; but the devil would have conquered, I fear, if you had not spoken. *We shall never meet again;* for within six months I shall have joined my wife in the land of the hereafter."

The presentiment was accurate. Within four months he died.

William Denton, the accomplished professor of geology, says, "There are forces coming out from all forms of matter: we cannot see them with the material eye, hear them with the material ear, know them by the material senses; but the soul has faculties by which to grasp them."

Reichenbach discovered that from every magnet, in proportion to its length, flowed forth luminous rays, and that some individuals are so susceptible as to be able to see these, while men generally have not the slightest idea of their existence. Some of the persons he experimented upon, were enabled to perceive the presence of a magnet, even when twenty to fifty feet distant. The luminous aura emitted from shells, minerals, magnets, the human body, and each of its organs, and, indeed, from all objects around us, serves, in some subtle way, to retain and convey to the seer an impression of their past history and surroundings. Mr. Denton's experiments have demonstrated that there are certain *sensitives* who can receive influences from the fossil remains of the far-off ages.

In that able contribution to the literature of Spiritualism, "The Soul of Things," he presents facts showing that the soul of man has power to read, even in the inorganic substances of nature, their eternal record. Thus, in a spiritual sense, whatever *has* been, *is* now. No mountain ever stood that stands not now; no human being ever shed an influence, who sheds it not now. Only the spiritual is the abiding.

"The air," says Professor Babbage, "is one vast library, on whose pages are for ever written all that man has ever said, or woman whispered."

The spirit, when it passes on, takes with it every thing necessary for the continuance of its individuality. Deprive a man at once of his good or his bad tendencies, and you rob him of his identity: he becomes somebody else at once. God is very patient, since he has an eternity in which to deal with us. He can put up with very slow gradations of progress; even with retrogressions. The loss will be our own; and the effort must be our own if we would make up for the loss.

Psychometry tells us, that the soul is what Aristotle calls an entelechy, or actuality, unlimited by this enclosure of flesh. "We ought not," says Dr. Bertrand, "to consider our body as containing our soul in the manner in which a thing material contains another; but only as limiting the extent of the matter in which it is given it to act and feel."

What an incentive to a scrupulous morality* would the facts of psychometry be, if rightly pondered! They show that every act and thought of our existence are for ever reproducible for ourselves and all spiritual intelligences to scan at pleasure; that the warp and woof of our spiritual substance include all that we have desired, done, and thought; that God's judgments are

* "If nothing could be Evil, nothing would be Good,
But all things whatsoever would be indifferent and unmoral.
The possibility of Vice is the condition of Virtue.
So likewise is Evil the revelation of Good,
And Human weakness of Divine strength.
If we had no lower impulses, no meaner passions,
No drawings toward the worse, no susceptibility of temptation,
Never should we distinguish God's voice in Conscience,
Nor know that God is moral, nor frame moral judgments."
Theism, by Francis W. Newman

recorded against us in the very structure of our being, as fast as our sins are committed.

There is no waiting for rewards and punishments. Poor conceptions of a heavenly reward must he have who regards it as something outside of the state of his own soul. Foretastes of heaven may be had even here by every righteous, loving, and aspiring spirit. All the good we do, all the pure happiness we enjoy, are happiness and good for ever. All our acquisitions in knowledge, in art, in virtue, are made for ever, and shall be the vantage-ground of ever new attainments.

On the other hand, the hell of the evil-doer yawns for him even now; and, in one sense, it is eternal; for, as we have seen, though the sinner may forsake his sin (and in every soul there is a redeeming principle antagonistic to everlasting wrong), the sin will not forsake *him*. Its record, which is itself, is for ever plain to the psychometrist of the spirit-world, and the sinner's own memory will not let it go.

The day of judgment, when is it, if not now? Shall He to whom the universe is a very small thing, need the forms of our poor human assizes for his purposes in the creation of man? The pressure of his laws is upon us every moment, spiritually as well as physically. We can no more violate his law of right, without a simultaneous penalty, than we can thrust our finger in the fire without injury. The spiritual, like the physical, offence, carries its punishment. We have but imperfect conceptions of the powers of our own souls. Clairvoyance, and the facts of Spiritualism, give us, here and there, a glimpse of them.

There will be no more awful tribunal than that of the awakened conscience; no more dreadful sentence than that which the roused and clear-seeing mind of man shall some day, in some stage of being, near or remote, pronounce, according to the degree of his development and his intelligence, against himself.

God's pardon! Can God arbitrarily or vicariously pardon? Yes: in all the ways by which we may truly seek it, God's pardon may be had, arbitrarily or freely, directly or vicariously;

through our own merits, or through another's, or through no merits at all; through reverence for a Saviour or saint of old time, or through heart-crushing affection for a poor little dying infant of to-day. Though our sins are as scarlet, his pardon goes with the asking.

But the soul's own pardon, — what of that? God, in his infinite mercy, may let the waters of Lethe serve us for a time; but, by the inexorable laws of our spiritual constitution, the soul's day of judgment must come, sooner or later, and the later the more terrible. The fearfulest judgment-seat will be that which in some moment of illumination, of expansion of our natural powers, we shall find established within the domain of our own intellectual being. Judge, jury, witnesses, will be there, —

> " *There* is no shuffling; there the action lies
> In his true nature; and we ourselves compelled,
> Even to the teeth and forehead of our faults,
> To give in evidence."

In that day of the soul, we can no more escape the ineffaceable brand which conscience will put upon us, than we can run from our own shadow in the sunlight.

Such are the teachings of psychometry.

CHAPTER XIV.

COGNATE FACTS AND PHENOMENA.

All life is Thy life, O Infinite One, and only the religious eye penetrates to the realm of True Beauty." — *J. G. Fichte.*

NO one who has carefully examined the facts of modern Spiritualism, can fail of being struck by the analogy they bear to many of the miraculous incidents recorded in the Bible. Nothing can be more certain than that the Bible distinctly recognizes a class of phenomena, rejected by modern skepticism as contrary to the order of nature, but the possibility of which is clearly proved in the attestations of thousands of intelligent contemporaries to similar occurrences.

Instances of the exercise of the prophetic faculty, by somnambulists and others, have been not unfrequent during the present century. The prophet Hosea represents God as saying, "I have spoken by the prophets, I have multiplied visions."

What clearer recognition of some of the higher experiences of somnambulism and trance can we have than the following: "God speaketh once, yea, twice, yet man perceiveth not; in a dream, in a vision of the night, when deep sleep falleth upon men, in slumberings upon the bed, and sealeth their instruction, that he may withdraw man from his purpose, and hide pride from man."

Among the earliest spiritual manifestations of the Old Testament are the spirit-voices. The Lord spake face to face with Adam and Eve (Gen. ii. 16, and again, Gen. iii. 9–22); again, he spake with Cain (Gen. iv. 6), and also spake and walked with Enoch.

What a life of spiritual experiences was that of Abraham! In Gen. xviii. is related the memorable visit of the three angels to him, and afterwards their visit to Lot, — "Be not forgetful to entertain strangers, for thereby some have entertained angels unawares." Angels of the Lord met Jacob on his return from Padanaram (Gen. xxxii. 1.); also at Peniel an angel met and wrestled with Jacob: refusing to give his name, he wrestled all the night, until he said, "Let me go, for the day breaketh." Moses was evidently in constant communication with the spirit-world.

An angel appeared to Hager (Gen. xvi.); and two to Lot (Gen. xix.). One called to Hagar (Gen. xxxi.); and to Abraham (Gen. xxii.); one spake to Jacob in a dream (Gen. xxxi.); one appeared to Moses (Exod. iii.); one went before the camp of Israel (Exod. xiv.); one spake to all the children of Israel (Judges ii.); one spake to Gideon (Judges vi.); and to the wife of Manoah (Judges xiii.); one appeared to Elijah (1 Kings xix.); one stood by the threshing-floor of Ornan (1 Chron. xxi.); one talked with Zachariah (Zach. i.); one appeared to the two Marys at the sepulchre (Matt. xxviii.); one foretold the birth of John the Baptist (Luke i.); one appeared to the Virgin Mary (*ibid.*); to the shepherds (Luke ii.); one opened the door of Peter's prison (Acts v.); two were seen by Jesus, Peter, James, and John (Luke ix.). It will not do for scriptural objectors to say these angels were a distinct order of beings from man; for those seen by the apostles were Moses and Elias, and that seen by John (Rev. xxii.), though called by him an angel, avowed himself to be his fellow-servant, and "one of his brethren, the prophets."

The instances of miraculous cures are numerous. Read Lev. xv. and xvi., Num. v., 1 Kings xiii., 1 Kings xvii., 2 Kings ii. 4; iv. 5; xix. 20; Josh. x., &c. Hundreds of such cases could be cited from the Old Testament, hundreds from the New Testament. Christ said this power would continue, and that these signs should *always* follow those that believe: "In my name shall they cast out devils; they shall speak with new tongues; they shall take up serpents; and if they drink any deadly thing, it shall not hurt them; they shall lay hands on the sick, and they shall recover" (Mark xvi. 17, 18).

The reported cures of Dr. Newton, the Zouave Jacob, and many others at the present day, are certainly not unworthy of investigation, if we are to believe passages like the above.

Modern skepticism accounts those persons fatuous who say, "We have seen writing that could never have been done by mortal hand;" or who say, "Our hands were moved to write involuntarily." And yet spirit-writing appeared on Belshazzar's palace-wall; and Ezekiel (ii. 9) says, "And when I looked, behold a hand was sent unto me; and lo, a roll of a book was therein, and he spread it before me, and it was written within and without."

There we have two distinct instances of spiritual manifestation, very similar to those coming under our own notice in the present day. Spirit-hands and spirit-writing were seen, without the seers being either mad, dreaming, or even entranced.

"All this," said David, "the Lord made me understand in writing, by his hand upon me, even all the marks of this pattern" (1 Chron. xxvii. 19). See, also, 2 Chron. xxi. 12, where it is stated that "There came a writing to Jehoram from Elijah the prophet;" and this must have been some years after Elijah's death; though some of the commentators quietly assume, in a marginal note, that the said writing was written *before* the prophet's death!

We have accounts of visions and trances, such as those of Balaam, the son of Beor, who heard the words of God, saw the vision of the Almighty; falling into a trance, having his eyes open, — a state accurately described, and which is familiar to those acquainted with certain forms of somnambulism; of Isaiah, the son of Amos, which he saw concerning Judah and Jerusalem; of Ezekiel, the priest, by the river Chebar, when the heavens were opened, and he saw visions of God; of Daniel, in the palace of Shushan, and by the great river Hiddekel; of Peter, at Joppa, who, when he had gone upon the house-top to pray, fell into a trance, and saw heaven opened; of Paul, who was in a trance while praying in the temple at Jerusalem; of John, the divine, in the isle that is called Patmos, and who was commanded by a voice from the heavens, "What thou seest write in a book;" and who, at the conclusion of his Apocalypse, tells us, "And I John saw and heard these things."

That spirits can move material objects, or manifest themselves materially to the touch of mortals, is clearly implied in such narratives as those of the angel who delivered Peter out of prison; of the angel who rolled away the stone from the door of the sepulchre; of the apostle Philip whom "the Spirit of the Lord caught away" and bore from Gaza to Azotus; and of Ezekiel's experiences, almost literally like those of some of our contemporaries, as mentioned in this volume: "So the Spirit lifted me up, and took me away. . . . And he put forth *the form of an hand*, and took me *by a lock of mine head*, and the Spirit *lifted me up* between the earth and the heaven."

Until within the last few years, who was more fit for a lunatic asylum than the man who would believe that a spirit could lift a table, "thus violating the law of gravitation"? Yet axes of iron were made to swim, and men were carried through the air, so often, indeed, that Obadiah was afraid lest the Spirit should carry away Elijah, after he had announced his presence to the king (1 Kings xviii.).

Of spiritual apparitions, it may be sufficient to refer to that of Samuel the prophet, who spoke to Saul, and foretold the impending fate of the king and of his sons.

Seership, in the earlier periods of Hebrew history, was a distinctive and honorable office. Thus we have Iddo, the seer; Gad, the king's seer; Jeduthun, the king's seer; and many more, whose sayings were written down and placed in the Jewish archives. We read of the time of Samuel, "He that is now called a prophet, was before-time called a seer;" and that, "The word of the Lord was precious in those days, there was no open vision;" or, as De Witte translates it, "The word of the Lord was rare, in those days visions were not frequent.

Besides these instances, so circumstantially related, and others of a like kind with which the Scriptures abound, exemplifying various modes of spirit influx and operation, there is the long series of miracles, prophecies, and revelations, running through and indissolubly blended with the sacred history; and the varied "spiritual gifts" concerning which St. Paul, writing to the Church of Corinth, says, "I would not have you ignorant."

Nor does the Church, in succeeding times, appear to have been ignorant.

Augustine asserts that miracles were so frequent and extraordinary in his time (the fourth century), that accounts of them were read in the churches. Some are said to have been done before many witnesses, and some in his own presence.

Evodius, a bishop in Africa, and a friend of Augustine, corresponded with the latter concerning spirit-manifestations. Of the reality of these, Evodius was well persuaded from his own experience. He says, "I remember well that Profuturus, Privatus, and Servitius, whom I had known in the monastery here, appeared to me, and talked to me, after their decease; and what they told me, happened. Was it their souls which appeared to me, *or was it some other spirits who assumed their forms?*" He also inquires, "If the soul on quitting its (mortal) body does not retain *a certain subtile body* with which it appears, and by means of which it is transported from one spot to another?" Augustine, in reply, acknowledges that there is a great distinction to be made between true and false visions, and that he could wish that he had some sure means of distinguishing them.

It is a common notion among Protestants, that all alleged supernatural occurrences in the Catholic Church are either the delusions of ignorant enthusiasts, or the inventions of priestcraft. There can be no greater mistake. Whether the miracles are genuine or not, the Catholic Church admits them only after a most thorough investigation. "I should not be a good Catholic," said Cardinal Wiseman, "if I did not believe in spiritual manifestations."

The working of miracles is a condition absolutely necessary in the canonization of saints; and it is only after a most careful scrutiny of facts that the Church allows canonization.

"In the scholastic ages," says Fleming, "the belief in return from the dead, in apparitions and spirits, was universal." Mr. Morison, in his "Life of St. Bernard," observes, "Miracles, ghostly apparitions, divine and demoniac interference with sublunary affairs, were matters which a man of the twelfth century would less doubt of than of his own existence."

St. Theresa, of whose experiences we have already made mention, writes in her account of her life, "Sometimes my whole body was *carried with my soul, so as to be raised from the ground;* but this was seldom. When I wished to resist these raptures, there seemed to be somewhat of such mighty force under my feet, which raised me up, that I knew not what to compare it to. All my resistance availed little."

A modern Spiritualist believes all this without difficulty.

Ernest Renan, in his "Life of Christ," makes light of the phenomena of the Bible, as well as of Spiritualism. He calls for "a miracle at Paris, for instance, before experienced *savans;*" one which would put an end to all doubt. Elsewhere, too, he explains more exactly what would suit him as to a miracle; that it should be wrought under conditions as to time and place, in a hall, and before a commission of physiologists, chemists, physicians, and critics; and that, when it had been done once, it should, on request, be repeated.

Well does William Mountford, in his "Anti-Supernaturalism of the Age," reply to expectations like these: "Are earthquakes, as reports, accounted incredible, as not occurring at a time and a place known beforehand, and submissive to the directions of men with clocks and spirit-levels, and with magnetic and other machines all ready for use? And, indeed, a miracle coming to order would scarcely be a miracle. For, coming to order patiently, punctually, and as a scientific certainty, it would, by that very fact, have parted probably with something essential to its nature as commonly understood."

The belief in guardian angels was common in the earliest historic times. According to Plato, a peculiar tutelary demon is allotted to every man, — an unseen, yet ever-present witness of his thoughts and conduct. Both Greeks and Romans had their genii. Plutarch says, "One Supreme Providence governs the world; and genii participate with him in its administration." That each individual has his guardian angel, has always been a favorite tenet of the Catholic Church; and its prayer for children recognizes the belief.

Instances in which persons have spoken in a language which was unknown to them in their normal state, are not infrequent in modern Spiritualism.

In Edward Irving's church, in England (1831), the utterances were sometimes in foreign languages as well as in "the unknown tongue."

Colquohon, in his "Isis Revelata," remarks, "Many authors have noticed this phenomenon of speaking a language unknown to the individual in his ordinary state; and it will very frequently be found coupled with the prophetic faculty, as arising out of the same or similar conditions."

Not only in Judea, but throughout the Orient, has the belief in spirit-communion prevailed from the earliest times. Mahomet was what would be called in our days a medium. He was subject to trances and ecstasies. He was a thorough Spiritualist. When he followed the mortal remains of his son Ibrahim to the grave, he invoked the child's spirit to hold fast to the foundations of the faith; the unity of God, &c. So Irving says.*

According to Huc, the Catholic missionary, table-rapping and table-turning were in use in the thirteenth century among the Mongols, in the wilds of Tartary. The Chinese recognize spiritual intervention as a fact, and it is an element in their religious systems. At the rites in honor of Confucius, Huc tells us that the spirit of Confucius is addressed as present.

Dr. Macgowan, in the "North-China Herald," tells us how writing is performed by the agency of spirits; from which we may infer that a form of Planchette is no novelty among the Chinese. He says, "The table is sprinkled equally with bran,

* From Irving, we learn that Columbus, too, was a Spiritualist; believing that a spirit-voice spoke to him, to comfort him in his troubles, in Hispaniola.

flour, dust, or other powder; and two mediums sit down at opposite sides, with their hands on the table. A hemispherical basket, eight inches in diameter, is now reversed, and laid down with its edges resting on the tips of one or two fingers of the two mediums. This basket is to act as penholder; and a reed, or style, is fastened to the rim, or a chopstick thrust through the interstices, with the point touching the powdered table.

"The ghost, meanwhile, has been duly invoked; and the spectators stand round, waiting the result. This is not uniform. Sometimes the spirit summoned is unable to write; sometimes he is mischievously inclined, and the pen — for it always moves — will make either a few senseless flourishes on the tables, or fashion sentences that are without meaning, or with a meaning that only misleads. This, however, is comparatively rare. In general, the words traced are arranged in the best form of composition, and they communicate intelligence wholly unknown to the operators. These operators are said to be not only unconscious, but unwilling, participators in the feat."

The same writer tells us that in Ningpo, in 1843, there was scarcely a house in which this mode of getting messages from the spirits was not practised. So it would seem that, some five years before the phenomena at Hydesville, Planchette, or a substitute for it, was common in China! *

* In the New-York "Round Table" of Dec. 12th, 1868, we find the following remarks upon the subject of Planchette: "Mr. Kirby is said to have sold over two hundred thousand planchettes, at a profit of fifty cents, cash, each. It need not surprise us that Mr. Kirby thinks well of planchette. Now what does so knowing a young lady as Miss Field think of it? In this neat little volume ('Planchette's Diary'), she tells her own experiences, and, as a conclusion of the whole, admits that she has no theory, is perplexed; and, finally, 'from the sensations undergone while using planchette, I am inclined to believe myself under the influence of a wonderfully subtle magnetic fluid.' To find a *name* to call a thing by, seems to satisfy most minds; but a name is nothing, — 'electricity,' 'magnetism,' 'odic force,' 'vital current,' and so on and on, and we are as much in the dark as ever about planchette, table-movings, hysteria, Spiritualism, demonism, witchcraft, possession of devils, &c. Are these any thing at all but 'derangement' of the normal forces of human nature, or a strange and unhealthy action? or are they, in some subtle way, the action of spiritual forces *outside* of ourselves? Science has not yet settled the question, and we commend it to the attention of our new school of positivists."

Seneca compares the birth of man into this world to his birth from the womb of Nature, into "another beginning, another state of things that expects us."

"It will be just as natural for you," says one, claiming to be a spirit, "to become suddenly conscious of the spirit-world, as it is for the infant to be ushered into the material world without consciously experiencing any unusual degree of excitement from the occurrence."

"A form which vanishes," says Gustave Aimand, "is the creation of a new form, a transformation of being. What we call death is a movement in advance, a progressive evolution, an aggrandizement of life. Our past furnishes us a double proof of this assertion; for it is through a double death, a double destruction of anterior forms, that we arrive at our present life.

"Suppose that the ovule which is to one day be a man, had sensibility and intelligence: would it not take for symptoms, premonitory of its end, the painful rendings of its ovulary organization? Error! Vain fears! The ovule becomes a fœtus; that is to say, passes from an inferior life to a superior; for the fœtus has an organization and a life distinct from those both of the ovule and of the infant.

"Suppose now that the fœtus, also sensitive and intelligent, approaching the end of its fœtal life, began to experience the sufferings of child-birth. Would not it, too, believe that the convulsive claspings of the uterus were the very embrace of death and the utter annihilation of life? Error again! Vain fears! For that which it took for its death-rattle of agony, and its last adieu to existence, is the first wailing of a new-born child, its salutation to a new and higher life.

"And so the end of one life is the commencement of another life less imperfect. It is in this manner, beyond a doubt, that by an endless series of evolutions or of deaths, we shall realize more and more the divine destiny which is revealed to us and promised by our aspirations, our infinite desires.

"Unless man is eternal in his substance, immortal in his personality, infinite in his destiny, even as he is in his desires, then there is neither Being of beings, nor Omnipotent Goodness, nor Infinite Love, nor Eternal Justice: God does not exist."

We know with what suddenness the prevalent fanatical notions in regard to witchcraft passed away from the civilized world. Mr. Lecky has described it in some striking sentences. It was as if people had awakened all at once from a dreadful nightmare. One day witchcraft seemed a fixed fact, and the next day it was spurned and gone. Unquestionably, with what there was in it fanatical and false, much that was true was repudiated. It will be the work of Spiritualism to point out and re-confirm the true. But the time is not far back, when, to deny witchcraft, and the construction put on it by the authority of the Old Testament, was regarded as a sort of atheism.

May it not be that our theological systems and creeds, widely but somewhat passively accepted as they now may be, are destined to a winnowing not unlike that which witchcraft has undergone? May not some of our professional religious teachers wake up some bright morning to find that their hearers have very generally outgrown a certain style of appeal to their lazy preferences, their self-indulgent hopes, their nervous fears, or their sordid calculations? Should such a change come, — and the signs are threatening, — we may be sure that Spiritualism, pure and undefiled, will be the unfailing conservator of all that is good and true in human beliefs on the subject of the relations of man to time and to eternity, to the universe and to its Author.

THE END.